Green Technologies for the Environment

ACS SYMPOSIUM SERIES **1186**

Green Technologies for the Environment

Sherine O. Obare, Editor
Western Michigan University
Kalamazoo, Michigan, United States

Rafael Luque, Editor
Universidad de Córdoba
Córdoba, Spain

Sponsored by the
ACS Division of Environmental Chemistry

American Chemical Society, Washington, DC

Distributed in print by Oxford University Press

Library of Congress Cataloging-in-Publication Data

Green technologies for the environment / Sherine O. Obare, editor, Western Michigan University, Kalamazoo, Michigan, United States, Rafael Luque, editor, Universidad de Córdoba, Córdoba, Spain ; sponsored by the ACS Division of Environmental Chemistry.
pages cm. -- (ACS symposium series ; 1186)
Includes bibliographical references and index.
ISBN 978-0-8412-3018-7 (alk. paper)
1. Green technology. 2. Green chemistry. 3. Environmental protection. I. Obare, Sherine O., editor. II. Luque, Rafael, editor. III. American Chemical Society. Division of Environmental Chemistry.
TA170.G747 2014
628--dc23
2014046394

The paper used in this publication meets the minimum requirements of American National Standard for Information Sciences—Permanence of Paper for Printed Library Materials, ANSI Z39.48n1984.

Distributed in print by Oxford University Press

Foreword

The ACS Symposium Series was first published in 1974 to provide a mechanism for publishing symposia quickly in book form. The purpose of the series is to publish timely, comprehensive books developed from the ACS sponsored symposia based on current scientific research. Occasionally, books are developed from symposia sponsored by other organizations when the topic is of keen interest to the chemistry audience.

Before agreeing to publish a book, the proposed table of contents is reviewed for appropriate and comprehensive coverage and for interest to the audience. Some papers may be excluded to better focus the book; others may be added to provide comprehensiveness. When appropriate, overview or introductory chapters are added. Drafts of chapters are peer-reviewed prior to final acceptance or rejection, and manuscripts are prepared in camera-ready format.

As a rule, only original research papers and original review papers are included in the volumes. Verbatim reproductions of previous published papers are not accepted.

ACS Books Department

Contents

1. **Green Technologies for the Environment** 1
   Sherine O. Obare and Rafael Luque

Solvent Systems for Green Chemistry Processes

2. **Carbon Dioxide Solvent Applications in a Biorefinery** 9
   Lindsay Soh

3. **Deep Eutectic Solvents: Environmentally Friendly Media for Metal-Catalyzed Organic Reactions** 37
   Joaquín García-Álvarez

4. **Amino Acid-Derived Imidazolium Zwitterions: Building Blocks for Renewable Ionic Liquids and Materials** 53
   Sarah Kirchhecker and Davide Esposito

5. **Opportunities for the Replacement/Minimization of Selective Hazardous Solvents: Applications, Concerns and Approaches To Identify Alternatives** 69
   Buchi Reddy Vaddula, Swathi Yalla, and Michael A. Gonzalez

Advances in Materials and Chemical Reactions

6. **Greener Oxidation of Alcohols, Glycosides and Sulfides** 117
   Mo Hunsen

7. **An EcoScale Comparison of Mechanochemistry and Solution Based Reactions** 129
   Kendra Leahy, Anthony M. Mack, and James Mack

8. **Some Strategies in Designing Highly Efficient Photocatalysts for Degradation of Organic Pollutants in Water** 139
   Changlin Yu, Wanqin Zhou, Gao Li, and Rongchao Jin

9. **Environmentally-Friendly Designs and Syntheses of Metal-Organic Frameworks (MOFs)** 161
   Tomislav Friščić, Patrick A. Julien, and Cristina Mottillo

Biomass and Waste Valorization to Chemicals, Materials and Fuels

10. Food Waste Valorisation for High Value Chemicals and Energy Production 187
Wan Chi Lam and Carol Sze Ki Lin

11. Lignin Based Carbon Materials for Energy Storage Applications 203
Sabornie Chatterjee, Tomonori Saito, Orlando Rios, and Alexander Johs

12. Synthesis of Nanoparticles for Biomass Conversion Processes 219
Jared T. Wabeke, Hazim Al-Zubaidi, Clara P. Adams, Liyana A. Wajira Ariyadasa, Setare Tahmasebi Nick, Ali Bolandi, Robert Y. Ofoli, and Sherine O. Obare

13. Agricultural Residues as Feedstocks for Lactic Acid Fermentation 247
Daniel Pleissner and Joachim Venus

Editors' Biographies 265

Indexes

Author Index 269

Subject Index 271

Chapter 1

Green Technologies for the Environment

Sherine O. Obare[1,*] and Rafael Luque[2,*]

[1]Department of Chemistry, Western Michigan University, Kalamazoo, Michigan, United States
[2]Departmento de Quimica Organica, Universidad de Cordoba Campus de Rabanales Edificio Marie Curie (C-3) Ctra NNal IV-A, Km 396, Cordoba, Spain E-14014
[*]E-mail: sherine.obare@wmich.edu (S.O.O.); q62alsor@uco.es (R.L.)

Green chemistry and engineering principles are central to advancements in fundamental research and development in academic, government and industrial sectors. Innovations in chemical design, development of new solvents or reducing/eliminating the use of solvents, and the processes that lead to sustainable practices, are essential toward a safe and clean environment. This chapter outlines important topics covered in this book relevant to recent advances in green technologies for the environment. It further provides a general outlook on topics for future research and education.

Introduction

In recent years tremendous research efforts have been devoted toward finding innovative and effective ways to develop important technologies, while ensuring that the processes are 'green'. The drive toward 'green' or sustainable processes arises from several aspects not only scientifically, but rather due to the need to establish a better quality of life, improve human health, and prevent damage to the environment. Coupled with this need, is the desire to avoid disasters caused by the chemical industry, for example, the 2014 chemical spill in West Virginia, USA that shut down the water supply of more that 300,000 residents (*1*), the explosion that occurred at the fertilizer company storage and distribution facility in West, Texas in 2013 killing 14 people and leaving more than 160 injured (*2*). In 2010, an alumina plant accident occurred in Hungary leading to the release of highly toxic and alkaline sludge that flooded several settlements and caused hundreds of injuries (*3*). Not to be forgotten is the 1984 disaster of Bhopal, India caused by a pesticide spill resulting from work at Union Carbide that led to the death of 3000 people and injuring hundreds of thousands of civilians (*4*).

The ideas of 'green chemistry' have been practiced for many years but were formalized in the 1990s with the implementation of the Twelve Principles of Green Chemistry by Paul T. Anastas and John C. Warner (*5*). The seminal work of Anastas and Warner has provided guidelines that the chemical and pharmaceutical companies, the food industry, government scientists and academic researchers can adopt for sustainability practices (*6–8*). Several journals have been established that focus on green chemistry, the three most notable ones being: *Green Chemistry* published by the Royal Society of Chemistry, *ChemSusChem* published by Wiley, *ACS Sustainable and Chemical Engineering* published by the American Chemical Society.

Overview

The book '*Green Technologies for the Environment*' was inspired by three symposia organized by Professors Rafael Luque (University of Cordoba, Spain) and Sherine Obare (Western Michigan University, USA), at the American Chemical Society National Meetings and Exposition in Fall 2012 (Philadelphia, PA), Spring 2013 (New Orleans, LA), and Fall 2013 (Indianapolis, IN). The symposia brought together experts in the field of biotechnology, chemistry, chemical engineering, environmental engineering and toxicology from both academia and industry, to discuss green processes for the environment. The topics included finding replacements for crude oil to meet both our energy needs as well as the supply of chemicals for the production of essential products, advances in chemical processing, waste valorization, alternative solvents, and developments in homogeneous and heterogeneous catalysis as well as enzyme-based processes for chemical transformations.

Advances in green chemistry concepts will further enhance the field through the design of new chemicals and solvents. In addition, obtaining a better understanding of the mechanistic pathways involved in various reactions is essential toward advances in the field. The goal of the work described in each of the chapters is to address the need for best practices for chemical processes and for the production of chemicals, while promoting sustainability.

This monograph is organized into three sections: (1) solvent systems for green chemistry processes, (2) advances in materials and chemical reactions, and (3) biomass and waste valorization to chemicals, materials and fuels. The first section focuses on innovative green technologies where solvent systems have been developed to replace toxic organic chemicals. Generally, the solvents need to be selected to satisfy the criteria of low cost, high solute solubility, environmental compatibility, low volatility and a wide potential window (*9*). Soh's work in Chapter 2 provides a description of the use of compressed carbon dioxide (CO_2) as a solvent for the biorefinery. The chapter describes important examples for the direct application of CO_2 for the processing of biomass for the production of fuels and chemicals. There is also a good outline of some of the advantages and limitations to the use of CO_2. Chapter 3 by Garcia-Alvarez *et al.* describes metal catalyzed organic chemical reactions that are carried out in deep eutectic solvents (DES). The use of DES was developed as an alternative to ionic liquids. DES consist of a mixture of two or more components, which may be solid or liquid and that at a particular composition present a high melting point depression becoming liquids at room temperature (*10*). Davide Esposito's work continues along the same line of finding alternative solvents. In Chapter 4, Esposito and coworkers describe the preparation of imidazolium zwitterions derived from amino acids and carbohydrates that serve as building blocks for ionic liquids. The section is summarized with a review in Chapter 5 by Gonzalez and coworkers, which focuses on the important characteristics of common solvents, their hazards, and methods that have been developed to eliminate the use of toxic solvents or replace them with environmentally-favorable alternatives.

The second section focuses on chemical processes that are fundamental for advancing green technologies. In Chapter 6, Hunsen describes greener reactions for preparation of aldehydes, ketones, and carboxylic acids using alcohols as precursors. These reactions are catalyzed using pyridinium chlorochromate in the presence of periodic acid, a co-oxidant. The chemistry relies on the selective oxidative ring-opening of glycosides and oxidation of sulfides to sulfones. Chapter 7, by Mack *et al.* describes the advantages of mechanochemical reactions over solvent-based methods that have been used to carry out chemical reactions. These reactions use a milling process and are emerging as an excellent environmentally friendly alternative to using solvents.

Photocatalytic reactions in which a photocatalyst is used to produce oxidative species to drive specific reactions have been recognized as environmentally friendly processes for aqueous-based remediation of toxic pollutants. In Chapter 8, Jin *et al.* describe the design of new photocatalysts that are hetero-structured or microspherical. The synthesis and characterization of the photocatalysts are described. The novelty of the new materials is their ability to minimize electron-hole recombination while improving the overall catalytic efficiency. Chapter 9 by Friscic *et al.* demonstrates an environmentally-friendly design and syntheses of metal-organic frameworks (MOFs), which are essential for fuel and gas storage. The general principles of green and sustainable chemistry have not been well established for the synthesis of inorganic compounds such as MOFs. In the chapter, Friscic outlines some of the challenges and opportunities that will enable advancement in green inorganic synthetic methods.

Understanding the fundamental chemistry involved for the utilization of biomass feedstocks and waste for the biorefinery to produce chemicals and fuel is paramount toward green technologies (*11*, *12*). An excessive amount of food waste is generated worldwide on a daily basis. One of the major concerns for food disposal is the biological hazard that arises due to bacterial and fungal growth, which leads to detrimental effects on human health and the environment. Furthermore, food waste is usually disposed in landfills leading to the release of toxic green house gases. In Chapter 10, Lin and co-workers provide an excellent overview of effective methods to convert food waste into energy and high value added chemicals. They also describe the use of fungi and microbial strains for driving important processes for food valorization.

Lignin is an important biomass component that has potential to be an alternative and very attractive feedstock for several commodity chemicals. However, several technical challenges must be overcome for lignin to become a practical and user-friendly resource. In Chapter 11, Chatterjee *et al.* show the processing and use of lignin to produce a new class of anode materials that have good electrochemical characteristics. These electrodes could potentially be used in current lithium ion batteries.

The last two chapters focus on biomass conversion. Chapter 12 by Ofoli *et al.* provides a review on the use of metallic and bimetallic nanoparticles for the catalytic conversion of biomass derived compounds to commodity chemicals with important aspects in methods for nanoparticle synthesis and applications in various reactions (for example, oxidation, hydrogenation and deoxygenation). Last, but not least, chapter 13, by Venus *et al.* outlines the use of agricultural residues as feedstocks in biotechnological processes. The chapter focuses on fermentative lactic acid production using straw, rapeseed residues and grass press juice as substrates.

Future Outlook

Future work in green technologies will focus on fulfilling the principles of green chemistry and engineering including advanced processes such as flow chemistry and continuous processing, which help make synthetic procedures more sustainable (*13*). It is paramount that technologies are established that bridge the gap between academic scientists and industrial researchers in order to ensure safe, reproducible, efficient and scalable processing.

One of the important aspects to focus on as we advance green technologies is to educate future generations and make society aware of the need for better and more sustainable practices. Several initiatives have been implemented toward addressing this important task (*14–18*) and should be utilized as a support system for future work. Furthermore, several organizations including, the Green Chemistry Education Network (GC Ed Net) that serves as a catalyst for integrating green chemistry in chemical education at all levels (*19*), the American Chemical Society Green Chemistry Institute (*20*), and the MIT Green Alternative Wizard are useful resources for education. The Green Chemistry Education Webinar Series launched in the fall of 2014 highlights relevant 'green chemistry' topics for chemistry educators and students to incorporate into their courses, labs and programs.

We need to redouble our efforts in seeking more sustainable alternatives for innovation and production of the type of products needed by consumers, with comparative or even improved performance to those currently available from petroleum-derived resources. Products that will enable efficient work to be conducted, products that will improve healthcare as well as improve the life standards of international citizens, must be generated using green technologies. Renewable resources are the key in the transition to a biobased economy and such efforts should also encompass an appropriate education to future generations to understand the importance of being "greener" to ensure a sustainable environment for the future of the planet.

Acknowledgments

S.O.O. is grateful to the National Science Foundation for funding under award CHE 1005456 and DMR 0963678. S.O.O. also thanks for Western Michigan University Office of the Vice President for Research for a Faculty Research and Creative Activities Award to support work on this monograph. R.L. gratefully acknowledges the Spanish MINECO for financial support under project CTQ2011 28954-C02-02 and Consejeria de Ciencia e Innovacion, Junta de Andalucia for funding project P10-FQM-6711.

References

1. Gutman, D. Freedom Industries execs are long time colleagues. *the Charleston Gazette*, January 10, 2014. http://www.wvgazette.com/News/201401100119 [accessed November 10th, 2014].
2. Krauss, C.; Santos, F. Report: As many as 35 killed in Texas plant explosion. *The New York Times*, retrieved April 19, 2013. http://www.nytimes.com/2013/04/20/us/texas-explosion.html?hp&_r=0. [accessed November 10th, 2014].
3. Gelencser, A.; Kovats, N.; Turoczi, B.; Rostasi, A.; Hoffer, A.; Imre, K.; Nyiro-Kosa, I.; Csakberenyi-Malasics, D.; Toth, A.; Czitrovszky, A.; Nagy, A.; Nagy, S.; Acs, A.; Kovacs, A.; Ferincz, A.; Hartyani, Z.; Posfai, M. *Environ. Sci. Technol.* **2005**, *45*, 1608–1615.
4. Lepkowski, W. *Chem. Eng. News* **1984**, *62*, 19–20.
5. Anastas, P. T.; Warner, J. C. *Green Chemistry: Theory and Practice*; Oxford University Press: New York, 1998; p 30.
6. Anastas, P. T.; Zimmerman, J. B. *Environ. Sci. Technol.* **2003**, *37*, 94A–101A.
7. Garcia-Serna, J.; Perez, B. L.; Cocero, M. J. *Chem. Eng. J.* **2007**, *133*, 7–30.
8. Boks, C.; Stevels, A. *Int. J. Prod. Res.* **2007**, *45*, 4021–4039.
9. Pena-Pereira, F.; Namiesnik, J. *ChemSusChem* **2014**, *7*, 1784–1800.
10. Paiva, A.; Craveiro, R.; Aroso, I.; Martins, M.; Reis, R. L.; Duarte, A. R. C. *ACS Sustainable Chem. Eng.* **2014**, *2*, 1063–1071.
11. Clark, J. H.; Luque, R.; Matharu, A. S. *Ann. Rev. Chem. Biomol. Eng.* **2012**, *3*, 183–207.
12. Clark, J. H.; Deswarte, F. E. I.; Farmer, T. J. *Biofuels, Bioprod. Biorefin.* **2009**, *3*, 72.
13. Newman, S. G.; Jensen, K. F. *Green Chemistry* **2013**, *15*, 1456–1472.
14. Murphy, C. F.; Allen, D.; Allenby, B.; Crittenden, J.; Davidson, C. I.; Hendrickson, C.; Matthews, H. S. *Environ. Sci. Technol.* **2009**, *43*, 5558–5564.
15. Aurandt, J. L.; Butler, E. C. *J. Prof. Iss. Eng. Educ. Prac.* **2011**, *137*, 102–106.

16. Kevern, J. T. *J. Prof. Iss. Eng. Educ. Prac* **2011**, *137*, 107–112.
17. Zimmerman, J. B.; Vanegas, J. *Int. J. Eng. Educ.* **2007**, *23*, 242–253.
18. Davidson, C. I.; Hendrickson, C. T.; Matthews, H. S. *Int. J. Eng. Educ.* **2007**, *23*, 287–293.
19. Green Chemistry Education Network Mission: The University of Oregon. http://cmetim.ning.com [accessed November 10th, 2014].
20. The American Chemical Society Green Chemistry Institute: http://www.acs.org/content/acs/en/greenchemistry.html [accessed November 10th, 2014].

Solvent Systems for Green Chemistry Processes

Chapter 2

Carbon Dioxide Solvent Applications in a Biorefinery

Lindsay Soh*

Department of Chemical and Biomolecular Engineering, Lafayette College, 740 High Street, Easton, Pennsylvania 18042
***E-mail: sohl@lafayette.edu**

In order to sustainably produce fuels and materials from renewable feedstocks, biorefineries must be designed for and operated with sustainable processes. Compressed carbon dioxide – subcritical, supercritical, and CO_2 expanded liquids – can potentially be used in a biorefinery as a green solvent. Key processes that may be improved via the use of CO_2 are extraction (including fractionation and refinement) and reaction. Numerous examples exist for the direct application of CO_2 for the processing of biomass for the production of fuels and other chemical outputs. While pitfalls exist for the application of compressed CO_2 in commercial production, this versatile compound holds great promise for utilization as a green solvent in a biorefinery setting.

Introduction

According to the International Energy Agency a biorefinery is "the sustainable processing of biomass into a spectrum of marketable products (food, feed, materials, chemicals) and energy (fuels, power, heat)" (*1*). This inclusive definition highlights the potential of biomass to be used for a number of different applications and the production of energy in different forms. As with petroleum refining, the idea of a biorefinery is to produce multiple products from a given feedstock with the goal of maximizing output as well as minimizing waste. In terms of economics, this concept produces high-value products from otherwise low-value material that might be otherwise discarded. From a sustainability perspective, a biorefinery minimizes waste while utilizing renewable feedstocks for products that generally have low inherent toxicity and are biodegradable.

Additionally, the emergence of integrated biorefineries provides an opportunity for the application of green chemistry (*2*) and engineering principles (*3*) to provide sustainable processes for the production of fuels and chemicals. Specifically, these principles serve as a guide to lower energy consumption, environmental impact, and waste. The use of solvents in chemical processing has a huge impact in all of these categories (*4*, *5*) as excess amounts are often needed to produce small volumes of product. The solvent must further be recycled or disposed of – both options imposing large impacts associated with either downstream separations or end-of-life. Additionally, many of the current organic solvents used are inherently toxic and/or flammable and are produced from non-renewable sources. The focus of this chapter will be the use of compressed carbon dioxide (CO_2) as an alternative and more sustainable process solvent with particular application to biorefineries.

Compressed CO_2 in either the liquid or supercritical state can be used as a effective solvent due to its favorable properties at elevated temperature and pressure. Namely sub- and supercritical CO_2 offer enhanced mass transfer and solvating properties due to its relatively low viscosity and high diffusivity, as compared to a liquid, and high density as compared to a gas (*6*). Furthermore, CO_2 is considered environmentally benign in that it is non-flammable, non-toxic, and readily abundant. The application of CO_2 for extractions and reactions has great promise, but the relatively high costs associated with high pressure processes bars its exclusive use for the production of low-value outputs such as fuels. The biorefinery provides a potential setting to utilize the beneficial properties of CO_2 while producing multiple products thus allowing for economical and sustainable fuel and chemical production.

Biomass

Biomass is uniquely valuable compared to other sources of renewable energy in its ability to provide liquid transportation fuels. Furthermore in comparison with the petroleum industry, biomass is the only renewable carbon source for the production of chemicals (*7*). Essentially, biological material is synthesized from a combination of sunlight, carbon dioxide, and nutrients and is comprised of a complex mixture of molecules whose functions range from structure and storage to energy production via photosynthesis. The major chemical constituents of biomass include carbohydrates/sugars, lipid, lignin, and proteins (*8*), each of which vary in proportion depending on the particular feedstock and growth conditions. For example, woody biomass mainly consists of cellulose, hemicellulose, and lignin (*9*), whereas microalgae lack lignocellulose altogether (*10*). Additionally, the lipid, starch, and protein content of microalgae can vary significantly depending on the phase of growth as well as the nutrients provided in the growth medium (*11*). Simplistically, biomass can be divided into subcategories of lipid-rich, sugar-rich, or lignocellulosics each of which require differences in processing to produce very different fuel and value-added products as discussed below.

In addition, biomass feedstocks can be classified according to how they are sourced. First generation biofuels are made from edible crops such as grains and seeds and thus directly compete with food production (*9*). Second generation biofuels on the other hand are made from sources that are not consumed as food. Examples of these sources include lignocellulosic materials, plant waste materials, and microalgae (sometimes named "third" generation). While first generation biomass feedstocks are generally easier to produce and process, the competition of their use for food versus fuel/material is controversial and thus the usage of second generation fuels is considered more sustainable (*12*). In order to utilize this biomass, significant processing must be done to generate fuel and other value-added products.

Biorefinery

While the main driver for establishing integrated biorefineries is to provide a sustainable source of transportation fuels, the generation of co-products improves the economic feasibility and decreases environmental impact (*1*). The importance of these multi-product biorefineries has led the International Energy Agency to establish the IEA Bioenergy Task 42 Biorefinery. Their report (*1*) identifies several global biorefineries in various states of commercialization and maps the potential conversion technologies that may be used with different feedstocks to produce a diverse set of chemicals and materials. The report further highlights that biorefineries will only be viable if biomass conversion efficiency is maximized through the use and production of multiple output streams.

These industrial products can be divided into three categories – energy, molecules, and materials (*13*). Energy products may be generated by direct combustion or conversion via thermal or chemical means (*8*). Direct combustion is the simplest of the three and can be used directly as a heat source or to generate steam for electricity production. Shortcomings of combustion include the inability to produce liquid transportation fuels and the inability to utilize valuable feedstock portions. Thermochemical conversion includes liquefaction, pyrolysis, and gasification. These conversions often use whole biomass fractions to produce biocrude and/or biogas. The biocrude that is produced can be further fractionated to various hydrocarbons and syngas as is the case with petroleum refineries (*9*, *14*). The high temperatures and pressures often associated with these processes (*15*) as well as the requirement for further product refinement necessitates high capital and operating expenditures. In chemical conversion (including biochemical) processes, selected biomass fractions are converted to fuel products. For example, ethanol is produced from starch and biodiesel from lipids. These processes require multiple steps – including extraction, conversion, and separations – and are costly with current technologies (*15*). In terms of fuel output, at least one of the energy products should be in the form of fuel as opposed to heat or electricity (*9*). Thus chemical and/or thermochemcial conversions are most promising for fuel production though there is currently no clear best technology. Developments in technology and process modeling will aid in biorefinery design for particular feedstocks and desired products (*8*).

In order for a biorefinery to be economically viable, at least one high-value product should be obtained along with low-value (though possibly high volume) products such as fuel and/or feed (*9*). Molecules extracted directly from biomass can potentially be used as chemical solvents or as building blocks for the production of other platform or commodity chemicals (*13*). In terms of materials, extracted compounds can be used to produce polymers or biomass fibers can be used directly, taking advantage of inherent properties such as strength and biodegradability. Of the three energy production routes, chemical conversion directly allows for conservation and potential valorization of non-fuel compounds. For example, vitamins, pigments, and nutraceuticals may be obtained from biomass post extraction of fuel precursor compounds (*16*). Similar processing may also be performed prior to thermochemical conversion, and value-added chemical products may be obtained from the resultant crude.

The processing steps required to obtain chemical and material products include extraction (mechanical or physicochemical) and/or conversion reactions. These steps require solvents such as hexane, DMSO, or methanol all of which have inherent hazards associated with high toxicity and/or flammability. Furthermore, separation of these solvents requires further unit processes and subsequent energy and capital costs (*17*). As such, improved technologies, particularly pertaining to the use of green solvents and efficient processing, are required to optimize costs and minimize environmental impacts associated with biorefineries (*1*).

Carbon Dioxide

Compressed CO_2 as a solvent can potentially be applied to a broad spectrum of biomass processes in biorefineries. CO_2 is non-toxic, non-flammable, easy to separate from a reaction mixture, and can improve process life cycle energy use and impacts (*18*). The enhanced mass transport properties and tunability make compressed CO_2 a promising solvent choice for both extractions and reactions. The low viscosity and high diffusivity facilitates increased transport of CO_2 into biomass substrates, potentially enabling improved extraction yields and kinetics. Furthermore, these same properties allow for enhanced reaction rates and high turnover frequencies for both homogenous and heterogeneous catalysts in CO_2 rich media (*19*). The tunability of CO_2 may enhance extraction and/or reaction selectivity (*6*). CO_2 may also be used to change the melting properties of compounds and increase the availability of reactants in immiscible phases (*6*). As the fourth most abundant gas in the air and a by-product of respiration, CO_2 is readily abundant. Furthermore, rising atmospheric CO_2 levels mean that CO_2 utilization could be environmentally advantageous. To straighten a common misconception, the use of CO_2 as a solvent would not directly generate CO_2, but in fact could utilize waste CO_2 which may otherwise be released as a greenhouse gas. In this work, both liquid and supercritical CO_2 are explored for their potential use as an extraction solvent and reaction medium.

The critical point of CO_2 is 304 K and 7.4 MPa, well within the limits of conventional chemical engineering processes (*20*). Neat supercritical carbon dioxide ($scCO_2$) is non-polar in nature and is effective for the dissolution of non-polar compounds or as an anti-solvent for more polar compounds (*21*). Adding polar co-solvents such as methanol or ethanol extends the range of solutes in $scCO_2$ to moderately polar compounds. Co-solvents are generally added at temperatures and pressures above the critical point of carbon dioxide and at concentrations so as to maintain a single supercritical phase. If excess modifier is added, above the solubility in CO_2, a multiple phase system occurs. In this instance, there exists a liquid phase that is expanded by the supercritical (vapor) phase, and is called a CO_2 expanded liquid (CXL). Typically, CXL's are used at lower temperatures and/or pressures and also exhibit enhanced mass transport properties (*i.e.* increased solute diffusivity compared to neat organic solvent (*22*)) while functioning as a more polar solvent and denser medium at lower pressures compared to $scCO_2$. CXL applications also include anti-solvent systems where the presence of CO_2 causes the dissolution of particular components allowing for efficient separations (*21*).

While CO_2 is a promising solvent, processes utilizing compressed CO_2 generally entail high capital and operating costs associated with pressure-rated equipment and compression. While potentially expensive, these costs are not prohibitive to commercial utilization – there are currently greater than 150 plants worldwide using supercritical fluid extraction at production volumes of greater than 500 L (*23*). These applications are mainly limited to extractions of high-value products using $scCO_2$. Examples range from decaffeinated coffee and natural product extracts to enriched food and nutraceuticals. These processes all take advantage of the inherent properties of CO_2 – enhanced mass transfer and non-toxicity allowing for use in edibles – while offering a high-value product and thus providing the economic incentive for use. As current utilization of CO_2 for the manufacture of biological products is limited to extractions of non-polar compounds, increased versatility of this solvent is required for its adoption into a multi-use scenario such as a biorefinery.

As mentioned above, the use of co-solvents and CXLs increases the range of applications for CO_2, though the presence of additional solvents may not be appropriate for certain purposes and may necessitate further separation processes downstream. Even when organic solvents are necessary, CO_2 may greatly reduce the amount needed and/or reduce the processing requirements in terms of time, waste, and downstream separations which can hence reduce energy use and the economic bottom line as seen in anti-solvent applications. An economic and environmental impact analysis by Fang et al. (*24*) showed that the use of CXL significantly reduced estimated environmental indices and cost for the hydroformylation of olefins. As seen in Figure 1, both $scCO_2$ and CXLs present green and efficient process alternatives for extractions and reactions with a variety of potential applications within the context of a biorefinery.

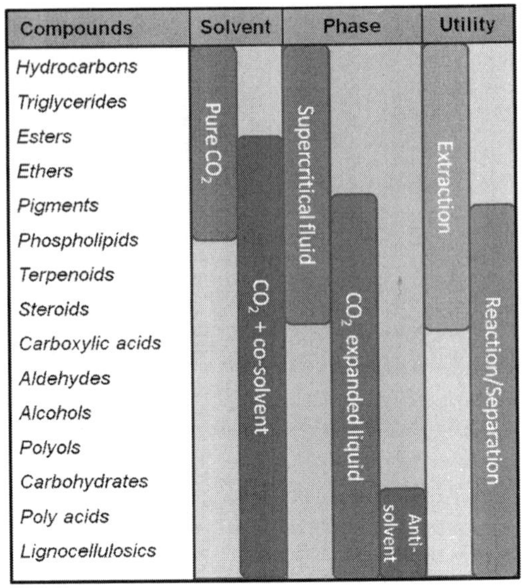

Figure 1. Solvent uses for carbon dioxide with biological compounds

CO_2 and the Biorefinery

Sub-, super-critical CO_2, and CXLs can be used as solvents in a biorefinery for extraction (including fractionation and refinement) and reactions. In order to use CO_2 in these settings, high purities must be attained though some processes have proved to be resilient to contaminants such as nitrogen or water (*25*). Proposed sources include by-product CO_2 from ethanol fermentation (*26*, *27*) and pressurized CO_2 from carbon capture and storage (*25*).

Utilization of CO_2 as a solvent in a biorefinery setting could provide not only an effective means for biomass processing but also a substitute for inherently hazardous solvents. For instance, CO_2 can replace hexane – a highly flammable, toxic, and volatile solvent – in many oil extractions. Commercial hexane use has been estimated at 151 million kg (1992) with environmental releases of over 32 million kg from 534 facilities in the United States alone (*28*). Supplanting this demand not only mitigates physical hazard and chemical exposure, but also avoids costs associated with downstream separations of hexane from the product. From sourcing to end-of-life, CO_2 has great potential to mitigate life cycle impacts associated with solvent use and unit operations.

Much of the state-of-the-art work in CO_2 solvent utilization is at the proof-of-concept phase where reactions and extractions are shown to be effective at laboratory scale. The steps towards commercial adoption would include further system definition and scale-up. For reactions, this development would entail optimization in terms of catalyst choice and phase behavior (*29*). Both reactions and extractions require a fundamental understanding of phase behavior which will impact both the system design as well as operating conditions. Given sufficient delineation, subsequent plant design will entail equipment choice, rating, and sizing as well as process synthesis including the incorporation of efficient product harvesting and solvent separation/recycle systems (*30*).

The application of CO_2 for fuel production is promising due to the selective solubility of low-polarity compounds (*21, 31*) and its ability to meditate conversion reactions such as transesterification for biodiesel production (*32*). Coupled with its ability to extract and produce value-added components, CO_2 has great potential for biorefinery applications that take advantage of its inherent properties to promote more efficient, greener processes. It is also worth mentioning that CO_2 may have other uses, such as its use as a working fluid (*33*) or in supercritical fluid chromatography (*34*), which are beyond the scope of this work, but are highly interesting fields of study. The research applications for the utilization of compressed CO_2 as a solvent are numerous and highly varied due to CO_2's high versatility. The following discussion, as summarized in Figure 1, is aimed at highlighting pertinent research applications so as to illustrate the potential use for CO_2 in the context of a biorefinery.

Extraction

The principal benefits of utilizing CO_2 for extraction are enhanced mass transfer and tunability. The rate and efficacy of extraction is a function of the ability of the fluid to penetrate the solid biomass matrix, the concentration gradient of the compound between the fluid and the matrix, and also of the ability of the solute to be incorporated into the fluid. In order to best utilize CO_2 for a given application, several models for phase prediction and process design have been proposed. General theories including film and penetration theory can be used to model transport of solute into the CO_2 and/or CO_2 into the biomass matrix (*35*). Del Valle et al. (*36*) provide a comprehensive overview of models useful for biomass extraction using $scCO_2$ taking into account internal and mass transfer limitations. In general, these models incorporate both kinetic and equilibrium properties influenced by solute diffusivities and solubilities respectively.

In terms of efficacy, the ability of CO_2 to solvate particular compounds is dependent on their mutual solubilities which can be modeled as a function of each substance's solubility parameter (δ) (*23*). Defined as the square root of a substance's internal energy divided by its molecular volume ($\delta = (E/V)^{1/2}$), the solubility parameter is unique to each compound at a given pressure and temperature (*37*). Generally, minimizing the difference between two component's δ values increases their mutual solubility with high solubilities attained when the difference is less than 5.11 $MPa^{1/2}$. The tunability of CO_2 stems from the range of δ_{CO2} that may be attained by varying pressure and temperature. At a given temperature, δ_{CO2} increases by over 4.5 times when raising the pressure from near critical to 30 MPa, following a sigmoidal relationship. Lipophilic molecules typically have lower solubility parameters (δ_{hexane} = 14.90 $MPa^{1/2}$) whereas hydrophilic compounds' are higher (δ_{water} = 47.81 $MPa^{1/2}$). Generally, raising CO_2 pressure increases the solubility parameter non-linearly while elevating temperature decreases δ_{CO2}. These differences allow CO_2 to be used for selective extraction and fractionation. At lower supercritical pressures, relatively large changes in δ_{CO2} results in significant differences in the solubility of different components based on small, tunable changes in pressure. Another consideration for solubility is the relative density of the CO_2 – as temperature decreases and pressure increases, the density of the supercritical fluid phase may increase significantly from densities close to a gas, approaching those of a liquid. The solvating power of the fluid varies greatly with the density thus influencing substrate solubility (*38–40*).

Biomass contains a range of compounds that are extractable using $scCO_2$ (*41–61*). Examples of components extracted from biomass using pure $scCO_2$ at varying pressures and temperatures include various lipophilic compounds such as triglycerides (*41*), tocopherols (*42*), and carotenoids (*43*). Organic solvents non-selectively extract lipid fractions consisting of mixtures of these components while the selectivity and tunability of CO_2 can allow for enriched product fractions of desired components. In terms of fuel production, triglycerides are the desired precursor for the production of biodiesel. Neat $scCO_2$ has been used for the selective extraction of triglycerides ($\delta_{sunflower\ oil}$ = 16.70) (*44*) versus other components such as carotenoids ($\delta_{\beta\text{-carotene}}$ = 17.80) (*45*) from biomass feedstocks (*41*). Solubility parameters of several biological extract classes can be seen in Table I. Increasing the system pressure and/or the use of co-solvents can increase the solubility parameter of CO_2 and thus allow for the extraction of carotenoids as well as tocopherols. Tocopherols (*42*) have vitamin E activity and carotenoids (*43, 46*) (*ex.* β-carotene, lycopene) are desireable for their use as antioxidants. These types of nutraceuticals have high retail value, several orders of magnitude larger than triglycerides. Furthermore, as these products are comestibles, the complete removal of conventional solvents via distillation is necessary in order to achieve high purities. Since these compounds are bio-active and sensitive to the high temperatures associated with distillation, the use of $scCO_2$ is highly pertinent for efficient extractions that allow for the compounds' functionalities to be retained (*23*).

As seen in Table II, a variety of biomass feedstocks have been tested for the ability of $scCO_2$ to extract lipophilic compounds. The feedstocks have been divided into general categories with select examples for each corresponding with pertinent extracts. The list is by no means exhaustive but provides a cross-section of the range of feedstocks and products that may be obtained.

Table I. Solubility parameters of biological extract classes

Compound	*δ, $MPa^{1/2}$*
Hydrocarbons (*39*)	17.06
Carotenoids (*39*)	17.84
Tocopherols (*39*)	18.12
Triglcyerides (*39*)	18.23
Fatty Acids (*39*)	18.61
Diglycerides (*39*)	19.33
Sterols (*39*)	19.47
Monoglyericdes (*39*)	20.86
Glycerol (*40*)	36.16

Table II. Biomass extraction by supercritical carbon dioxide

Feedstock	*Examples*	*Extract*
Algae	C. vulgaris (*43*) B. braunii (*47*) S. dimorphus (*41*)	**Lipids:** Triglycerides, Hydrocarbons, Polyunsaturated Fatty Acids **Pigments:** Lutein, Carotenoids, Chlorophyll
Fish	Mackerel (*48*) Salmon (*49*)	Triglycerides Polyunsaturated fatty acids
Oil seeds	Jatropha Seed (*50*) Palm kernel (*51*) Soybeans (*52*)	**Lipids:** Triglycerides, Polyunsaturated Fatty Acids Carotenoids
Leaves	Chamomile (*53*) Citronella (*54*) Sage (*55*)	"Essential oils"
Other plant matter	Chili pepper (*56*) Turmeric (*57*)	
Wood	Pine (*58, 59*) Spruce (*59*)	Terpenoids Sterols
	Eucalyptus (*60*)	Phenolics
Waste	Olive (*42*)	Beta carotene, lycopene
	Tomato (*61*)	Tocopherols
	Wine (*35*)	Ethanol

Table III. Solubility parameters of common extraction solvents and CO_2 co-solvents. (data source: Laurie (*40*))

Compound (40)	*δ, MPa½*
Hexane	14.90
Ethyl Acetate	18.15
Toluene	18.16
Benzene	18.51
Chloroform	18.95
Acetone	19.94
Dichloromethane	20.20
Ethanol	26.52
Methanol	29.61
Water	47.81

Polar co-solvents may also be used to enhance the range of extractable compounds using scCO_2. Adding a miscible co-solvent such as methanol, ethanol, or ethyl acetate (*62*) increases the effective polarity of the solvent mixture and allows for extraction of molecules of intermediate polarities (*21*). Solubility parameters for common extraction solvents and CO_2 co-solvents can be seen in Table III. Generally, the mixed solvent solubility parameter can be estimated by a weighted volume fraction contribution from each component (*63*). Additionally, the co-solvent may increase the mixed fluid density and thus enhance the volumetric carrying capacity of the fluid. Potential disadvantages of co-solvent utilization stem from the incompatibility of the solvents with downstream processes and utilization. For instance, methanol is inherently toxic and contamination of edible feedstocks is undesired. Further separation processes such as distillation may be applied to the extract in order to recover the co-solvent and isolate the desired product. However, such steps may increase processing complexity leading to additional cost and energy use. While co-solvent addition is a potential means to augment the range of extractable compounds using scCO_2, the comparative advantages of using a supercritical fluid must be definite, as succinctly underscored by King (*23*): "The question then becomes whether another compressed fluid might be better for the extraction/fractionation of these more polar solutes." The unique value of compressed CO_2 as compared to conventional solvents stems from its high mass transport, low temperature, selectivity, and environmental benign behavior – it is non-flammable, relatively inert, and non-toxic. These properties have been successfully exploited for effective extractions of a diverse set of compounds including high value applications such as in the food and pharmaceutical industries where CO_2's non-toxic behavior is highly pertinent.

More references highlighting the use of $scCO_2$ for biomass extraction can be found in published works by Mantell et al. (*35*) and Marriott and Sin (*27*). Further details on the use of $scCO_2$ extraction in various areas of applications including the food industry (*23, 62*), algae (*64*), and other commercial products (*65*) are also available.

Fractionation and Refinement

In addition to extraction from complex matrices, CO_2 can be used for modification of previously or concurrently extracted materials through fractionation or refinement. These techniques take advantage of variable component solubilities in CO_2, allowing for mixed feedstocks to be separated into fractions for either product enrichment or contaminant removal. Fractionation, or rather the production of enriched product portions, may be done with compounds that are similar in nature but have slight polarity differences allowing for tunable solubilities in CO_2. For example, ethyl ester mixtures produced from fish oils can be fractionated based mainly on the length of the alkyl chains (*66*). Starting with a mixture of commercial fish oil derived ethyl esters, Peretti et al. (*67*) demonstrated the enrichment of polyunsaturated fatty acids and even specific ω-3 fatty acids given appropriate optimization of pressure, temperature, and CO_2 flow rate. While fractionation of ethyl and methyl esters is possible, the fractionation of underivatized triglycerides is ineffective (*68*). In terms of fuel products, Brunner (*69*) summarizes results of lipid mixture fractionation including free fatty acids and tocopherols.

Refinement employs the differences in solubility of components in CO_2 to allow for purification or isolation of particular products. For example, Brunner (*69*) utilized CO_2 for the separation of methyl oleate from squalene as well as squalene from tocopherols and sterols. Additionally, free fatty acids were separated from triglycerides due to the far greater solubility of the acids in $scCO_2$. This deacidification as well as other contaminant removal is an important step prior to biodiesel production. Such contaminants, together with carotenes and other impurities, were also effectively removed from crude palm oil and palm-oil-esters with $scCO_2$.

CO_2 has also been used for the purification of crude biodiesel, which consists of methyl esters (biodiesel) as well as unreacted triglyceride, intermediates (di- and monoglycerides), and glycerol (*70*). As shown in Figure 2 for the oleate family of glycerides, the solubility of the methyl esters in $scCO_2$ is far higher than the other components (*71*). The high solubility of methyl esters allows for its dissolution into a supercritical phase, leaving behind components above their solubility capacity. According to Corazza et al. (*70*), ternary phase behavior measurements of this system (*72*) were used to predict conditions separation between 6-12 MPa and 298 to 323 K. Addition of CO_2 to crude biodiesel at these conditions forms a binary phase system where the lighter phase contains methyl esters that meets and exceeds specifications as set by the National Agency of Petroleum, Natural Gas and Biofuels in Brazil.

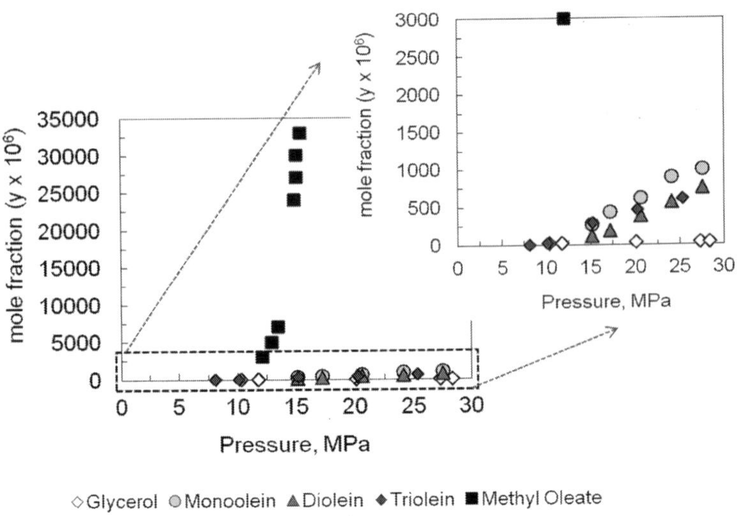

Figure 2. Solubility of oleate species in CO_2 at 323K. (data source: Gupta (71))

CO_2 as an Anti-Solvent

The relatively insoluble nature of certain components in CO_2 may also be utilized for harvesting of products. The use of CO_2 as an anti-solvent entails the solubilization of a mixed feedstock into an organic solvent. Adding CO_2 changes the solvent properties including acidity and polarity and can lead to the precipitation of insoluble components from the carbon dioxide expanded liquid (CXL). Depending on the desired product and its interaction with the CXL, the product may be in the liquid phase or the precipitated solid. Table IV illustrates the use of CO_2 as an anti-solvent for a range of starch, lignocellulosic, and oil-rich feedstocks.

Eckert et al. (*73*) precipitated lignin from a solution of crude lignin and methanol using very slight partial pressures of CO_2. Due to the decreased solubility of lignin in CO_2 expanded methanol, lignin precipitated out of the expanded methanol phase effectively extracting vanillin, syringol, and syringaldehyde in the fluid phase. The authors also postulate that the precipitation is attributed to the insoluble nature of lignin in acidic media (*73*). Another example is the separation of phospholipids and triglyceride oils from hexane post crude soy lecithin extraction (*74*). CO_2 addition causes the precipitation of the phospholipids and oils both of which may be used for other purposes such as emulsification or fuel respectively. While this example shows the effectiveness of using CO_2 as an anti-solvent, hexane would be better substituted with a greener solvent.

Table IV. Uses of CO_2 as an anti-solvent

Feedstock	*Solvent*	*Temp*	*Pressure*	*Product phase*	*Product*
Lignin (*73*)	Methanol	298K	1.3 MPa	Liquid	Vanillin, syringol, syringaldehyde
Crude soy lecithin (*74*)	Hexane	298K	6.5 MPa	Liquid Solid	Oil Phospholipids
Hemi-cellulose (*75*)	DMSO, 10% water	313K	15 MPa	Solid	Xylans and mannans
Microalgae (*46*) *(N. oculata)*	Ethanol	313K	20 MPa	Solid	Carotenoids (Zeaxanthin)
Soybean flakes (*81*)	Amidine	298K	N/A	Switchable solvent	Oil
Microalgae (*82*) *(B. braunii)*	DBU-octanol	313K	N/A	Switchable solvent	Hydrocarbons

In the previous examples, the products remain in the liquid solvent phase whereas other applications utilize the precipitated (solid) phase as seen in the collection of carotenoids from the ethanol soluble fraction of microalgae (*46*). Precipitation can potentially also be controlled to produce solid particles of tunable size and porosity. Haimer et al. (*75*) reported the production of purified, uniform hemicellulose particles consisting of xylans and mannans for potential application in coating materials. Other work has established the utility of CO_2 for adding value to biologically derived components. For example, the production of polymer particles laden with extracted pigments takes advantage of the mutual insolubility of both polymer and colorant in the CXL. For example, lutein laden poly-ethylene glycol beads have been produced by precipitation from dichloromethane (*76*). Additionally poly-lactic acid particles have also been imbued with lutein and precipitated from ethyl acetate (*77*). The authors have shown that for both pure lycopene and polymeric beads, CO_2 may be used to tune particle size (*77*, *78*). Specifically, larger particles were formed at lower temperatures and higher pressures above the critical point, but that particle size increased as pressure decreased from the critical pressure (*78*). CO_2 serves to plasticize and liquefy these mixtures at moderate temperatures, allowing for this value-adding process to be used with temperature labile compounds (*79*).

The use of CO_2 as an anti-solvent is also demonstrated in switchable solvent systems. Switchable polarity solvents take advantage of a reversal in solvent polarity due to a change the system conditions (*80*). CO_2 can act as this trigger and has been used for the extraction of oils from biomass and subsequent solvent separation (*81*, *82*). Phan et al. (*81*) found that an amidine/carbonate water system was effective to extract oil from soybean flakes and allow for separation of the extracted oil and aqueous/amidine rich phases after the trigger was applied. The authors did note that this system is not yet optimal as water amidine mixtures are difficult to separate. Samori et al. (*82*) used 1,8-diazabicyclo-[5.4.0]-undec-7-ene (DBU) mixed with octanol to extract hydrocarbons from *Botryococcus braunii.* CO_2 was bubbled through the system to allow for phase splitting with a yield of up to 16 wt% from freeze-dried samples (compared to 7.8 wt% with n-hexane) representing 81% of available hydrocarbons.

Reactions

In addition to efficient extraction, the favorable mass transfer and tunable properties of compressed CO_2 can enhance reaction kinetics and selectivity (*22*). Reaction efficiency is critical to the conversion and production of fuels, platform molecules, and value-added chemicals, as underscored by Bozell and Petersen (*83*): "Biobased chemical production is challenged by a lack of conversion technology. Conversion of renewable carbon to chemicals is the least developed and most complicated of all biorefinery operations, especially when compared to conversion processes available for nonrenewable hydrocarbons." As such the application of promising technologies such as the use of compressed CO_2 should be explored.

The increased diffusivity and decreased viscosity offered by CO_2, can lead to faster reaction kinetics due to the higher rate of mass transfer, which allows for greater contact between immiscible phases and catalyst efficacy (*84*). Furthermore, gases have increased solubility in both sccCO_2 and CXL's as compared to conventional solvents leading to improved availability of gaseous reactants in the liquid phase (*85*, *86*). Other advantages of using CO_2 as a solvent for reaction mediation are that CO_2 cannot be oxidized, is generally immune to free radical chemistry, and is miscible with gases and many organic materials (*87*). In terms of reagent availability, CO_2 may change the melting point properties of certain compounds thus allowing for reactions to occur at closer to ambient temperatures (*i.e.* liquid-liquid versus solid-liquid reactions at temperatures well below the substrate melting point) (*85*). CO_2 may also serve as an anti-solvent in reaction systems as illustrated by Organic Aqueous Tunable Solvents (OATS) where CO_2 facilitates solvent-catalyst separations post-reaction as with switchable solvent systems (*88*).

The use of catalysts is fundamental to green chemistry in order to facilitate reactions using less materials, heat, and time, thus minimizing energy use and waste (*2*). CO_2 may increase catalyst efficiency by enhancing reaction rates and turnover frequencies (*84*). Particular advantages exist when employing CO_2 with heterogeneous catalysts – namely CO_2 enables their use in situations where excessively slow reaction rates would require extreme processing conditions or the use of homogeneous catalysts. Heterogeneous catalysts are generally less hazardous and easier to separate/reuse than their homogenous analogs (*84*). The increased diffusivities afforded by CO_2 decreases internal and external mass transfer resistances in the solid matrix, allowing for greater catalyst activity (*19*). In some cases, namely with polymer supported catalysts, polymer swelling by CO_2 can allow for greater availability of internal catalytic sites and greater contact between the substrate and active sites (*89*). Additionally, CO_2 may be used with immobilized enzymes as long as temperatures are moderated to prevent denaturing of the enzyme (*22*).

In addition to kinetics, the use of CO_2 may improve reaction selectivity and potentially allow for a given reaction to be tuned for a particular product distribution (*85*). Even with the use of solvents, CO_2 may decrease the negative impacts associated with a particular reaction by minimizing the need for hazardous solvents, decreasing the flammability envelope of a system, improving the usage of a particular feedstock, and reducing energy usage through process intensification. Comprehensive reviews for the utilization of CO_2 in reactions are available, though none are specifically geared towards biomass processing (*22*, *84–87*, *90*). In terms of catalysts, Seki et al. (*91*) offer a critical review on the use of supercritical fluids for use in heterogeneous catalyst applications. Marriott and Sin (*27*) mention several potential reactions that would be pertinent to biorefining.

The production of biodiesel from oils illustrates several advantages of CO_2-mediated reactions pertaining to the transesterification of triglycerides into fatty acid alkyl esters. Biodiesel is traditionally produced using a homogeneous base catalyst (sodium or potassium hydroxide) and methanol. However, this process may cause problems with downstream separations of catalyst, glycerol, and unreacted intermediates/reactants in the two product phases (*92*). Supercritical transesterification has been shown to effectively transesterify triglycerides without the use of a catalyst (*93*, *94*). Unfortunately, this method requires high temperatures to form a supercritical phase (*95*), requiring cost-prohibitive amounts of energy for heating (*96*). Additionally, large excesses of alcohol are required, leading to challenges in downstream separations and potential hazards associated with toxicity and flammability. The use of CO_2 as a co-solvent has been explored to moderate these reaction conditions with transesterification using both supercritical ethanol (*97*) and methanol (*98*). CO_2 at percentages up to 10% was found to lower the required reaction temperature, but heating to at least 550K is still required to reach a 98% yield in this uncatalyzed reaction (*98*).

CO_2-mediated transesterification with heterogeneous catalysts takes advantage of the increased mass transport properties to allow for the effective transesterification at decreased reaction temperatures with the benefit of minimized downstream separations associated with the use of a solid catalyst (*32*, *81*, *82*). This reaction can either take place in a single supercritical phase or multiphase system where the alcohol is expanded by the CO_2. Maçaira, et al. (*99*) achieved 88% conversion of sunflower oil at a temperature of 473 K and 25 MPa using Nafion SAC-13 in $scCO_2$ and methanol. Similarly, Jackson and King transesterified corn oil in a continuous supercritical CO_2 and methanol system using immobilized lipase at 17.2 MPa and 323 K with a yield greater than 98%. Though the yields and processing conditions using enzymatic catalysis are potentially better, the high cost of enzymatic catalysts could hinder their employment (*100*).

Multi-phase systems have also been utilized effectively for transesterification with methanol over acidic catalyst resins, taking advantage of the properties of CO_2 while requiring moderate temperatures and pressures as compared to those needed to form a single supercritical phase (*32, 101*). Furthermore, greater yields were obtained in the CXL versus the supercritical system. It is postulated that high yields are achieved because CO_2 increases the contact between the immiscible phases (substrate-substrate and substrate-catalyst) and also swells the catalyst resin. Product distribution between the separate phases in a CXL system may potentially lead to easier downstream separations (*70*) as illustrated in the refining section of this discussion.

There is clear potential for CO_2-mediated reactions within a biorefinery. In addition to transesterification of triglycerides to fatty acid methyl esters, CXLs have been shown to effectively esterify free fatty acids to their alkyl ester counterparts (*102–106*). Other examples may be seen below in Table III and will be detailed in the following section.

Biorefinery Products Utilizing CO_2

Sustainability of fuels, molecular feedstocks, and chemicals production is enhanced by targeting biorefining towards appropriate compounds. Bozell and Petersen (*83*) have identified some of the top chemical opportunities for biorefineries with the greatest potential for high impact. These compounds include ethanol, furans, glycerol and its derivatives, hydrocarbons, carboxylic acids (lactic acid and succinic acid), aldehydes/ketones (hydroxypropionic aldehyde and levulinic acid), and polyols (sorbitol and xylitol). These compounds were chosen due to their abundance in the literature, multi-functionality, ability to replace existing petroleum-based platform and commodity chemicals, potential for high volume production and scale up, as well as ability to be sourced from renewable sources (*83*). The use of compressed CO_2 has been established in the literature for the extraction and reactions associated with these compound classes. Table V illustrates this potential for both the use of CO_2 for production and separations of the listed compounds as well as a medium to convert platform chemical to value-added products.

As outlined in this chapter, CO_2 can be used as a solvent for extractions and reactions. For extraction, CO_2's strength lies in the selectivity of low-polarity molecules. In this context, CO_2 can be utilized for extraction of hydrocarbons such as diolefins produced by *B. braunii* (*47*). Extraction yields were comparable with hexane, a more hazardous organic solvent, at pressures of 30 MPa. As detailed above, hydrocarbon extractions with *B. braunii* have also been demonstrated with switchable solvents using CO_2 as a trigger for solvent polarity reversal (*82*).

Table V. CO_2 utilization for products in a biorefinery

Compound(s)	*Source*	*Role of CO_2*
Aldehydes/ Ketones	Biosynthetic sugar isomerization	Levulinic acid conversion to and separation of γ-valerolactone (*116*)
Bio-hydrocarbons	Biosynthesis and extraction	Extraction from producing organisms (*47*, *77*)
Carboxylic Acids	Sugar fermentation	Succinic acid and γ-butyrolactone synthesis from maleic acid (*114*) Polymerization and PLA/PLGA particle production (*114*, *115*)
Ethanol	Starch fermentation	Separation from aqueous solutions (*35*) Extractions with ethanol as a co-solvent (*84*)
Furans	Carbohydrate extraction	Acid hydrolysis and extraction (*107*, *108*) Derivatization with ionic liquids (*107*)
Glycerol derivatives	Biodiesel by-product	CO_2-mediated biodiesel production (*32*) Telomerization of diene/terpenes (*111*, *112*)
Polyols	Chemical synthesis from cellulose	Synthesis of retinol glycosides (*113*)

The high solubility of ethanol and low solubility of water in $scCO_2$ makes it useful for the separation of ethanol rich aqueous mixtures (*35*). Furthermore, the use of ethanol as a polar co-solvent with CO_2 has been used for many applications for the extraction of mildly polar compounds such as phospholipids and pigments (*64*, *65*). The use of ethanol has several advantages in that it is also considered a green solvent due to low toxicity, and availability from a renewable resource.

The high solubility of furans – including furfural, HMP, and FDCA – in $scCO_2$ allows for effective extraction from mixed media (*107*). For example, furfural is formed and extracted with high selectivity from both the acid hydrolysis (*108*) and hydrothermal decomposition (*109*) of lignocellulosic materials. This one-step process was shown to improve furfural yields over the typical two-step conversion and dehydration with yields around 90% of the theoretical maximum in rice husks (*108*). In addition to the solubility of furans, the insolubility of sugar precursors and ionic liquids in $scCO_2$ allows for synthesis of compounds such as 5-hydroxymethylfuran in the ionic liquid followed by recovery with $scCO_2$ (*107*). Reactions such as these employ the ability of CO_2 to act either as a solvent or anti-solvent.

Other examples of CO_2 utilization for the production of chemical building blocks and value-added products include derivatization, synthesis, and polymerization. As detailed above, glycerol is produced as a by-product of the biodiesel production process which can be mediated by both $scCO_2$ and CXLs (*32, 101*). The low solubility of glycerol in CO_2 may facilitate separations of the fuel product, glycerol, and unreacted substrates. As glycerol comprises roughly 20 wt% of the biodiesel product, high volumes of biodiesel production will result in a large glut unless uses for glycerol and its derivatives are found (*110*). CO_2 may be used for reactions involving glycerol, for example for the catalytic telomerization of glycerol with diene (*111*) and myrcene (*112*). These products can be utilized as building blocks for use in the production of value-added products such as surfactants. Another example of derivatization of polar molecules in $scCO_2$ is the production of retinol glycosides from polyols such as sorbitol, mannose, fructose, and glucose (*113*). Charles et al. (*113*) have synthesized retinol glycosides from the enzymatic glycosylation of retinol with polyols such as mannose and sorbitol as mediated by CO_2. The authors showed high regioselectivity with certain sugars allowing for selective production of these particular vitamin A derivatives, which exhibit greater stability than the parent compound. Additionally, the derivatization causes the fat soluble vitamin to be water soluble and hence allows for its greater usage as a pharmacological agent.

Lactic acid and succinic acid are two carboxylic acids that are of particular interest for a biorefinery utilizing the fermentation of sugars (*83*). Succinic acid and/or γ-butyrolactone can be synthesized from the hydrogenation of maleic acid in CO_2 using a heterogenous catalyst (1% Pd/Al_2O_3) (*114*). Maleic acid was found to be completely miscible in CO_2 at 11 MPa. Selectivity towards either product could be controlled by reaction temperature, pressure and reaction time. Conversions with the catalyst were found to be complete, though γ-butyrolactone was preferred at increased temperature, pressure, and reaction time. Lactic acid polymerization to poly-lactic acid (PLA) is advantageous in CO_2 due to the high solubility of the esterification catalysts (dicyclohexyldimethylcarbodiimide and 4-dimethylaminopyridine) producing high molecular weight polymers (up to 13,500 g/mol) with a 95% yield at ~24 MPa. As the polymer is formed, the PLA will precipitate from solution allowing for facile separation and reuse of the CO_2-catalyst mixture. This method was also shown to produce particles of with narrow and tunable size distributions (*79*). As mentioned above, these polymers may also be imbued with drugs for specialized delivery applications (*115*).

The case of aldehyde/ketone production is well exemplified by the conversion of levulinic acid to γ-valerolactone where using CO_2 has the dual benefits of increased reaction rates and γ-valerolactone separation. Bourne et al. (*116*) have demonstrated yields of γ-valerolactone over 99% through the hydrogenation of levulinic acid using compressed CO_2 and water as a co-solvent over a heterogeneous acid catalyst. In addition to high yields of γ-valerolactone, the authors have shown that favorable system phase behavior provides an isolated product stream of γ-valerolactone with less than 0.4% water and no detectable amounts of starting material. The authors also note that a number of hydrogenation reactions may be possible using this technique (*116*).

These examples have demonstrated the versatility of compressed CO_2 as a solvent and showed how the unique properties can be used strategically for optimizing solubility, tunability, mass transport, separations, and overall process efficiency. When designing systems for the utilization of CO_2, these properties must be fully exploited and rational process design must be levied in order for the successful application of this technology in a chemical or biological refinery (*117*).

Sustainable Design

In 2002, Thomas Swan & Co, Ltd., in collaboration with the University of Nottingham, built a demonstration plant for the scale up and production of trimethylcyclohexanone from the hydrogenation of isophorone with compressed CO_2 (*29*). Despite the initial success, the plant ultimately had to shut down due to overly high operating costs resulting from the large energy consumption associated with CO_2 re-compression post product isolation (*90*). While a setback for the field of CO_2-mediated reaction chemistry, the plant was successful at demonstrating appropriate scale-up and provided valuable lessons for future design considerations. While much promise surrounds the use of CO_2 for a variety of applications, drawbacks – including relatively high operating pressures, low dielectric constant, limited solvating power, incompatibility with certain substrates, and Lewis acid behavior also attributing to its low pH in water – may limit the use of CO_2 in certain situations (*87*).

Integration of CO_2 into Biorefinery Processes

In order to avoid these pitfalls, Beckman (*87*) suggests a set of design heuristics for operations with compressed CO_2:

- Maximize substrate concentrations
- Minimize operating pressure
- Avoid pressure drops during product recovery
- Maximize the recovery and reuse of catalysts and CO_2-philes
- Utilize continuous reactions if possible

These heuristics are aimed at increasing process efficiency and guiding research and technological innovation towards processes that will take advantage of the inherent properties of CO_2. Commercial adoption of CO_2 as a solvent for both extractions and reactions requires scale-up, optimization, and process synthesis. At the current juncture, most of the reviewed uses for CO_2 in this chapter are at the concept stage of process design. Moving forward, fundamental knowledge of high pressure systems as well as thorough understanding of theory and phase behavior will be essential for determining feasibility and operating conditions necessary (*29*, *87*).

Operating parameters include pressure, temperature, substrate ratios, and concentrations, all of which are determined by the system effectiveness and phase behavior. Further understanding of the extraction/process mechanism - whether the process is solubility or diffusion limited (or both) - is also necessary for the choice, design, and sizing of equipment (*30*). Other considerations include choice of packing materials, distributors, and catalysts.

The use of catalysts, as illustrated earlier, is a highly promising and pertinent research area that couples elegantly with compressed CO_2. Catalysts have the potential to improve atom economy and energy efficiency in biomass conversions (*85*) but will not prove effective unless these often valuable materials are recovered and re-used. For reaction systems, experimental optimization is necessary not only to obtain appropriate operating conditions but also to understand and optimize the trade-offs associated with selectivity versus yield (*118*).

Current utilization of CO_2 as a solvent for biological feedstocks is limited to extractions of high-value products. Preliminary design for extraction from solids generally utilizes at least three extraction vessels to attain a semi-continuous batch process. Typically, these extractions are operated with counter-current solvent flow (*23*). Further gains in efficiency may be gained by moving from batch-wise to continuous configurations, particularly for reaction systems (*27, 90*). Combining multiple unit processes using extractions with and without co-solvents as well as reactions can also reduce the costs and negative impacts associated with both CO_2 and added chemicals (*26*). While the integration of multiple processes has not yet been adopted in industry, there is great potential to couple multiple extractions/fractionations and reactions in CO_2 or with other solvents (including other supercritical fluids) that are generally regarded as safe (*23*).

Other process considerations include the design of solvent and product separation and recycle systems. While depressurization is an effective way to obtain pure product, practically these pressure swings are not efficient and alternative product recovery methods should be used such as liquid-liquid extraction (*6*), membrane separations (*26*), or temperature induced product precipitation (*119*). These separation systems are integral for the economic utilization of pressurized CO_2. Although compressing CO_2 can be expensive, costs may be recovered by that fact that a number of downstream separation processes (i.e. distillation) can be eliminated. Also negative externalities, such as those associated with the environmental ramifications of utilizing hazardous solvent, are mediated. While these rules are useful guidelines, more comprehensive economic and environmental impact assessments are essential to better assess process viability (*5*).

Economic and Environmental Impact

Few economic studies and life cycle assessments (LCA) have been published to evaluate the viability of compressed CO_2 processing. Lisboa et al. (*100*) performed an economic analysis of enzyme catalyzed biodiesel production and found that the major costs of processing were actually attributed to the high enzyme price. The compressor and reactor comprised a large part of the capital expenses. Electricity, mainly associated with heating and compression, accounted for less than a quarter of direct-costs. Decreasing the estimated catalyst price from 800€ to 8€ per kg more than halved the estimated cost of biodiesel. Use of alternative heterogeneous catalysts may help to realize these cost reductions. In order to perform analyses such as these, several aspects must be considered in the design including the influence on mass transfer on column sizing – both in terms of HETP (height equivalent of a theoretical plate) and diameter – as well as how to recover products and regenerate unspent methanol and catalyst (*69*).

In addition to economic viability, environmental impacts over the production life cycle must be assessed. These impacts include energy use, global warming potential, acidification potential, water use, eutrophication, etc. An LCA by Brentner et al. (*96*) of the algae to biodiesel production process found that, from an energy perspective, $scCO_2$ extraction was favorable over extraction with conventional solvents. For both extraction and conversions the process requiring the least energy input was found to be supercritical methanol direct-transesterification as it can theoretically combine the two process steps. It should be noted that the assessment used assumed values based on laboratory results and that the energy requirements for the production of algal biodiesel still provided a net negative energy ratio over the entire life cycle. Potential processing improvements would include lowering the energy costs associated with heating by using CO_2 mediated transesterification (*32*). This example illustrates the benefit of combining multiple unit processes as well as moderating process operating conditions.

In a comprehensive economic and environmental assessment, Fang et al. (*24*) compared the catalytic hydroformylation of olefins with CXL to conventional processes. The CXL process had lower impact from both economic and environmental perspectives. The overall cost of the CXL-mediated hydroformylation was lower in both capital and operating cost, with major reductions in reactor design and operation due to the decreased retention times (higher turnover frequencies) and mild operating conditions. The CXL process had elevated catalyst and energy costs for solvent recycle, highlighting where further optimization would make the greatest impact. All nine environmental impacts categories evaluated showed lower repercussions with the use of CXL in all process stages. The major impacts arose from discharges due to the catalyst regeneration step. The analysis assumed almost complete catalyst recycle rates, which, given appropriate process design, was considered feasible by the authors. Reactions in CXL are a highly promising though case-by-case impact assessments must be made in order to realize the best alternative.

Conclusion

While current usage of CO_2 in a biorefinery setting is limited and its effectiveness is yet to be proven, much promise surrounds the use of this alternative solvent for effective and environmentally benign production of a variety of products. Projects are aimed at the adoption of CO_2 include the use of CO_2 for the initial processing and wax extraction from wheat straw in the

UK (*118*). Another scheme involves the use of $scCO_2$ for the production of defatted-rice bran with future plans to utilize multiple supercritical fluids for the further processing of the extracted lipids (*23*).

As summarized in Figure 3, this chapter focused on the potential usage of compressed CO_2 as a solvent in an advanced biorefinery setting for the co-production of fuels and chemicals and subsequent minimization of process inputs. Though there are many promising applications using compressed CO_2 as a solvent for improved processing, engineering limitations exist. Thoughtful consideration of the advantages and limitations of integrated CO_2 usage in the biorefinery setting can aid to avoid potential pitfalls associated with using this alternative solvent. Sustainable design necessitates the use of green solvents such as CO_2 that have low hazard, are renewable sourced, and/or are easily regenerable to cover the range of applications required by industrial processing (*21*). Informed solvent choice will be key for the design of multiple unit process systems utilizing sustainable practices and technologies in a biorefinery (*5*).

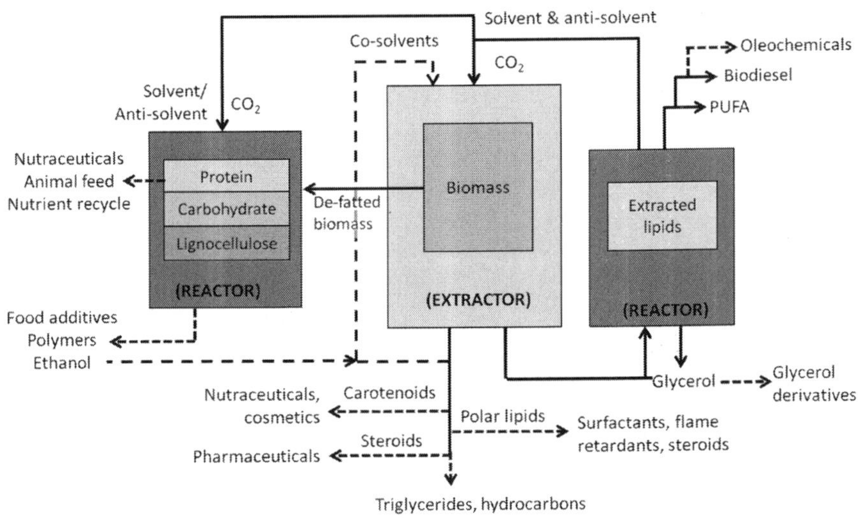

Figure 3. Biorefinery applications for CO_2 as a solvent

Acknowledgments

Much thanks to Dr. Julie Zimmerman for her support, guidance, and knowledge as well as the Center for Green Chemistry and Engineering at Yale. Further thanks to Dr. Ngai Yin Yip for his editorial insights.

References

1. Sonnenberg, A.; Baars, J.; Hendrickx, P. IEA Bioenergy Task 42 Biorefinery. http://www.iea-bioenergy.task42-biorefineries.com (accessed Oct. 10, 2014).

2. Anastas, P. T.; Warner, J. C. *Green chemistry: theory and practice*; Oxford University Press: New York, 2000.
3. Anastas, P. T.; Zimmerman, J. B. *Environ. Sci. Technol.* **2003**, *37*, 94A–101A.
4. Sheldon, R. A. *Green Chem.* **2005**, *7*, 267–278.
5. Capello, C.; Fischer, U.; Hungerbühler, K. *Green Chem.* **2007**, *9*, 927–934.
6. Beckman, E. J. *J. Supercrit. Fluids* **2004**, *28*, 121–191.
7. Demirbaş, A. *Energy Convers. Manage.* **2001**, *42*, 1357–1378.
8. Demirbas, M. F. *Appl. Energy* **2009**, *86*, S151–S161.
9. Cherubini, F. *Energy Convers. Manage.* **2010**, *51*, 1412–1421.
10. Singh, J.; Gu, S. *Renewable Sustainable Energy Rev.* **2010**, *14*, 2596–2610.
11. Soh, L.; Montazeri, M.; Haznedaroglu, B. Z.; Kelly, C.; Peccia, J.; Eckelman, M. J.; Zimmerman, J. B. *Bioresour. Technol.* **2014**, *151*, 19–27.
12. Naik, S. N.; Goud, V. V.; Rout, P. K.; Dalai, A. K. *Renewable Sustainable Energy Rev.* **2010**, *14*, 578–597.
13. Octave, S.; Thomas, D. *Biochimie* **2009**, *91*, 659–664.
14. Wang, B.; Gebreslassie, B. H.; You, F. *Comput. Chem. Eng.* **2013**, *52*, 55–76.
15. McKendry, P. *Bioresource Technol.* **2002**, *83*, 47–54.
16. Haveren, J. v.; Scott, E. L.; Sanders, J. *Biofuels, Bioprod. Biorefin.* **2008**, *2*, 41–57.
17. Ragauskas, A. J.; Williams, C. K.; Davison, B. H.; Britovsek, G.; Cairney, J.; Eckert, C. A.; Frederick, W. J.; Hallett, J. P.; Leak, D. J.; Liotta, C. L. *Science* **2006**, *311*, 484–489.
18. Leitner, W. *Acc. Chem. Res.* **2002**, *35*, 746–756.
19. Baiker, A. *Chem. Rev.* **1999**, *99*, 453–474.
20. Rana, M. S.; Samano, V.; Ancheyta, J.; Diaz, J. *Fuel* **2007**, *86*, 1216–1231.
21. Payne, S. M.; Kerton, F. M. *Green Chem.* **2010**, *12*, 1648–1653.
22. Jessop, P. G.; Subramaniam, B. *Chem. Rev.* **2007**, *107*, 2666–2694.
23. King, J. W. *Annu. Rev. Food Sci. Technol.* **2014**, *5*, 215–238.
24. Fang, J.; Jin, H.; Ruddy, T.; Pennybaker, K.; Fahey, D.; Subramaniam, B. *Ind. Eng. Chem. Res.* **2007**, *46*, 8687–8692.
25. Stevens, J. G.; Gomez, P.; Bourne, R. A.; Drage, T. C.; George, M. W.; Poliakoff, M. *Green Chem.* **2011**, *13*, 2727–2733.
26. King, J. W.; Srinivas, K. *J. Supercrit. Fluids* **2009**, *47*, 598–610.
27. Marriott, R.; Sin, E. In *The Role of Green Chemistry in Biomass Processing and Conversion*; Xie, H., Gathergood, N., Eds.; John Wiley & Sons, Inc.: Hoboken, NJ, 2012; pp 181−204.
28. Agency for Toxic Substances and Disease Registry (ATSDR), Toxicological profile for n-Hexane. U.S. Deptartment of Health and Human Services, 1999. http://www.atsdr.cdc.gov/toxprofiles/tp113.pdf (accessed Oct. 10, 2014).
29. Licence, P.; Ke, J.; Sokolova, M.; Ross, S. K.; Poliakoff, M. *Green Chem.* **2003**, *5*, 99–104.
30. Leboeuf, F.; Deschamps, F. In *Current Trends of Supercritical Fluid Technology in Pharmaceutical, Nutraceutical and Food Processing Industries*; Duarte, A. R., Duarte, C. M., Eds.; Bentham Science Publishers: Bussum, NL, 2010; pp 97−115.

31. Jessop, P. G. *Green Chem.* **2011**, *13*, 1391–1398.
32. Soh, L.; Curry, J.; Beckman, E. J.; Zimmerman, J. B. *ACS Sustainable Chem. Eng.* **2013**, *2*, 387–395.
33. Nekså, P. *Int. J. Refrig.* **2002**, *25*, 421–427.
34. Taylor, L. T. *Anal. Chem.* **2010**, *82*, 4925–4935.
35. Mantell, C.; Casas, L.; Rodríguez, M.; de la Ossa, E. M. In *Separation and Purification Technologies in Biorefineries;* Ramaswamy, S., Huang, H. J., Ramarao, B. V., Eds.; John Wiley & Sons, Inc.: Hoboken, NJ, 2013, pp 79–100.
36. del Valle, J.; de la Fuente, J.; Uquiche, E.; Zetzl, C.; Brunner, G. In *Separation and Purification Technologies in Biorefineries*; Ramaswamy, S., Huang, H. J., Ramarao, B. V., Ed.; John Wiley & Sons, Inc.: Hoboken, NJ, 2011; pp 393–470.
37. Hansen, C. M. *The three dimensional solubility parameter*; Danish Technical: Copenhagen, Denmark, 1967; Vol. 14.
38. Taylor, L. T. *Supercritical fluid extraction*; John Wiley & Sons, Inc.: New York, 1996.
39. King, J. W. *Grasas y Aceites* **2002**, *53*, 8–21.
40. Laurie, L. W. In *Hansen Solubility Parameters*; Hansen, C. M., Ed.; Taylor & Francis Group: Boca Raton, FL, 2007; pp 177–201.
41. Soh, L.; Zimmerman, J. *Green Chem.* **2011**, *13*, 1422–1429.
42. Gracia, I.; Rodríguez, J. F.; de Lucas, A.; Fernandez-Ronco, M. P.; García, M. T. *J. Supercrit. Fluids* **2011**, *59*, 72–77.
43. Mendes, R. L.; Fernandes, H. L.; Coelho, J.; Reis, E. C.; Cabral, J.; Novais, J. M.; Palavra, A. F. *Food Chem.* **1995**, *53*, 99–103.
44. Stortz, T. A.; Marangoni, A. G. *Green Chem.* **2014**, *16*, 3064–3070.
45. Favati, F.; King, J. W.; Friedrich, J. P.; Eskins, K. *J. Food Sci.* **1988**, *53*, 1532–1536.
46. Liau, B.-C.; Shen, C.-T.; Liang, F.-P.; Hong, S.-E.; Hsu, S.-L.; Jong, T.-T.; Chang, C.-M. J. *J. Supercrit. Fluids* **2010**, *55*, 169–175.
47. Mendes, R. L.; Fernandes, H. L.; Coelho, J. A.; Cabral, J. M.; Palavra, A. M.; Novais, J. M. *J. Appl. Phycol.* **1994**, *6*, 289–293.
48. Dunford, N. T.; Temelli, F.; LeBlanc, E. *J. Food Sci.* **1997**, *62*, 289–294.
49. Sprague, M.; Bendiksen, E. Å.; Dick, J. R.; Strachan, F.; Pratoomyot, J.; Berntssen, M. H.; Tocher, D. R.; Bell, J. G. *Br. J. Nutr.* **2010**, *103*, 1442–1451.
50. Min, J.; Li, S.; Hao, J.; Liu, N. *J. Chem. Eng. Data* **2010**, *55*, 3755–3758.
51. Md Zaidul, I.; Nik Norulaini, N.; Mohd Omar, A. *J. Sci. Food Agric.* **2006**, *86*, 1138–1145.
52. Friedrich, J.; List, G.; Heakin, A. *J. Am. Oil Chem. Soc.* **1982**, *59*, 288–292.
53. Vuorela, H.; Holm, Y.; Hiltunen, R.; Harvala, T.; Laitinen, A. *Flavour Fragrance J.* **1990**, *5*, 81–84.
54. Silva, C.; Moura, F.; Mendes, M.; Pessoa, F. *Braz. J. Chem. Eng.* **2011**, *28*, 343–350.
55. Reverchon, E.; Taddeo, R.; Porta, G. *J. Supercrit. Fluids* **1995**, *8*, 302–309.
56. Del Valle, J.; Jiménez, M.; De la Fuente, J. *J. Supercrit. Fluids* **2003**, *25*, 33–44.

57. Chassagnez-Mendez, A. L.; Machado, N. T.; Araujo, M. E.; Maia, J.; Meireles, M. A. A. *Ind. Eng. Chem. Res.* **2000**, *39*, 4729–4733.
58. Yesil-Celiktas, O.; Otto, F.; Parlar, H. *Eur. Food Res. Technol.* **2009**, *229*, 671–677.
59. Conde, E.; Díaz-Reinoso, B.; Moure, A.; Hemming, J.; Willför, S.; Domínguez, H.; Parajó, J. Extraction of phenolic and lipophilic compounds from Pinus pinaster knots and stemwood by supercritical CO_2. In *Proceedings of III Iberoamerican Conference on Supercritical Fluids*, Cartagena de Indias, Columbia, April 1−5, 2013; pp 1−7.
60. Pimenta, A.; Vital, B.; Bayona, J.; Alzaga, R. *Fuel* **1998**, *77*, 1133–1139.
61. Baysal, T.; Ersus, S.; Starmans, D. *J. Agric. Food Chem.* **2000**, *48*, 5507–5511.
62. Raventós, M.; Duarte, S.; Alarcón, R. *Food Sci. Technol. Int.* **2002**, *8*, 269–284.
63. Charles, M. H. In *Hansen Solubility Parameters*; Hansen, C. M., Ed.; Taylor & Francis Group: Boca Raton, FL, 2007; pp 27−43.
64. Herrero, M.; Cifuentes, A.; Ibanez, E. *Food Chem.* **2006**, *98*, 136–148.
65. Sahena, F.; Zaidul, I.; Jinap, S.; Karim, A.; Abbas, K.; Norulaini, N.; Omar, A. *J. Food Eng.* **2009**, *95*, 240–253.
66. Nilsson, W. B.; Gauglitz, E. J., Jr.; Hudson, J. K. *J. Am. Oil Chem. Soc.* **1989**, *66*, 1596–1600.
67. Perretti, G.; Motori, A.; Bravi, E.; Favati, F.; Montanari, L.; Fantozzi, P. *J. Supercrit. Fluids* **2007**, *40*, 349–353.
68. Corrêa, A. P. A.; Peixoto, C. A.; Gonçalves, L. A. G.; Cabral, F. A. *J. Food Eng.* **2008**, *88*, 381–387.
69. Brunner, G. *Eur. J. Lipid Sci. Technol.* **2000**, *102*, 240–245.
70. Corazza, M.; Ndiaye, P.; Escorsin, A. *Proceedings of the Workshop on Supercritical Fluids and Energy*, Campinas, Brazil, Dec. 8−11, 2013; pp 127−128
71. Gupta, R. B.; Shim, J. J. *Solubility in Supercritical Carbon Dioxide*; Taylor & Francis Group, LLC: Boca Raton, FL, 2007.
72. Pinto, L. F.; da Silva, D. I. S.; Rosa da Silva, F.; Ramos, L. P.; Ndiaye, P. M.; Corazza, M. L. *J. Chem. Thermodyn.* **2012**, *44*, 57–65.
73. Eckert, C.; Liotta, C.; Ragauskas, A.; Hallett, J.; Kitchens, C.; Hill, E.; Draucker, L. *Green Chem.* **2007**, *9*, 545–548.
74. Mukhopadhyay, M.; Singh, S. *J. Supercrit. Fluids* **2004**, *30*, 201–211.
75. Haimer, E.; Wendland, M.; Potthast, A.; Henniges, U.; Rosenau, T.; Liebner, F. *J. Supercrit. Fluids* **2010**, *53*, 121–130.
76. Martín, A.; Mattea, F.; Gutiérrez, L.; Miguel, F.; Cocero, M. J. *J. Supercrit. Fluids* **2007**, *41*, 138–147.
77. Miguel, F.; Martin, A.; Mattea, F.; Cocero, M. *Chem. Eng. Process.* **2008**, *47*, 1594–1602.
78. Miguel, F.; Martin, A.; Gamse, T.; Cocero, M. *J. Supercrit. Fluids* **2006**, *36*, 225–235.
79. Hao, J.; Whitaker, M. J.; Wong, B.; Serhatkulu, G.; Shakesheff, K. M.; Howdle, S. M. *J. Pharm. Sci.* **2004**, *93*, 1083–1090.

80. Jessop, P. G.; Mercer, S. M.; Heldebrant, D. J. *Energy Environ. Sci.* **2012**, *5*, 7240–7253.
81. Phan, L.; Brown, H.; White, J.; Hodgson, A.; Jessop, P. G. *Green Chem.* **2009**, *11*, 53–59.
82. Samorì, C.; Torri, C.; Samorì, G.; Fabbri, D.; Galletti, P.; Guerrini, F.; Pistocchi, R.; Tagliavini, E. *Bioresour. Technol.* **2010**, *101*, 3274–3279.
83. Bozell, J. J.; Petersen, G. R. *Green Chem.* **2010**, *12*, 539–554.
84. Scurto, A. M.; Hutchenson, K.; Subramaniam, B. In *Gas-Expanded Liquids and Near-Critical Media*; Hutchenson, K. W., Scurto, A. M., Subramaniam, B., Ed.; ACS Symnposium Series; American Chemical Society: Washington, DC, 2009; Vol. 1006, pp 3−37.
85. Subramaniam, B. In *Innovations in Green Chemistry and Green Engineering*; Anastas, P. T., Zimmerman J. B., Eds.; Springer: New York, 2013; pp 5−36.
86. Savage, P. E.; Gopalan, S.; Mizan, T. I.; Martino, C. J.; Brock, E. E. *AIChE J.* **1995**, *41*, 1723–1778.
87. Beckman, E. J. *J. Supercrit. Fluid* **s2004**, *28*, 121–191.
88. Pollet, P.; Hart, R. J.; Eckert, C. A.; Liotta, C. L. *Acc. Chem. Res.* **2010**, *43*, 1237–1245.
89. Fujita, S.-I.; Akihara, S.; Fujisawa, S.; Arai, M. *J. Mol. Catal. A: Chem.* **2007**, *268*, 244–250.
90. Han, X.; Poliakoff, M. *Chem. Soc. Rev.* **2012**, *41*, 1428–1436.
91. Seki, T.; Grunwaldt, J.-D.; Baiker, A. *Ind. Eng. Chem. Res.* **2008**, *47*, 4561–4585.
92. Meher, L. C.; Vidya Sagar, D.; Naik, S. N. *Renewable Sustainable Energy Rev.* **2006**, *10*, 248–268.
93. Pinnarat, T.; Savage, P. E. *Ind. Eng. Chem. Res.* **2008**, *47*, 6801–6808.
94. Sawangkeaw, R.; Bunyakiat, K.; Ngamprasertsith, S. *J. Supercrit. Fluids* **2010**, *55*, 1–13.
95. Hegel, P.; Andreatta, A.; Pereda, S.; Bottini, S.; Brignole, E. A. *Fluid Phase Equilibr.* **2008**, *266*, 31–37.
96. Brentner, L. B.; Eckelman, M. J.; Zimmerman, J. B. *Environ. Sci. Technol.* **2011**, *45*, 7060–7067.
97. Bertoldi, C.; da Silva, C.; Bernardon, J. P.; Corazza, M. L.; Filho, L. C.; Oliveira, J. V.; Corazza, F. C. *Energy Fuels* **2009**, *23*, 5165–5172.
98. Han, H.; Cao, W.; Zhang, J. *Process Biochem.* **2005**, *40*, 3148–3151.
99. Maçaira, J.; Santana, A.; Recasens, F.; Angeles Larrayoz, M. *Fuel* **2011**, *90*, 2280–2288.
100. Lisboa, P.; Rodrigues, A. R.; Martín, J. L.; Simões, P.; Barreiros, S.; Paiva, A. *J. Supercrit. Fluids* **2014**, *85*, 31–40.
101. Galia, A.; Scialdone, O.; Tortorici, E. *J. Supercrit. Fluids* **2011**, *56*, 186–193.
102. Gao, L.; Wu, W.; Hou, Z.; Jiang, T.; Han, B.; Liu, J.; Liu, Z. *J. Phys. Chem. B* **2003**, *107*, 13093–13099.
103. Hou, Z.; Han, B.; Zhang, X.; Zhang, H.; Liu, Z. *J. Phys. Chem. B* **2001**, *105*, 4510–4513.
104. Vermue, M.; Tramper, J.; De Jong, J.; Oostrom, W. *Enzyme Microb. Technol.* **1992**, *14*, 649–655.

105. Knez, Ž.; Laudani, C. G.; Habulin, M.; Reverchon, E. *Biotechnol. Bioeng.* **2007**, *97*, 1366–1375.
106. Wyatt, V.; Haas, M. *J. Am. Oil Chem. Soc.* **2009**, *86*, 1009–1016.
107. Moreau, C.; Belgacem, M.; Gandini, A. *Top. Catal.* **2004**, *27*, 11–30.
108. Sangarunlert, W.; Piumsomboon, P.; Ngamprasertsith, S. *Korean J. Chem. Eng.* **2007**, *24*, 936–941.
109. Gairola, K.; Smirnova, I. *Bioresour. Technol.* **2012**, *123*, 592–598.
110. Johnson, D. T.; Taconi, K. A. *Environ. Prog.* **2007**, *26*, 338–348.
111. Conceiçao, L.; Bogel-Łukasik, R.; Bogel-Łukasik, E. *Green Chem.* **2012**, *14*, 673–681.
112. Lopes, J. M.; Petrovski, Z.; Bogel-Łukasik, R.; Bogel-Łukasik, E. *Green Chem.* **2011**, *13*, 2013–2016.
113. Charles, R. E.; Ponrasu, T.; Sankar, K. U.; Divakar, S. *Appl. Biochem. Biotechnol.* **2009**, *159*, 199–207.
114. Pillai, U. R.; Sahle-Demessie, E. *Chem. Commun.* **2002**, *5*, 422–423.
115. Pini, R.; Storti, G.; Mazzotti, M.; Tai, H.; Shakesheff, K. M.; Howdle, S. M. *J. Polym. Sci., Polym. Phys.* **2008**, *46*, 483–496.
116. Bourne, R. A.; Stevens, J. G.; Ke, J.; Poliakoff, M. *Chem. Commun.* **2007**, *44*, 4632–4634.
117. Beckman, E. J. *Environ. Sci. Technol.* **2002**, *36*, 347A–353A.
118. Deswarte, F. E. I.; Clark, J. H.; Wilson, A. J.; Hardy, J. J. E.; Marriott, R.; Chahal, S. P.; Jackson, C.; Heslop, G.; Birkett, M.; Bruce, T. J.; Whiteley, G. *Biofuels, Bioprod. Biorefin.* **2007**, *1*, 245–254.
119. Brennecke, J. F.; Eckert, C. A. *AIChE J.* **1989**, *35*, 1409–1427.

Chapter 3

Deep Eutectic Solvents: Environmentally Friendly Media for Metal-Catalyzed Organic Reactions

Joaquín García-Álvarez*

Laboratorio de Compuestos Organometálicos y Catálisis (Unidad Asociada al CSIC), Red ORFEO-CINQA - Centro de Innovación en Química Avanzada, Departamento de Química Orgánica e Inorgánica (IUQOEM), Facultad de Química, Universidad de Oviedo, Oviedo, Principado de Asturias, E-33071, Spain
***E-mail: garciajoaquin@uniovi.es**

Deep Eutectic Solvents (*DESs*), also known in the literature as *Deep Eutectic Ionic Liquids* (*DEILs*), *Low Melting Mixtures* (*LMMs*) or *Low Transition Temperature Mixtures* (*LTTMs*), are defined as combinations of two or three safe and inexpensive components which are able to establish hydrogen bond interactions with each other to form an eutectic mixture, that is a liquid at the desired temperature, with a melting point lower than that of the each individual component. Due to the advantages of *DESs* (*i.e.*, low cost of components, easy to prepare, tuneable physicochemical properties, negligible vapour pressure, biorenewability and biodegradability) when compared with traditional ionic liquids (*ILs*), these eutectic mixtures have been used as *Green* and biorenewable solvents in different metal-catalyzed organic reactions like: *i*) dehydration of carbohydrates to 5-hydroxymethyl-furfural (*HMF*); *ii*) production of furfural (furan-2-carbaldehyde) from Xylose or Xylan; *iii*) cross-coupling reactions; *iv*) *Click Chemistry* processes; *v*) isomerization procedures; and *vi*) hydrogenation reactions.

Introduction

Due to both environmental and economic issues, *Chemistry* is driven to reduce waste and re-use materials in order to meet the standards of the *12 Principles of Green Chemistry* (*1*). These *Principles* have become a widely accepted set of criteria for the assessment of the *Greenness* of a given chemical transformation (*2–4*). In this sense, some of the crucial points in realizing a *Catalytic Green Chemical Process* (*5*) involve: *i*) the design of a chemical process with energy efficiency (if possible reaction should be conducted at ambient temperature); and *ii*) atom economy should be maximized (*6–10*). It is also desirable that the catalytic reactions proceed in high yields, as selective and specific processes, under aerobic conditions, in the absence of co-catalyst and the isolation of the resulting products should be straightforward.

Together with these important points in realizing a *Catalytic Green Chemical Process*, the choice of a safe, non-toxic, biorenewable and cheap solvent is a crucial goal in metal-catalyzed organic reactions (from a *Green Chemistry* point of view) (*1–4*). In this sense, one of the largest areas of consumption of petroleum based chemicals in a conventional metal-catalyzed organic transformation is the solvent used as a reaction medium (*i.e.*, solvents account for 80-90% of mass utilization in a typical pharmaceutical/fine chemical operational process) (*11*, *12*). In top of this, conventional volatile organic solvents (*VOCs*), commonly used as reaction media in metal-mediated organic procedures, can cause well-established environmental problems due to their: *i*) high toxicity; *ii*) non-biodegradability; *iii*) accumulation in the atmosphere (low boiling points); and *iv*) flammability (*13*, *14*). Despite these intrinsic drawbacks of *VOCs*, most of the metal-catalyzed organic transformations are performed in solution (using these hazardous solvents) to: *i*) achieve homogeneity of reagents and products; *ii*) efficiently control the heat flow; *iii*) ensure rapid and safe conversion; and *iv*) avoid undesired side products by dilution, thus enhancing the reaction rate (*15*).

To overcome the aforementioned drawbacks of traditional *VOCs* and still benefit from the *solvent effect*, remarkable research efforts have been focused on the replacement of traditional organic solvents by *Green Solvents* (*16*, *17*). Within this regard, ionic liquids (*ILs*, generally composed of an imidazolium or pyridonium cation in a stoichiometric combination with a weakly coordinating anion) have been intensively investigated as an alternative reaction media due to their: *i*) non-flammability; *ii*) thermal stability; *iii*) low vapour pressure; and *iv*) catalytic effect in many chemical transformations (*18–24*). However, the current consensus is that *ILs* cannot be generalized as either *Green* or toxic, but their environmental impact strongly depends on the type of cation and anion used to produce the *IL* (*25–27*). In this sense, recent pioneering work from several synthetic laboratories around the world has recognized the potential of *Deep Eutectic Solvents* (*DESs*, a related class of *ILs*) as superior *Green* and biorenewable solvents (*28–38*). Thus, this chapter is intended to cover the progress made in the application of *DESs* as environmentally-friendly reaction media in a variety of metal-catalyzed organic reaction, fundamental processes in modern organic synthesis.

General Description of Deep Eutectic Solvents

Deep Eutectic Solvents (*DESs*), also known in the literature as *Deep Eutectic Ionic Liquids* (*DEILs*), *Low Melting Mixtures* (*LMMs*) or *Low Transition Temperature Mixtures* (*LTTMs*), can be defined as molecular complexes typically formed between quaternary ammonium salts and hydrogen bond donors. The charge delocalization that occurs through hydrogen bonding between the halide anion and the hydrogen-donor moiety is responsible for the decrease in the melting point of the mixture relative to the melting points of the individual components (see Figure 1). The concept "*Deep Eutectic Solvent*" was firstly coined by Abbott *et al.* (*39*) to describe the formation of a liquid eutectic mixture (melting point 12 °C) starting from two solid materials with high melting points: *i*) choline chloride [*ChCl,* (2-hydroxyethyl)trimethylammoniun chloride, melting point 302 °C] (*40–42*); and *ii*) urea (melting point 133 °C) in a ratio 1:2 (1*ChCl*/2*Urea*) (*39, 43*). However, it is important to note that strictly speaking *Deep Eutectic Mixtures* were used before to: *i*) lower the melting point of salts by forming metal-halide complexes (*44, 45*); and *ii*) reduce the lattice energy of halide salts by the addition of urea (*46*).

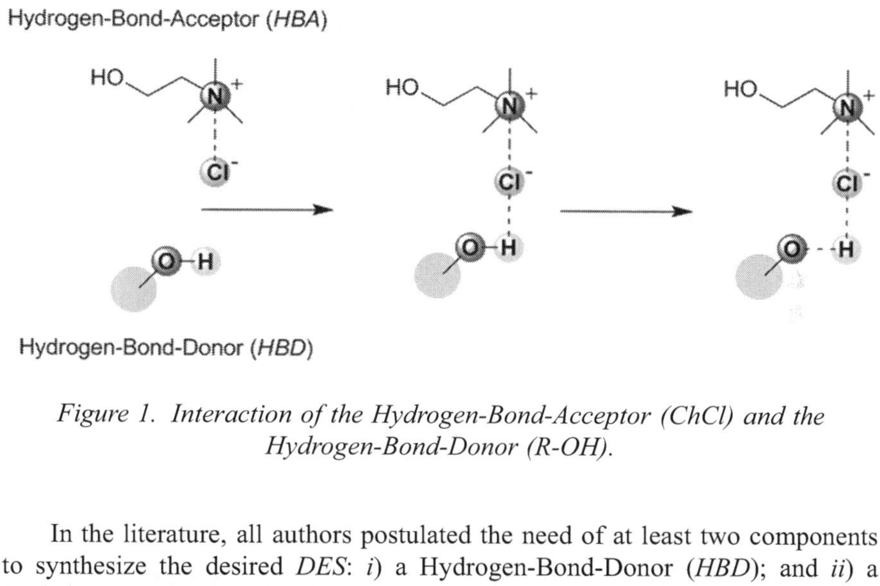

Figure 1. Interaction of the Hydrogen-Bond-Acceptor (ChCl) and the Hydrogen-Bond-Donor (R-OH).

In the literature, all authors postulated the need of at least two components to synthesize the desired *DES*: *i*) a Hydrogen-Bond-Donor (*HBD*); and *ii*) a Hydrogen-Bond-Acceptor (*HBA*) (*28–38*). Taking into account the large pool of inexpensive, safe, renewable and biodegradable components which are capable of forming eutectic mixtures, the possibility of creating new *Green* and biorenewable *Deep Eutectic Solvents* is enormous (see Figure 2). Examples of environmentally-friendly and low cost eutectic mixtures include *DESs* based on: *i*) *ChCl* described by Abbott and co-workers (*39, 47, 48*); *ii*) low melting eutectic mixtures of sugars, urea and salts first reported by König et *al.* (*49*); *iii*) natural deep eutectic solvents (*NADES*) described by Choi and co-workers (*50, 51*); and *iv*) low transition temperature mixtures (*LTTMs*) reported by Kroon *et al.* (*52*).

At this point, is important to note that, although most of these *DESs* are made from ionic species (*i.e.*, *ChCl*, NH_4Cl or NaCl) *DESs* cannot be considered as traditional *ILs* because: *i*) *DESs* are not entirely composed of ionic species; and *ii*) can also be obtained from non-ionic species. Also, and due to the nature of their components, these *DESs* gather many advantages as compared with traditional *ILs*: *i*) low price; *ii*) chemical inertness with water; *iii*) easy to prepare (since *DESs* are obtained by simply mixing two components, thus over-passing all problems of purification and waste disposal generally encountered with traditional *ILs*); and *iv*) most of them are biodegradable, biocompatible and non-toxic, reinforcing the *Greenness* of these media (*28–38*).

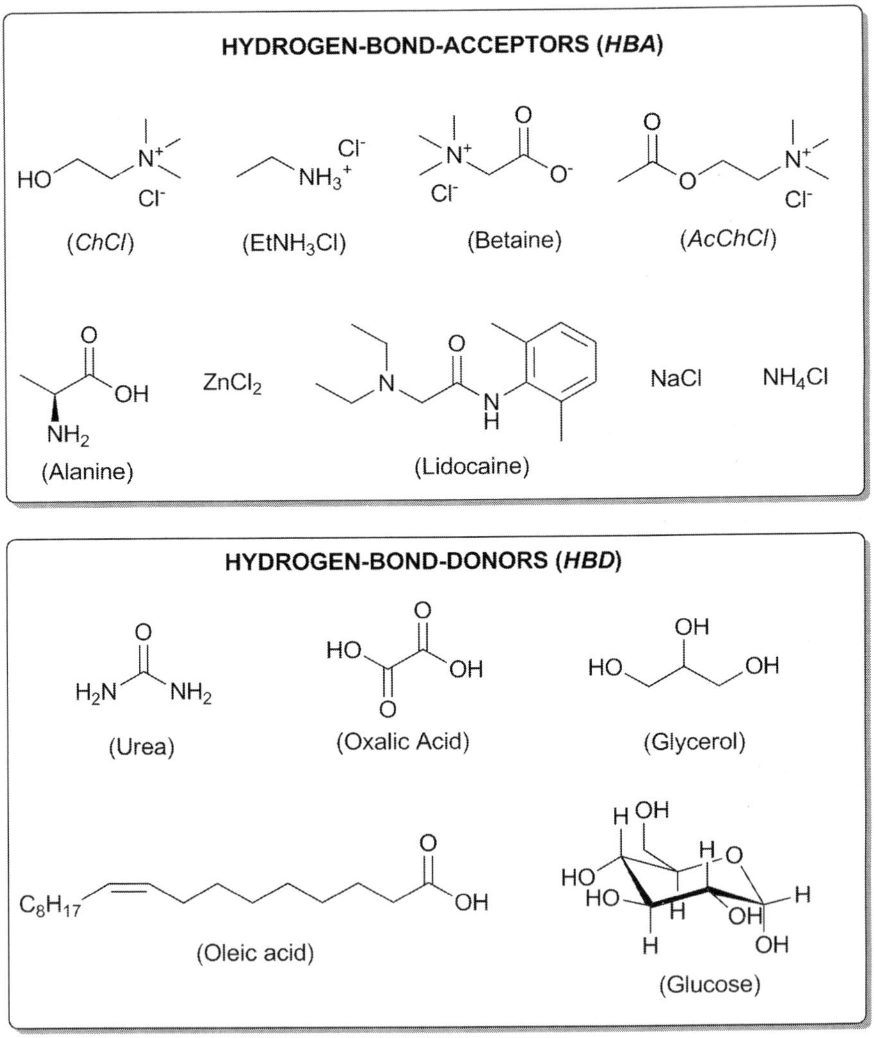

Figure 2. Structure of some Hydrogen-Bond-Acceptors (HBA) and Hydrogen-Bond-Donors (HBD) commonly used for DESs synthesis.

Applications of *DESs* as New *Green* and Biorenewable Reaction Media for Metal-Catalyzed Organic Reactions

Due to the aforementioned advantages of *DESs* (*i.e.*, low cost of components, easy to prepare, tuneable physicochemical properties, negligible vapour pressure, biorenewability and biodegradability), it is not surprising that these eutectic mixtures have found a wide variety of applications in different fields of modern chemistry (*28–38*), including: *i*) organic synthesis; *ii*) biocatalytic reactions; *iii*) dissolution of metal oxides; *iv)* electrodeposition of metals; and *iv*) material chemistry. In this section, our main objective is to cover the recent developments in metal-catalyzed organic reactions using *DESs* as *Green* and biorenewable solvents.

Metal-mediated organic reactions are one of the most frequently encountered methodologies in the synthetic chemist´s toolbox, providing powerful instruments to achieve elevated levels of chemo-, regio-, and stereoselectivity in the manufacture of highly functionalized organic products (*53–55*). In this sense, *DESs* have been used as environmentally-friendly reaction media for several metal-catalyzed organic reactions, like: *i*) dehydration of carbohydrates to 5-hydroxymethyl-furfural (*HMF*); *ii*) production of furfural (furan-2-carbaldehyde) from Xylose or Xylan; *iii*) cross-coupling reactions; *iv*) *Click Chemistry* processes; *v*) isomerization procedures; and *vi*) hydrogenation reactions.

Lewis Acid Catalyzed Dehydration of Carbohydrates to 5-Hydroxymethyl-furfural (HMF)

Efficient conversion of biorenewable carbohydrates (like Glucose, Sucrose or Fructose) to 5-hydroxymethyl-furfural (*HMF*) is a key step for using these saccharides to produce liquid fuels and value added chemicals (*56*). Thus, different metallic salts (*i.e.*, $FeCl_3$, $ZnCl_2$, $CrCl_2$, $Sc(OTf)_3$ or $AlCl_3$) have been employed in the Lewis acid catalyzed dehydration of carbohydrates to 5-hydroxymethyl-furfural (*HMF*), using *DESs* as environmentally-friendly solvents (*57–62*). These eutectic mixtures arc ideal solvents for this transformation as: *i*) most of *DESs* are able to dissolve enormous amounts of saccharides; *ii*) 5-hydroxymethyl-furfural can be easily isolated from the reaction media (by liquid-liquid extraction); *iii*) the eutectic mixture is able to dilute the water generated in the dehydration process (avoiding the rehydration of *HMF* to levulinic and formic acid, see Scheme 1); and *iv*) *DESs* are not toxic, biodegradable and biocompatible (toxic and non-biodegradable *ILs* have been previously used in this dehydration reaction) (*63–67*).

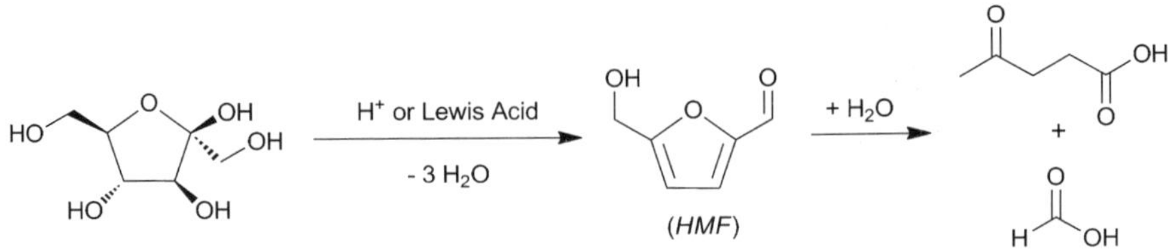

Scheme 1. Protic- or Lewis acid-catalyzed dehydration of Fructose to HMF and undesired rehydration reaction.

In this sense, Han and co-workers were the first to describe the catalytic activity of several Lewis acids ($ZnCl_2$ or $CrCl_3$) in the aforementioned dehydration reaction of Fructose to *HMF* in various *ChCl*-based eutectic mixtures (*68*). However, this catalytic system displayed only moderate conversions and selectivities. Similar fruitless results were latter reported by König and co-workers (*69*), employing in this case, the eutectic mixture 1*ChCl*/2*Urea* as solvent in the presence of catalytic amounts of different metal salts ($ZnCl_2$, $CrCl_2$, $CrCl_3$ and $FeCl_3$). The authors assumed that the low yields observed (< 30%) in this eutectic mixture (1*ChCl*/2*Urea*) can be cause by a side reaction between urea and Fructose. Thus, König *et al.* reported in the same article (*69*), that higher conversion in *HMF* (up to 60%, see Scheme 2) in this metal-catalyzed dehydration reaction can be easily reached just by avoiding the presence of urea in the eutectic mixture and using instead the low melting mixture 6*ChCl*/4Fructose. At this point, it is important to note that, under this catalytic conditions and using their corresponding low melting mixtures with *ChCl* as solvents, other monosaccharides (Glucose), disaccharide (Sucrose) and the polyfructan Inulin can be successfully converted into *HMF*. These precedents clearly indicate that *DESs* will illuminate the way to follow to succeed in the conversion of biorenewable raw materials into value-added chemicals.

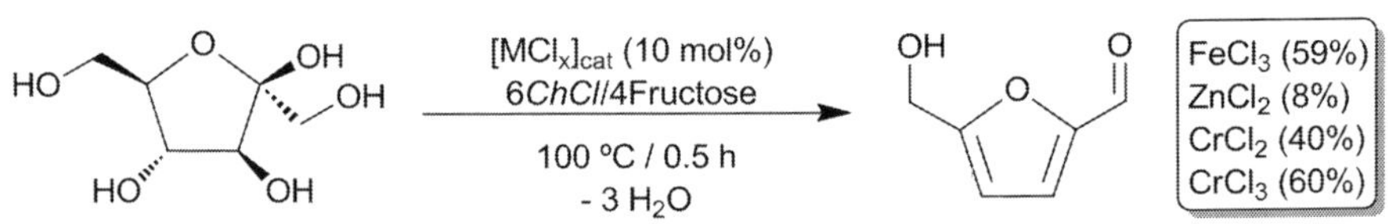

Scheme 2. Metal-catalyzed dehydration of Fructose to HMF in the low melting mixture 6ChCl/4Fructose.

Lewis Acid Catalyzed Production of Furfural (Furan-2-carbaldehyde) from Xylose or Xylan

Furfural, derived from lignocellulosic biomass, is a very versatile and key bio-based platform chemical used for the production of important non-petroleum-derived chemicals and the development of new generation of bioplastics (*70*). Besides, it is a biofuel precursor and also in high demand in industries including petroleum refining, agrochemical and pharmaceutical industries (*71*). In this sense, Yu and co-workers reported the metal-catalyzed furfural production from Xylose or Xylan under mild reaction conditions (50 to 100 ºC) and using the eutectic mixture *ChCl*/oxalic acid as biorenewable solvent (*72*). In this case, the addition of a Lewis acid ($AlCl_3$) to the reaction media allowed to achieved high values of Xylose conversion (92%). However, at this point, it is important to note that this carboxylic acid-containing eutectic mixture is able to catalyze the conversion of Xylose by its own and without the help of metal salts (60%). A similar experimental fact was previously reported independently by Han (*68*) and Jérôme (*73*), in the study of the conversion of both Fructose and Inulin to *HMF* in *DESs* containing carboxylic acids (*i.e., ChCl*/citric acid) (*68*), or betaine hydrochloride (*BHC*) (*73*).

Metal-Catalyzed Cross-Coupling Reactions

Metal-catalyzed cross-coupling reactions between an organic electrophile, typically an organic halide, and an organic nucleophile have developed into a standard component of the synthetic chemistry toolbox for the formation of C-C and C-heteroatom bonds (*74–76*). Palladium-catalyzed reactions, which can be generally carried out under milder conditions and with a wider range of substrates than reactions promoted by other metals, clearly dominate the field. The organic halide can be a sp-, sp^2-, or sp^3- hybridized carbon with any halogen or pseudo-halogen leaving group. In this regard, *DESs* have been already employed as environmentally-friendly solvents in these transformations. Thus, König and co-workers firstly reported the Pd-catalyzed Suzuki coupling of phenyl boronic acid and a variety of aryl bromides in different sugar-based eutectic mixtures (see Table 1) (*77*). In all the cases, quantitative conversion into the biaryl products was observed. Moreover, the addition of water and pentane to the reaction crude allows the easy layer separation of the desired organic product. It is important to note that, hydrodeborination as a competing reaction was not observed under these optimized reaction conditions.

After this seminal work described by König *et al.*, further efforts devoted to the development of other palladium-catalyzed cross-coupling reactions performed in the aforementioned sugar-based *DESs* as biorenewable solvents and employing other nucleophiles were reported. In this sense, the same research group described the Pd-catalyzed: *i*) Stille coupling of aryl bromides and iodides with alkyl and aryl stannates (*78*); *ii)* Heck cross-coupling with *n*-butyl acrylate and several iodobenzenes (*79*); and *iii*) Sonogashira coupling of *in situ* generated acetylide anions and aryl bromides (*79*). Apart from C-C coupling reactions, other palladium catalyzed organic reactions have been successfully performed in different *DESs*. In this sense, Jérôme, Tilloy and co-workers reported in 2014 the palladium-catalyzed cleavage of allylcarbonates (Tsuji-Trost reaction) using new low melting mixtures based on *β*-cyclodextrines and *DMU* (*80*). The cleavage of the desired allylcarbonate was carried out at 90 ºC under nitrogen with diethylamine as an allyl scavenger. Finally, it should be mentioned that in the same article the authors reported the use of these *β*-cyclodextrine-based eutectic mixtures as *Green* and biorenewable solvents for the rhodium-catalyzed hydroformylation of 1-decene (*80*).

Table 1. Pd-catalyzed Suzuki reaction in different sugar-based *DESs*.

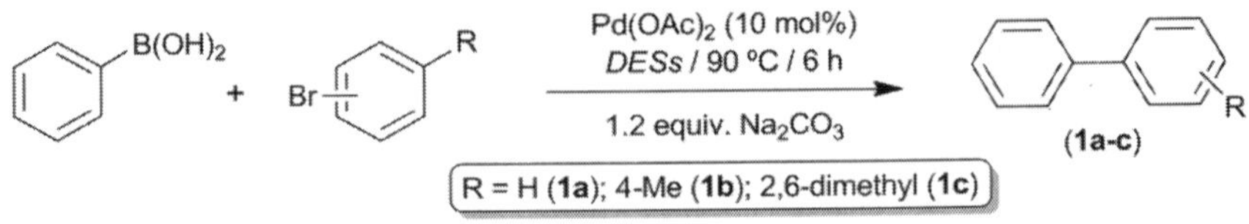

Entry	Carbohydrate	Urea	Salt	Yield (**1a-c**)
1	Fructose (70%)	Urea (20%)	NaCl (10%)	78-82%
2	Maltose (50%)	*DMU*[a] (40%)	NH_4Cl (10%)	81-84%
3	Mannose (30%)	*DMU*[a] (70%)	-	88-99%
4	Lactose (60%)	*DMU*[a] (30%)	NH_4Cl (10%)	86-91%
5	Mannitol (50%)	*DMU*[a] (40%)	NH_4Cl (10%)	96-97%
6	Sorbitol (70%)	*DMU*[a] (20%)	NH_4Cl (10%)	94-98%
7	Glucose (60%)	Urea (30%)	NH_4Cl (10%)	80-86%

[a] *DMU* = *N,N'*-dimethylurea.

Cu(I)-Catalyzed "Click Chemistry" Reactions (CuAAC)

The copper-catalyzed 1,3-dipolar cycloaddition of azides and terminal alkynes (*CuAAC*) reported independently in 2002 by Meldal (*81, 82*) and Sharpless (*83*) (performed in *Green* and biorenewable solvents); fulfils the aforementioned principles of *Green Chemistry* (*1*) as usually proceeds: *i*) at ambient temperature (with energy efficiency); *ii*) under atom economy (*6–10*); *iii*) in the presence of air; and *iv*) in the absence of co-catalyst. Surprisingly, and despite that the application of this Cu(I)-catalyzed organic reaction has grown tremendously (*84–89*), there is only one example in the literature that reports the use of *DESs* in this transformation. In this context, the aforementioned eutectic mixture Sorbitol/Urea/NH_4Cl was employed by König and co-workers in the CuI-catalyzed 1,3-dipolar cycloaddition of benzylazide and phenylacetylene in the absence of base (see Scheme 3) (*79*). Although the three-component version (involving NaN_3, $PhCH_2Br$ and phenylacetylene) also takes place in the studied eutectic mixture, high catalytic loading (5 mol%) and temperatures (85 °C) were required to achieve complete conversion. The rate of the reaction was only slightly enhanced when *L*-Carnitine was employed in the synthesis of the eutectic mixture. According to the authors, this is due to the *in situ* formation of the organometallic complex [CuI·(*L*-carnitine)], avoiding the intrinsic instability of Cu(I) species.

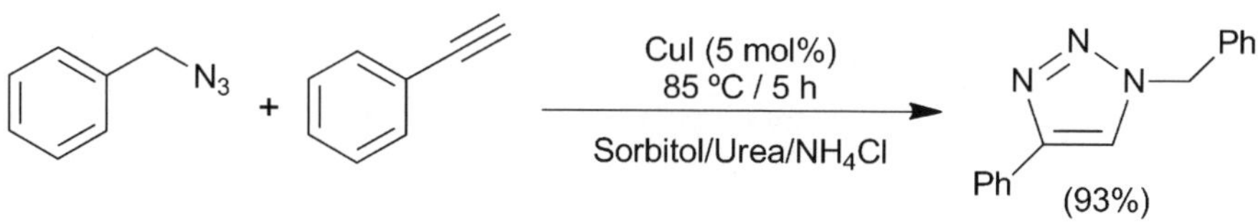

Scheme 3. CuI-catalyzed 1,3-dipolar cycloaddition of azides and terminal alkynes performed in the eutectic mixture Sorbitol/Urea/NH_4Cl.

Metal-Catalyzed Isomerization Reactions

Currently, the search for novel organic transformations which proceed with high levels of efficiency, selectivity and atom economy has emerged as one of the major challenges in synthetic chemistry (*6–10*). Isomerization reactions are typical examples of atom economic processes in which no by-products are generated. Thus, the redox isomerization of readily accessible allylic alcohols catalyzed by transition-metal complexes is a useful and straightforward synthetic route to saturated carbonyl compounds, which are very important raw materials in organic synthesis (*90–97*). In this sense, García-Álvarez *et al.* described the unique example reported to date for the redox isomerization of allylic alcohols in *DESs* catalyzed by the Ru(IV)-complex [Ru(η^3:η^3-$C_{10}H_{16}$)Cl_2(κ^1-*N*-benzimidazole)] (**2**) (*98*). After solvent and reaction conditions screening, the authors found that catalyst **2** was highly efficient for the transformation of 1-octen-3-ol into octan-3-one in the eutectic mixture 1*ChCl*/2*Gly*. The heating at 75 °C using 0.2 mol% of **2** in absence of base conducted to the isolation of octan-3-one in 98% after only 10 min. The modification of the *ChCl*/*Gly* ratio and/or the substitution of one of those two components dramatically decreased the reaction rate. Remarkably, the catalytic activity of **2** in the eutectic mixture 1*ChCl*/2*Gly* is much higher than the reported for ionic liquids (*99*), thus suggesting a relevant role of the hydrogen interactions with the solvent. Catalyst **2** was also efficient for the isomerization of a family of mono- and di-substituted allylic alcohols (see Scheme 4). As previously observed in this metal-catalyzed isomerization (*90–97*), a strong dependence upon the substitution grade on the olefin was observed, thereby higher ruthenium charges (5-10 mol% Ru) and reaction time (8-24 h) were required to promote the isomerization of di-substituted allylic alcohols. Finally, the mentioned conditions allowed for the recycling of the catalytic system during 4 consecutive cycles, with no significant observation of loss of activity during the two first runs.

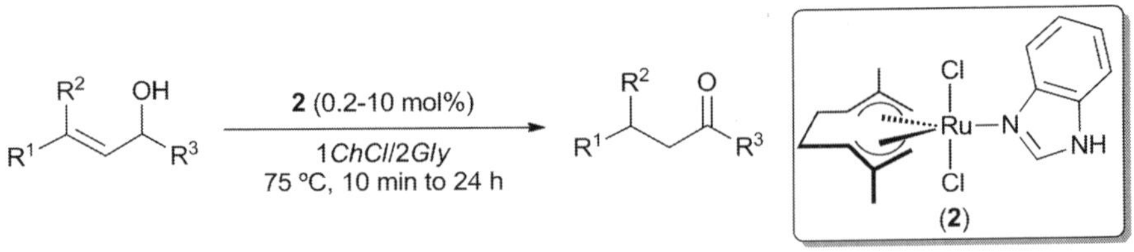

Scheme 4. Ru(IV)-catalyzed redox isomerization of allylic alcohols in the eutectic mixture 1ChCl/2Gly.

The metal-catalyzed rearrangement of aldoximes, a process closely related to the well-known Beckmann rearrangement of ketoximes (*100, 101*), has emerged in recent years as an appealing tool for the generation of primary amides (*102–105*). In this sense, Nagarkar and co-workers reported the synthesis of primary amides in the eutectic mixture 1*ChCl*/2$ZnCl_2$, *via in situ* generation of the corresponding aldoxime intermediates (starting directly from aldehydes) and concomitant isomerization reaction of the aldoxime into the desired amide (see Scheme 5) (*106*). The authors proposed that coordination of the $ZnCl_2$ to the nitrogen atom of the aldoxime is one of the key steps in this isomerization reaction.

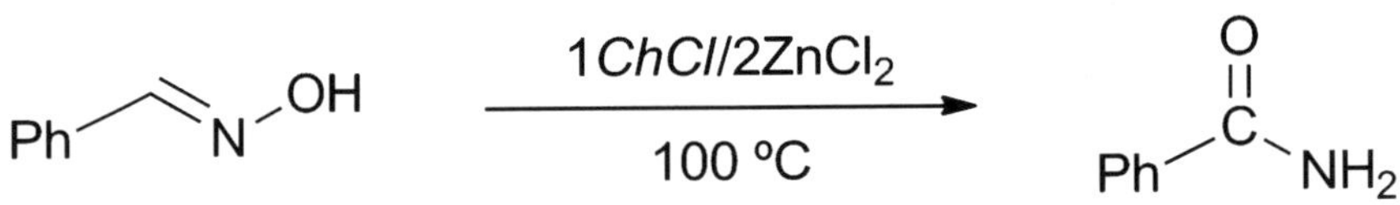

Scheme 5. Rearrangement of aldoximes catalyzed by the metal-containing eutectic mixture 1ChCl/2ZnCl₂.

Metal-Catalyzed Hydrogenation of Olefins

Metal-catalyzed hydrogenation reactions are among the most frequently encountered transformations in laboratories and industry, appearing as a core technology for the production of a large variety of commodity and fine chemicals, as well as pharmaceutical substances. Concerning the homogeneous version of these reactions, it is now well-accepted that high rates and selectivities are only attainable by using pre-isolated and well-defined organometallic complexes (*107, 108*). In this sense, König and co-workers reported the hydrogenation of methyl α-cinnamide catalyzed by the Wilkinson's complex $[RhCl(PPh_3)_3]$ (**3**) in the carboxylic acid-based eutectic mixture citric acid/*DMU*, at 90 °C and 1 atm of H_2 (see Scheme 6) (*77*). Trying to achieve an enantiomeric excess in this hydrogenation reaction, the authors employed chiral eutectic mixtures (*i.e.*, Sorbitol/Urea/NH_4Cl) as reaction media. However, only racemic mixtures were obtained in all cases. More recently, Carrier and co-workers reported the hydrogenation of butadiene in the eutectic mixture formed by *ChCl* and urea; catalyzed by gold nanoparticles supported on TiO_2 (*109*).

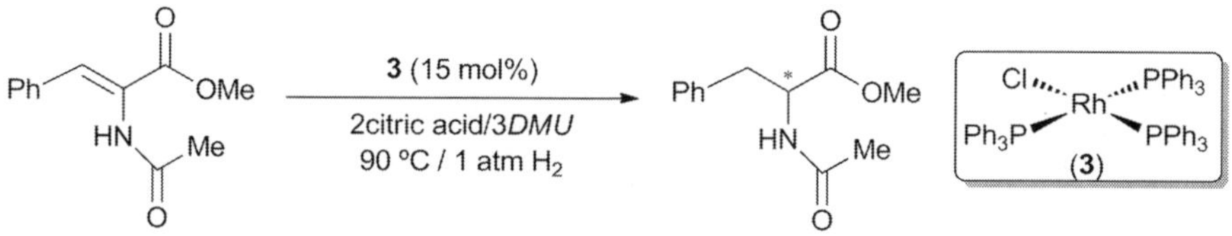

Scheme 6. Rh(I)-catalyzed hydrogenation of methyl α-cinnamide in the eutectic mixture citric acid/DMU.

Conclusions

This chapter clearly exemplifies the reliability gained by *Deep Eutectic Solvents* as new *Green* and biorenewable solvents in different metal-catalyzed organic reactions. These eutectic mixtures have been employed as new reaction media in a variety of relevant transformations in synthetic organic chemistry, such as hydrogenations, C-C couplings, rearrangements or cycloaddition reactions; and satisfactory results have been in most cases obtained. Despite all these promising applications, much effort is still needed in order to widen the utilization of *DESs* as *Green*, low toxic and biocompatible reaction medium in metal-catalyzed organic reactions.

The almost infinite variety of *Deep Eutectic Solvents* available, obtained through the adequate selection and combination of components [Hydrogen-Bond-Donor (*HBD*) and Hydrogen-Bond-Acceptor (*HBA*)], opens a new gate that will certainly lead in the near future to the design of new metal-mediated catalytic reactions in these solvents. We hope that this chapter will serve as inspiration for future work in this promising research field, which is clearly open with many opportunities for new discoveries.

Acknowledgments

Financial support by the MICINN (project CTQ2010-14796/BQU) of Spain and COST action Smart Inorganic Polymers (SIPs-CM1302) is gratefully acknowledged. JG-A thanks MICINN and the European Social Fund for the award of a "Ramón y Cajal" contract.

References

1. Anastas, P. T.; Warner, J. C. In *Green Chemistry: Theory and Practice*; Oxford University Press: New York, 1998; p 30.
2. Matlack, A. S. In *Introduction to Green Chemistry*; Marcel Dekker: New York, 2001.
3. Poliakoff, M.; Fitzpatrick, J. M.; Farren, T. R.; Anastas, P. T. *Science* **2002**, *297*, 807–810.

4. Lancaster, M. In *Green Chemistry: An Introductory Text*; RSC Publishing: Cambridge, 2002.
5. Sheldon, R. A.; Arends, I. W. C. E.; Henefeld, U. In *Green Chemistry and Catalysis*; Wiley-VCH: Weinheim, 2007.
6. Trost, B. M. *Science* **1991**, *254*, 1471–1477.
7. Trost, B. M. *Angew. Chem., Int. Ed. Engl.* **1995**, *34*, 259–281.
8. Sheldon, R. A. *Pure Appl. Chem.* **2000**, *72*, 1233–1246.
9. Trost, B. M.; Frederiksen, M. U.; Rudd, M. T. *Angew. Chem., Int. Ed.* **2005**, *44*, 6630–6666.
10. Sheldon, R. A. *Green Chem.* **2007**, *9*, 1273–1283.
11. Constable, D. J. C.; Jiménez-González, C.; Henderson, R. K. *Org. Process Res. Dev.* **2007**, *11*, 133–137.
12. Kemeling, G. M. *ChemSusChem* **2012**, *5*, 291–2292.
13. A recent editorial in *Organic Process Research and Development* discourages chemists to use solvents that are either known to be toxic, dangerous for large scale preparations or expensive to dispose as waste. Laird, T. *Org. Process Res. Dev.* **2012**, *16*, 1–2.
14. Jessop, P. G. *Green Chem.* **2011**, *13*, 1391–1398.
15. Reichard, C. In *Solvents and Solvents Effects in Organic Chemistry*, 3rd ed.; Wiley-VCH: Weinheim, Germany, 2003.
16. Anastas, P. T. In *Handbook of Green Chemistry*, Vols. 4, 5 and 6; *Green Solvents: Volume 4, Supercritical Solvents* (Leitner, W., Jessop, P. G., Eds.); *Volume 5: Reactions in Water* (Li, C.-J., Ed.); *Volume 6: Ionic Liquids* (Wasserschied, P., Stark, A., Eds.); Wiley-VCH: Weinheim, 2010.
17. In theory, an ideal *Green Solvent* should be safe for both human beings and the environment, with its use and manufacture being also sustainable. Therefore, to be qualified as a *Green* medium, these solvents have to meet different criteria such as availability, non-toxicity, biodegradability, recyclability, non-flammability and low price. Moity, L.; Durand, M.; Benazzouz, A.; Pierlot, C.; Molinier, V.; Aubry, J.-M. *Green Chem.* **2012**, *14*, 1132–1145.
18. Welton, T. *Chem. Rev.* **1999**, *99*, 2071–2083.
19. Wasserscheid, P.; Keim, W. *Angew. Chem., Int. Ed.* **2000**, *39*, 3772–3789.
20. Sheldon, R. *Chem. Commun.* **2001**, *2001*, 2399–2407.
21. Dupont, J.; de Souza, R. F.; Suárez, P. A. Z. *Chem. Rev.* **2002**, *102*, 3667–3691.
22. Parvulescu, V. I.; Hardacre, C. *Chem. Rev.* **2007**, *107*, 2615–2665.
23. Wassercheid, P.; Welton, T. In *Ionic Liquids in Synthesis*; Wiley-VCH: Weinheim, 2008.
24. Hallet, J. P.; Welton, T. *Chem. Rev.* **2011**, *111*, 3508–3576.
25. Reinhardt, D.; Ilgen, F.; Kralisch, D.; König, B.; Kreisel, G. *Green Chem.* **2008**, *10*, 1170–1181.
26. Deetlefs, M.; Seddon, K. R. *Green Chem.* **2010**, *12*, 17–31.
27. Welton, T. *Green Chem.* **2011**, *13*, 225–225.
28. Avalos, M.; Babiano, R.; Cintas, P.; Jiménez, J. L.; Palacios, J. C. *Angew. Chem., Int. Ed.* **2006**, *45*, 3904–3908.

29. Abbott, A. P.; Harris, R. C.; Ryder, K.; d'Agostino, C.; Gladden, L.; Mantle, M. D. *Green Chem.* **2011**, *13*, 82–90.
30. Ruβ, C.; König, B. *Green Chem.* **2012**, *14*, 2969–2982.
31. Carriazo, D.; Serrano, M. C.; Gutiérrez, M. C.; Ferrer, M. L.; Del Monte, F. *Chem. Soc. Rev.* **2012**, *41*, 4996–5014.
32. Zhang, Q.; de Oliveira Vigier, K.; Royer, S.; Jérôme, F. *Chem. Soc. Rev.* **2012**, *41*, 7108–7146.
33. Zhao, H.; Baker, G. A. *J. Chem. Technol. Biotechnol.* **2013**, *88*, 3–12.
34. Francisco, M.; van den Bruinhorst, A.; Kroon, M. C. *Angew. Chem., Int. Ed.* **2013**, *52*, 3074–3085.
35. Tang, B.; Row, K. H. *Monatsh. Chem.* **2013**, *144*, 1427–1454.
36. Yu, G.; Jérôme, F. *Chem. Soc. Rev.* **2013**, *42*, 9550–9570.
37. Del Monte, F.; Carriazo, D.; Serrano, M. C.; Gutierrez, M. C.; Ferrer, M. L. *ChemSusChem* **2014**, *7*, 999–1009.
38. García-Álvarez, J. Deep Eutectic Solvents and Their Applications as New Green and Biorenewable Reaction Media. In *Handbook of Solvents: Use, Health, and Environment*, 2nd ed.; Wypych, G., Ed.; ChemTec Publishing: Toronto, 2014; Vol. 2.
39. Abbott, A. P.; Capper, G.; Davies, D. L.; Rasheed, R. K.; Tambyrajah, V. *Chem. Commun.* **2003**, 70–71.
40. *ChCl,* so-called vitamin B_4 (an essential micro- and human nutrient), is produced on the scale of million metric tons per year as an additive for chicken feed and many other applications. This ammonium salt is not only cheap and easy to obtain (can be extracted from biomass) but also non-toxic and biodegradable: Blusztajn, J. K. *Science* **1998**, *281*, 794–795.
41. Petkovic, M.; Ferguson, J. L.; Gunaratne, H. Q. N.; Ferreira, R.; Leitão, M. C.; Seddon, K. R.; Rebelo, L. P. N.; Pereira, C. S. *Green Chem.* **2010**, *12*, 643–649.
42. Hayyan, M.; Hashim, M. A.; Hayyan, A.; Al-Saadi, M. A.; AlNashef, I. M.; Mirghani, M. E. S.; Saheed, O. K. *Chemosphere* **2013**, *90*, 2193–2196.
43. Abbott, A. P.; Capper, G.; Davies, D. L.; Rasheed, R. K. *Inorg. Chem.* **2004**, *43*, 3447–3452.
44. Laitinen, H.; Ferguson, W.; Osteryoung, R. *J. Electrochem. Soc.* **1957**, *104*, 516–520.
45. Duke, F.; Iverson, M. *J. Phys. Chem.* **1958**, *62*, 417–418.
46. Gambino, M.; Gaune, P.; Nabavian, M.; Gaune-Escard, M.; Bros, J. *Thermochim. Acta* **1987**, *111*, 37–47.
47. Abbott, A. P.; Capper, G.; Gray, S. *ChemPhysChem* **2006**, *7*, 803–806.
48. Abbott, A. P.; Boothby, D.; Capper, G.; Davics, D. L.; Rasheed, R. K. *J. Am. Chem. Soc.* **2004**, *126*, 9142–9147.
49. Imperato, G.; Eibler, E.; Niedermaier, J.; König, B. *Chem. Commun.* **2005**, *2005*, 1170–1172.
50. Choi, Y. H.; van Spronsen, J.; Dai, Y.; Verberne, M.; Hollmann, F.; Arends, I. W. C. E.; Witkamp, G.-J.; Verpoorte, R. *Plant Physiol.* **2011**, *156*, 1701–1705.
51. Daia, Y.; van Spronsenb, J.; Witkampb, G.-J.; Verpoortea, R.; Choi, Y. H. *Anal. Chim. Acta* **2013**, *766*, 61–68.

52. Francisco, M.; González, A. S. B.; García de Dios, S. L.; Weggemans, W.; Kroon, M. C. *RSC Adv.* **2013**, *3*, 23553–23561.
53. van Leeuwen, P. W. N. M. In *Homogeneous Catalysis: Understanding the Art*; Kluwer Academic Publishers: Dordrecht, 2004.
54. Beller, M.; Renken, A.; van Santen, R. A. In *Catalysis: From Principles to Applications*; Wiley-VCH: Weinheim, 2011.
55. Steinborn, D. In *Fundamentals of Organometallic Catalysis*; Wiley-VCH: Weinheim, 2011.
56. *HMF* is a valuable molecule platform, especially for the fabrication of safer polymers and fuel additives. See for example: Karinen, R.; Vilonen, K.; Niemelä, M. *ChemSusChem* **2011**, *4*, 1002–1016.
57. The conversion of Fructose to *HMF* has been carried out in some polar organic solvents, such as dimethylsulfoxide, dimethylformamide, and sub-critical or high temperature water. For pioneering studies see references (*58*–*62*).
58. Nakamura, Y.; Morikawa, S. *Bull. Chem. Soc. Jpn.* **1980**, *53*, 3705–3706.
59. Seri, K.; Inoue, Y.; Ishida, H. *Chem. Lett.* **2000**, 22–23.
60. Seri, K.; Inoue, Y.; Ishida, H. *Bull. Chem. Soc. Jpn.* **2001**, *74*, 1145–1150.
61. Asghari, F. S.; Yoshida, H. *Cabohydr. Res.* **2006**, *341*, 2379–2387.
62. Asghari, F. S.; Yoshida, H. *Ind. Eng. Chem. Res.* **2006**, *45*, 2163–2173.
63. It has been reported that some *ILs* have strong solvent power for carbohydrates, see references (*64*–*67*).
64. Swatloski, R. P.; Spear, S. K.; Holbrey, J. D.; Rogers, R. D. *J. Am. Chem. Soc.* **2002**, *124*, 4974–4975.
65. Abbott, A. P.; Bell, T. J.; Handa, S.; Stoddart, B. *Green Chem.* **2005**, *7*, 705–707.
66. Moreau, C.; Finiels, A.; Vanoye, L. *J. Mol. Catal. A: Chem.* **2006**, *253*, 165–169.
67. Zhao, H. B.; Holladay, J. E.; Brown, H.; Zhang, Z. C. *Science* **2007**, *316*, 1597–1600.
68. Hu, S.; Zhang, Z.; Zhou, Y.; Han, B.; Fan, H.; Li, W.; Song, J.; Xie, Y. *Green Chem.* **2008**, *10*, 1280–1283.
69. Illgen, F.; Ott, D.; Kralisch, D.; Reil, C.; Palmberger, A.; König, B. *Green Chem.* **2009**, *11*, 1948–1954.
70. Mamman, A. S.; Lee, J. M.; Kim, Y. C.; Hwang, I. T.; Park, N. J.; Hwang, Y. K.; Chang, J. S.; Hwang, J. S. *Biofuels Bioprod. Biorefin.* **2008**, *2*, 438–454.
71. Dutta, S.; De, S.; Saha, B.; Alam, M. I. *Catal. Sci. Technol.* **2012**, *2*, 2025–2036.
72. Zhang, L.-X.; Yu, H.; Yu, H.-B.; Chen, Z.; Yang, L. *Chin. Chem. Lett.* **2014**, *25*, 1132–1136.
73. de Oliveira Vigier, K.; Benguerba, V. A.; Barrault, J.; Jérôme, F. *Green Chem.* **2012**, *14*, 285–289.
74. Miyaura, N. In *Cross-coupling Reactions: a Practical Guide*; Springer: Berlin, 2002.
75. de Meijere, A.; Diederich, F. In *Metal-catalyzed Cross-coupling Reactions*, 2nd ed.; Wiley-VCH: Weinheim, 2004.
76. Nishihara, Y. In *Applied Cross-coupling Reactions*; Springer: Berlin, 2013.

77. Imperato, G.; Höger, S.; Leinor, D.; König, B. *Green Chem.* **2006**, *8*, 1051–1055.
78. Imperato, G.; Vasold, R.; König, B. *Adv. Synth. Catal.* **2006**, *348*, 2243–2247.
79. Illgen, F.; König, B. *Green Chem.* **2009**, *11*, 848–854.
80. Jérôme, F.; Ferreira, M.; Bricout, H.; Menuel, S.; Monflier, E.; Tilloy, S. *Green Chem.* **2014**, *16*, 3876–3880.
81. Meldal, M.; Christensen, C.; Tornøe, C. W. *J. Org. Chem.* **2002**, *67*, 3057–3064.
82. Meldal, M.; Tornøe, C. W. *Chem. Rev.* **2008**, *108*, 2952–3015.
83. Sharpless, K. B.; Fokin, V. V.; Green, L. G.; Rostovtset, V. V. *Angew. Chem., Int. Ed.* **2002**, *114*, 2708–2711.
84. *CuAAC* has found wide application in pharmaceutical, combinatorial and material chemistry. For selected reviews see references (*85–89*).
85. Wu, P.; Fokin, V. V. *Aldrichimica Acta* **2007**, *40*, 7–17.
86. Moses, J. E.; Moorhouse, A. D. *Chem. Soc. Rev.* **2007**, *36*, 1249–1262.
87. Spiteri, A. D.; Moses, J. E. *ChemMedChem* **2008**, *3*, 715–723.
88. Holub, J. M.; Kirshenbaum, K. *Chem. Soc. Rev.* **2010**, *39*, 1325–1337.
89. Agalave, S. G.; Maujan, S. R.; Pore, V. S. *Chem. Asian J.* **2011**, *6*, 2696–2718.
90. van der Drift, R. C.; Bouwman, E.; Drent, E. *J. Organomet. Chem.* **2002**, *650*, 1–24.
91. Uma, R.; Crésvisy, C.; Greé, R. *Chem. Rev.* **2003**, *103*, 27–52.
92. Cadierno, V.; Crochet, P.; García-Garrido, S. E.; Gimeno, J. *Curr. Org. Chem.* **2006**, *10*, 165–183.
93. Cadierno, V.; Crochet, P.; Gimeno, J. *Synlett* **2008**, *2008*, 1105–1124.
94. Mantilli, L.; Mazet, C. *Chem. Lett.* **2011**, *40*, 341–344.
95. Lorenzo-Luis, P.; Romerosa, A.; Serrano-Ruiz, M. *ACS Catal.* **2012**, *2*, 1079–1086.
96. Alhsten, N.; Bartoszewicz, A.; Martín-Matute, B. *Daltons Trans.* **2012**, *41*, 1660–1670.
97. García-Álvarez, J.; García-Garrido, S. E.; Crochet, P.; Cadierno, V. *Curr. Top. Catal.* **2012**, *10*, 35–56.
98. Vidal, C.; Suárez, F. J.; García-Álvarez, J. *Catal. Commun.* **2014**, *44*, 76–79.
99. Suárez, F. J.; Vidal, C.; García-Álvarez, J. *Curr. Green Chem.* **2014**, *1*, 121–127.
100. Kumar, R. R.; Vanithan, K. A.; Balasubramanian, M. In *Name Reactions for Homologations - Part 2*; Li, J. J., Ed.; John Wiley & Sons: Hoboken, NJ, 2009; pp 274–292.
101. Gawley, R. E. *Org. React.* **1988**, *35*, 1–420.
102. Ghose, A. K.; Viswanadhan, V. N. *J. Comb. Chem.* **1999**, *1*, 55–68.
103. Singh, G. S. *Tetrahedron* **2003**, *59*, 7631–7649.
104. Albericio, F. *Curr. Opin. Chem. Biol.* **2004**, *8*, 211–221.
105. Shen, B.; Makley, D. M.; Johnston, J. N. *Nature* **2010**, *465*, 1027–1032.
106. Patil, U. B.; Singh, A. S.; Nagarkar, J. M. *RSC Adv.* **2014**, *4*, 1102–1106.
107. Chaloner, P. A.; Esteruelas, M. A.; Joó, F.; Oro, L. A. In *Homogeneous Hydrogenation*; Kluwer Academic Publishers: Dordrecht, 1994.

108. de Vries, J. G.; Elsevier, C. J. In *Handbook of Homogeneous Hydrogenation*; Wiley-VCH: Weinheim, 2007.
109. Oumahi, C.; Lombard, J.; Casale, S.; Calers, C.; Delannoy, L.; Louis, C.; Carrier, X. *Catal. Today* **2014**, *235*, 58–71.

Chapter 4

Amino Acid-Derived Imidazolium Zwitterions: Building Blocks for Renewable Ionic Liquids and Materials

Sarah Kirchhecker and Davide Esposito*

**Max Planck Institute of Colloids and Interfaces,
Department of Colloid Chemistry, 14424 Potsdam, Germany
*E-mail: davide.esposito@mpikg.mpg.de**

Imidazolium zwitterions were prepared as novel building blocks from amino acids and carbohydrate-derivable dicarbonyl fragments with a simple and sustainable method. These building blocks were converted into ionic liquids via hydrothermal decarboxylation in flow and into acidic ionic liquids, cross linking moieties and monomers for renewable polymers.

Introduction

With the diminishing of the world's oil reserves and the political and economic problems this entails, there is an ever growing demand for new synthetic strategies based on renewable resources, to ensure the provision of chemical building blocks for the future.

Heterocycles are a very important and diverse group of compounds, present in many types of fine chemicals. In the context of integrated biorefineries, the two furan derivatives furfural and 5-hydroxy methyl furfural (HMF) have received a great deal of attention (*1*, *2*). They can be derived via acid catalyzed dehydration from pentoses and hexoses, respectively, and especially HMF has now been established as a new platform molecule, which can be further converted to various value-added chemicals (*3*). Nitrogen – containing heterocycles, in turn, cannot be derived by direct pyrolysis or hydrothermal treatment of lignocellulosic biomass alone, as a second component is needed to introduce a nitrogen containing functionality. Nevertheless, nitrogen-containing heterocycles are very important molecules. Interestingly, with the increasing industrial use of biotechnology large amounts of protein waste are being generated which could be used as a cheap source of amino acids in combination with lignocellulosic biomass to generate N-heterocycles. In addition, sustainable sources of amino acids could be provided by the cultivation of protein-rich macroalgae, which do not compete with conventional food crops for land use and have much higher growth rates.

Among other N-heterocycles, imidazole and imidazolium compounds have received much attention in different research fields in recent decades, most notably in ionic liquids (*4*). But they are also important motifs used in a variety of other chemicals, including antifungal drugs (*5*, *6*), surfactants (*7*), N-heterocyclic carbenes (*8*, *9*) and as carbon material precursors (*10*) to name but a few.

In the following we will present our recent work towards imidazolium building blocks derivable from carbohydrate fragments and amino acids and their derivatization into different types of functionalized molecules and materials (Figure 1). For the transformations we placed an emphasis on simple one-pot procedures and continuous processing using water as a benign solvent.

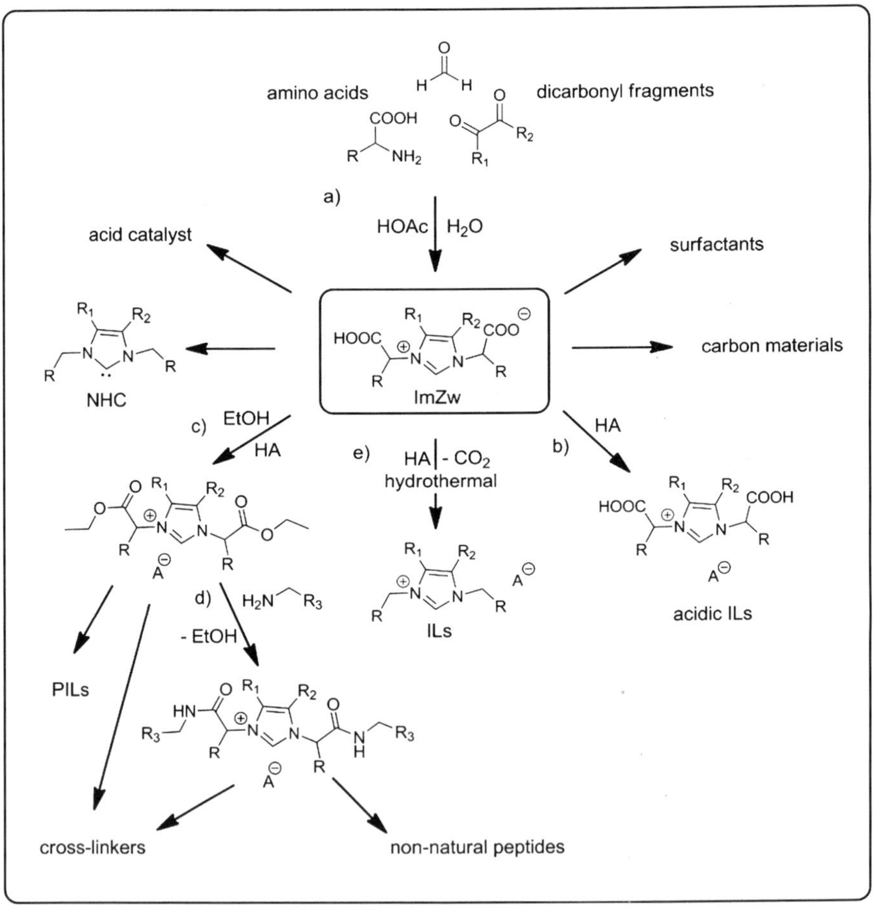

Figure 1. Bio-derived imidazolium building blocks as versatile scaffolds for the generation of functional molecules and different types of materials

Imidazolium Zwitterions as Building Block Molecules

The Debus-Radziszewski imidazole synthesis (*11*, *12*) is a one-pot reaction that affords substituted imidazoles via condensation of an amine and/or ammonia, a dicarbonyl and an aldehyde. In a bid to synthesize imidazolium compounds from renewable resources, we developed a modified version of this long-established reaction, using amino acids as the amine component and dicarbonyl compounds which can be derived from carbohydrates. Employing acetic acid as an environmentally benign catalyst, we obtained a library of differently functionalized imidazolium compounds in a simple one-pot reaction at ambient or very moderate temperatures (*13*). The reaction uses reagents in stoichiometric amounts and is very atom efficient, the only side product being water (Figure 1a). The different amino acid side chains conveniently add different functionalities as well as chirality to the final building block (*13*). All amino acid-derived imidazolium compounds formed as zwitterionic salts and were isolated as solids, which can be purified easily by removal of water (e.g. lyophilization), followed by washing of the residue. In the following, these compounds will be identified as ImZw (imidazolium zwitterions).

While optimizing our strategy, we initially focused on pyruvaldehyde (methylglyoxal) as the dicarbonyl component. Pyruvaldehyde is an intermediate in the alkaline conversion of glucose to lactic acid (*14*) and it can be generated via dihydroxyacetone from glycerol (*15*, *16*), available in large quantities as the side product of biodiesel production. Successively, the methodology was extended to glyoxal and 2,3-butadione, which are both also accessible from biomass via pyrolysis (*17–20*) or fermentation (*21*, *22*), respectively. The ImZw are presented in Table 1. Most ImZw derived from pyruvaldehyde were formed in good to very high yields, apart from serine-derived **7**, which was only obtained in 59% yield. The yields for several of the glyoxal- and butadione- derived ImZw were slightly lower than their pyruvaldehyde-derived counterparts using the same amino acids.

The methodology is also relatively versatile with respect to the nitrogen donor. When aliphatic amines were used, ionic liquids were obtained as products, with the counterion provided by the acetic acid catalyst (*13*), as shown also by other groups previously (*23*). Unsubstituted and hydroxyl-functionalized amines could also be converted with good yields. However, when more reactive functionalities were used such as alkynes or nitriles, the yields were low due to side reactions.

Acidic Properties of ImZw and Acidic ILs

Task specific ILs are a subsection of ILs that contain a functional group in their side chains, so that they can be used as solvent or catalyst or both at the same time (*24*). Acid-catalyzed reactions are ubiquitous and therefore acidic ILs are an important new class of molecules. There are many environmental problems associated with the use of strong liquid mineral acids, and recycling is difficult. In industry, solid acid catalysts are preferred due to the ease of separation, but they also have associated problems, and can often be deactivated quickly due to adsorption of byproducts. Acidic ILs, due to their negligible vapour pressure may be seen as a cross between liquid and solid catalysts. For example, the products of the reaction could be distilled off, leaving the liquid catalyst available for a further reaction (*25*).

The synthesis of acidic ionic liquids is usually performed by quaternization of *N*-alkyl imidazoles with sulfocontaining groups such as 1,3 propane sultone resulting in a zwitterion, followed by addition of stoichiometric amounts of a strong acid to protonate the zwitterion and provide the counterion (*25*). In a similar way, we considered the possibility of using ImZw as precursors for acidic ILs capitalizing on their internal acidic functionalities. In order to gain insights into the reactivity of the newly synthesized ImZw, the pKas of several representative compounds were determined (Table 2). The pKas were found to be similar to those of the corresponding amino acids, with methyl substituents on the imidazolium ring increasing the pKa slightly. Considering the acidity of these compounds, one could image their use as acid catalysts in their own right. As an example, the deprotection of benzaldehyde dimethyl acetal was performed in a biphasic mixture of chloroform, containing the reagent, and water containing 10 mol% of **1**. After 3 h at 50 °C the acetal was hydrolyzed in 95 % yield as assessed by NMR.

Furthermore, in an attempt to generate acidic ILs from ImZw, we reacted **1** with a series of strong acids. Such protonation reactions afforded different viscous oils, which were characterized by differential scanning calorimetry (Table 3). These acidic compounds can be imagined as an internally buffered system, where three anions, namely two carboxylates and the conjugate base of the strong acid, are available for each acid proton. In accordance, the measured pH of a 0.1 M aqueous solution of **12** (pH 1.29) lay in between those of the acid (pH 0.97) and the ImZw (pH 2.12) dissolved alone. As all ImZw contain the double carboxylic acid/carboxylate functionalities, a wide variety of these acidic compounds could be generated by choice of different acid and ImZw, whose different side chains would modulate their acid strength and confer different properties to the compound (Figure 1b).

Reactions of the Carboxylic Functionalities

The carboxylic functionalities of the ImZw offer several possibilities for functionalization and incorporation of the ImZw motif into different architectures. The most straightforward transformation that could be applied is transesterification, to introduce different alkyl functionalities. Surface active imidazolium ionic liquids (*26, 27*) which contain one long alkyl chain attached to the imidazolium ring or gemini surfactants (*7*) derived from two imidazole heads and two long alkyl chains are well established systems. We surmised that similar structures could be obtained by esterification of the carboxylic functionalities of ImZw with long chain alcohols. As an example, **1** was reacted with decanol using triflic acid as a catalyst, giving the expected diester in 99% yield (*13*).

Table 1. ImZw derived from amino acids and carbohydrate derivable dicarbonyl compounds.

entry	*amino acid*	*dicarbonyl*	*ImZw*	*conditions*	*Yield[1] [%]*
1	Gly	PRV	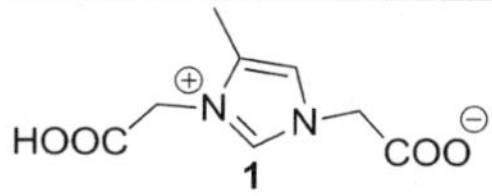 **1**	a	98
2	β-Ala	PRV	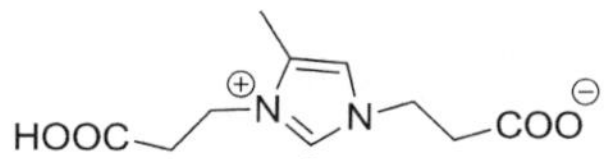 **2**	a	95
3	Ala	PRV	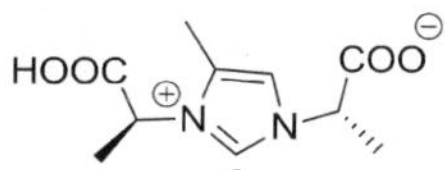 **3**	a	87
4	Phe	PRV	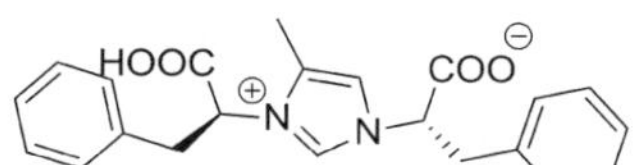 **4**	b	78
5	Leu	PRV	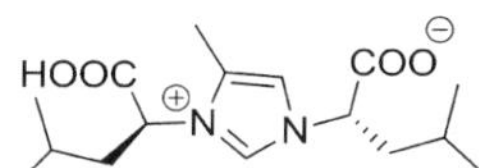 **5**	b	77
6	Boc-Lys	PRV	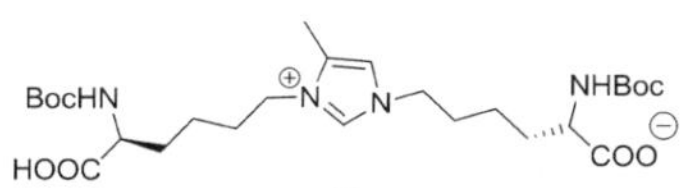 **6**	a	80
7	Ser	PRV	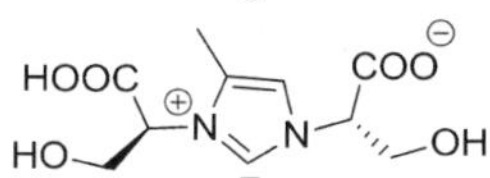 **7**	b	59
8	Gly	GX	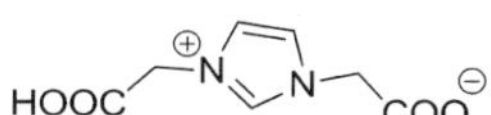 **8**	a	98
9	Ala	GX	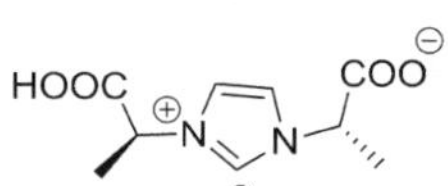 **9**	a	67
10	Phe	GX	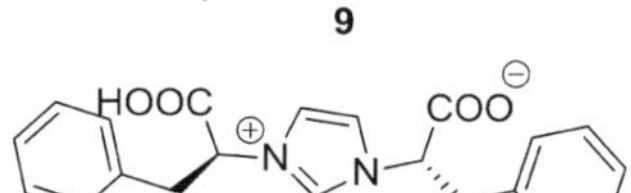 **10**	a	54
11	Gly	BD	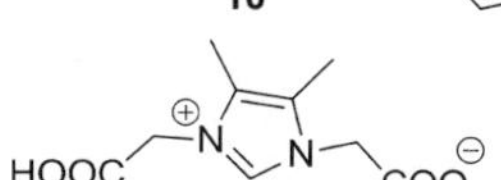 **11**	b	71

1) Yield determined by NMR using 2-methyl imidazole as internal standard. Reaction conditions: H_2O, AcOH (6 eq), *a)* RT, 1 h, *b)* 50 °C, 3 h. PRV: pyruvaldehyde, GX: glyoxal, BD: butadione. Adapted with permission from reference 13. Copyright 2013, Wiley.

Table 2. pKa values of some ImZw in comparison to the parent amino acid

entry	*ImZw*	*pKa*	*amino acid*	*pKa*
1	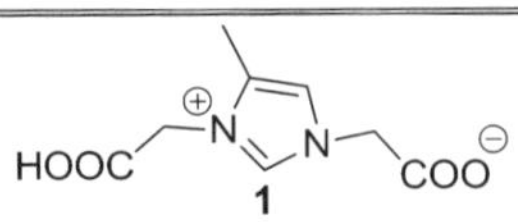 **1**	2.74	Gly	2.34
2	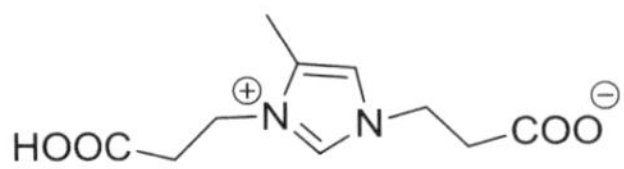 **2**	4.53	β-Ala	3.63
3	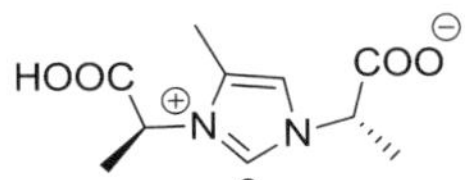 **3**	2.76	Ala	2.35
4	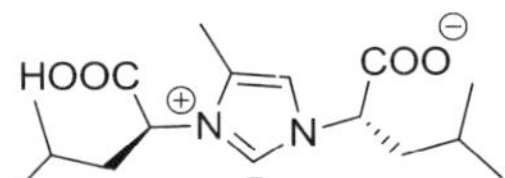 **5**	2.88	Leu	2.36
5	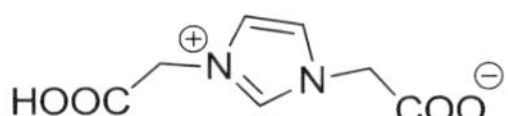 **8**	2.38	Gly	2.34
6	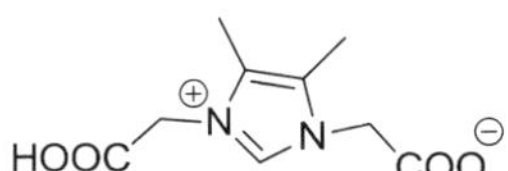 **11**	2.80	Gly	2.34

Table 3. Glass transition temperatures for adducts of 1 with different acids measured by differential scanning calorimetry. Reproduced with permission from reference (*13*). Copyright 2013 Wiley.

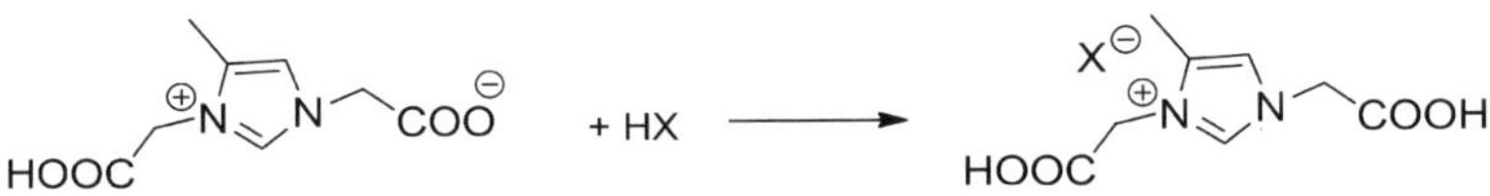

HX	*Product*	*Tg (°C)*	*σ (+/-)*
CF_3SO_3H	**12**	16.61	1.13
CH_3SO_3H	**13**	31.42	1.17
$CH_3C_6H_4SO_3H$	**14**	50.07	4.85
BF_4H	**15**	17.87	1.60

On the other hand, esterification of **1** with ethanol catalyzed by HCl produced a viscous ionic liquid (Figure 1c). Ethanol-esterified ionic liquids such as this one are in this way activated for condensation reactions, including transesterifications or amidations, during which ethanol is eliminated. These sustainable building blocks open the way for the preparation of polyesters and polyamides. Polyelectrolytes based on the imidazolium moiety have shown interesting properties for the absorption of CO_2 (*28*). Therefore we are currently developing novel renewable polyionic liquids (PILs) based on this moiety.

In addition, the ethyl esters can be converted into different types of ionic cross-linkers via exchange of the ester bond for an amide. The reaction could be performed with several differently functionalized amines by simply stirring the esterified compound with a slight excess of primary amine in ethanol at room temperature without the need for a catalyst (Figure 1d). For example, exchange with ethylenediamine proceeded with 82 % (*29*), introducing amine functionalities into the molecule, which could be employed for the synthesis of non-natural peptides.

Based on this kind of reactivity and considering their bifunctional nature ethanol esterified ImZw can also be envisaged as cross-linkers for the reaction with alcohol- or amino-functionalized polymers. Cellulose materials functionalized with these compounds will be discussed later on.

Ionic Liquids

The classical synthesis of imidazolium ionic liquids (ILs) generally relies on the preparation of 1-methyl-imidazole, followed by a second quaternization with an alkyl halide and final anion metathesis. These reaction pathways use petroleum-derived chemicals and often require long reaction times and high temperatures, followed by non-trivial purification procedures (*30*). Another potential problem is the contamination with halide salts after metathesis, which are difficult to remove completely and affect the physicochemical properties of ILs, with adverse effects especially for electrochemical applications (*31, 32*).

Recently there have been many efforts to synthesize ionic liquids based on renewable resources (*33–36*) and in many cases amino acids are utilized (*37, 38*). In these examples amino acids are used as anions or, with little derivatisation, for the synthesis of the cation. We, however, wanted to employ amino acids for the renewable synthesis of ionic liquids that are structurally similar to the well-known and widely used imidazolium ILs. In this respect ImZw can be regarded as useful synthons. Obtained via a straightforward one pot condensation, a simple decarboxylation of ImZw could be in principle used to convert such building blocks into ILs, as shown in the following retrosynthetic analysis (Figure 2).

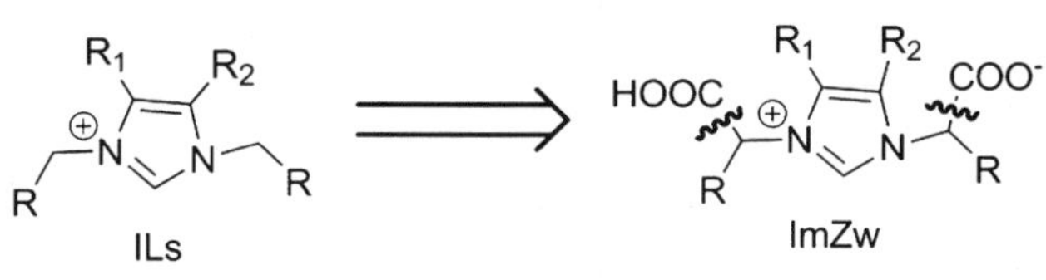

Figure 2. Retrosynthetic analysis for the synthesis of ILs from ImZw

Continuous processing in microreactors enables fast and efficient reactions due to the optimized heat transfer and mixing, while allowing reactions under near- or supercritical conditions. Superheating of water confers very special and useful properties to this benign solvent, which has been shown as a versatile medium for a variety of different transformations, including condensations, Diels-Alder reactions, rearrangements and oxidation reactions (*39*). We have recently shown how the amino acid-derived ImZw can be decarboxylated in the presence of two equivalents of acetic acid under hydrothermal conditions in a flow reactor, with good yields of the decarboxylated product (Figure 3) (*40*). Apart from butadione-derived **21,** all the compounds obtained following this approach are ionic liquids, with melting points or glass transitions well below 100 °C, and with the exception of **16** (Tm = 65.1 °C), they classify as room temperature ionic liquids. The decarboxylated compounds are shown in Table 4.

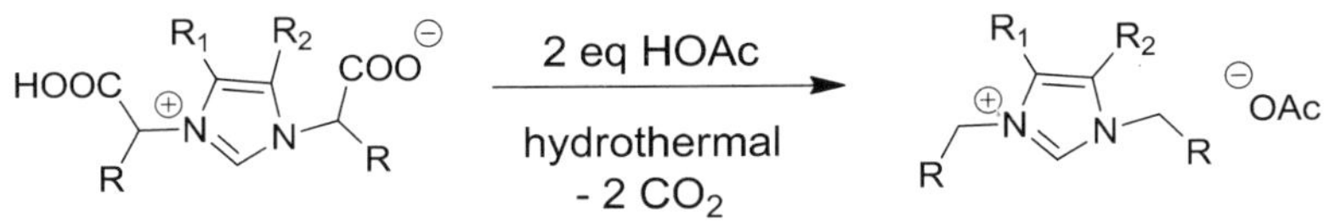

Figure 3. Hydrothermal decarboxylation of ImZw in the presence of acetic acid to form ionic liquids

In the final compound the counterion is provided by the conjugate base of the added acid, in this case, acetate. However, decarboxylation can also be performed with other acids and even salts. For example, the chloride IL of **4** was prepared by hydrothermal decarboxylation in the presence of two equivalents of NaCl. As for other ionic liquids, the anion can be exchanged by metathesis. The latter strategy can also be used as a simple and efficient method for the purification of the ionic liquids. In addition the ILs can be purified by dissolution in the appropriate solvent followed by precipitation. As this methodology avoids alkylation reactions with halides, ionic liquids can be easily prepared halide-free. In order to improve the efficiency of the process, we tried to reduce the number of steps for the synthesis of these new ILs and therefore, the direct hydrothermal processing of crude ImZw contained in the condensation reaction mixture of amino acids, dicarbonyl and formaldehyde was assessed for compounds **1**, **3** and **4.** Using such an approach, compounds **1** and **4** were obtained with slightly lower yields (less than 10 % difference) compared to the synthesis performed with pure ImZw. This appeared as a negligible difference, considering that some ImZw is lost during purification in a two-step process. In addition when the reaction mixture was injected directly, higher concentrations of ImZw, especially for the more hydrophobic **4,** could be used, increasing the efficiency of the synthesis even more. The yield for the direct decarboxylation of **3** however was lower (46 % compared to 69 % via the 2-step process).

Table 4. Ionic liquids derived from ImZw by hydrothermal decarboxylation in flow.

entry	*amino acid*	*dicarbonyl*	*ImZw*	*product*	*Yield [%]*	*T_g [°C]*	*T_m [°C]*
1	Gly	PRV	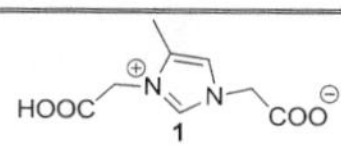 **1**	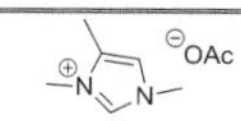 **16**	90.0		65.1
2	Ala	PRV	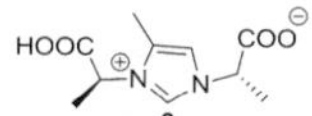 **3**	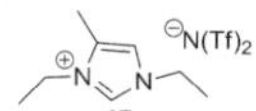 **17**	65.4[a]	-88.5	
3	Phe	PRV	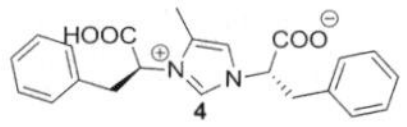 **4**	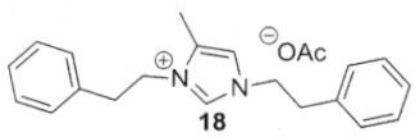 **18**	89.4	-20.0	
4	Leu	PRV	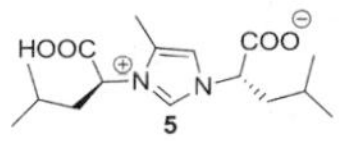 **5**	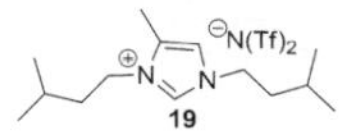 **19**	17.9[a]	n/d	n/d
5	Gly	GX	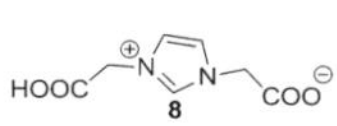 **8**	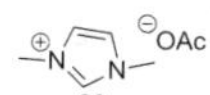 **20**	95.9	-51.7	
6	Gly	BD	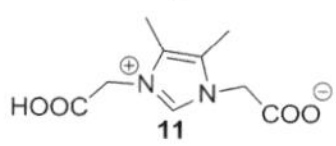 **11**	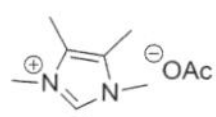 **21**	98.0		125.9
7	Ala	GX	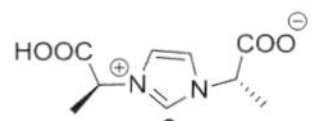 **9**	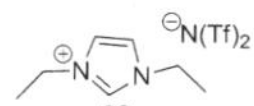 **22**	12.5		-13.4
8	Ala	PRV	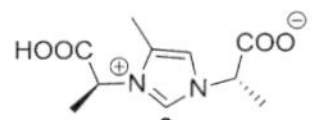 **3**	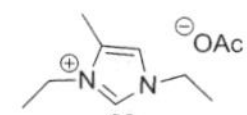 **23**	93.0[b]	-62.0	

Reaction conditions: H_2O, 300 °C, 150 bar, 4 mins, except entry 3: 50:50 H_2O:EtOH, 250 °C, 120 bar, 4 mins. a) after metathesis with $LiN(Tf)_2$, b) by direct Debus-Radzizsewski reaction with ethylamine. n/d: none detected down to -150 °C. PRV: pyruvaldehyde, GX: glyoxal, BD: butadione. Adapted with permission from reference (*40*). Copyright 2014 RSC.

As observed by control experiments, free amino acids cannot be decarboxlyated under these conditions, pointing to the beneficial effect of the imidazolium ring. The aromatic ring probably has a catalytic effect on the decarboxylation, similar to that of conjugated carbonyl compounds which have found application as catalyst electron sinks for decarboxylations due to the formation of a conjugate imine (*41*).The neighbouring positive charge could in principle stabilize the developing anion at the site of decarboxylation. A tentative mechanism for the decarboxylation is shown in Figure 4.

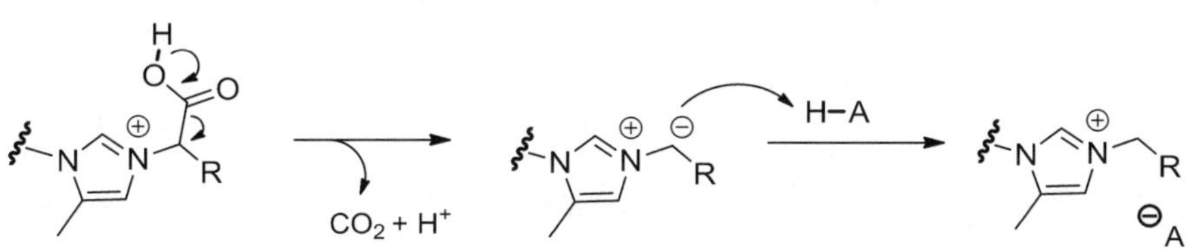

Figure 4. Proposed mechanism for the decarboxylation of ImZw and formation of ionic liquids.

Applications of the Ionic Liquids

To demonstrate the applicability of these novel renewable ionic liquids we chose to use the alanine-derived ILs **17** and **23**, due to their structural similarity with the commercially available 1-ethyl-3-methyl imidazolium (Emim) –based ionic liquids. Some physicochemical properties of these ILs are reported in Table 5.

Table 5. Some physicochemical properties of ILs used for solvent applications. Reproduced with permission from ref (*40*). Copyright 2014 RSC.

ionic liquid	*water content [%]*	*density [g cm-1]*		*viscosity [mPa s]*		*conductivity [mS cm-1]*	T_g/T_m *[°C]*	T_d *[°C]*
		25 °C	*90 °C*	*25 °C*	*90 °C*			
17	0.07	1.452	1.389	38.3	6.8	2.98	-88.5[a]	443
23	6.68	1.070	1.027	50.0	5.8	2.70	-62.0[b]	242

[a] T_g. [b] T_m.

Two areas where ionic liquids have a high potential for increasing the sustainability of synthetic procedure applied in industry - the Heck reaction and the dissolution of cellulose – were chosen to demonstrate that these novel bio-based ionic liquids can be used with the same efficiency of other commercially available ILs.

Heck Reaction

The Heck coupling is a well-known and very important Pd catalyzed carbon-carbon cross coupling reaction (*42*, *43*). While initially demonstrated using unstabilized Pd salts, variations that make use of highly engineered ligands based on phosphine or N-heterocyclic carbenes are nowadays widespread in the literature for reactions involving more recalcitrant substrates. A major drawback of these approaches is to be seen in the issues of cost and recycling of the expensive catalysts. Ionic liquids have the ability to immobilize and stabilize metal nanoparticles, thereby improving the lifespan and recyclability of the catalyst (*44*). To assess the ease of recycling of catalyst in our ionic liquid **17,** we performed a standard Heck reaction between iodobenzene and methyl acrylate catalyzed by 0.2 mol% of Pd acetate with respect to iodobenzene. The isolated yields of the product for five consecutive runs are shown in Figure 5a. After each reaction the product was extracted with toluene and the catalyst-IL phase was reused. In the 3rd run the yield of the reaction was lower due to the buildup of byproducts in the reaction mixture. Washing of the ionic liquid with water and drying in vacuo fully restored the catalytic ability (run 4). TEM analysis of the IL showed the formation of Pd nanoparticles of about 5 nm during the reaction

(Figure 5b). These results confirmed that the renewable IL **17** can be used for the Heck coupling in the same way as commercially available, non-renewable ILs.

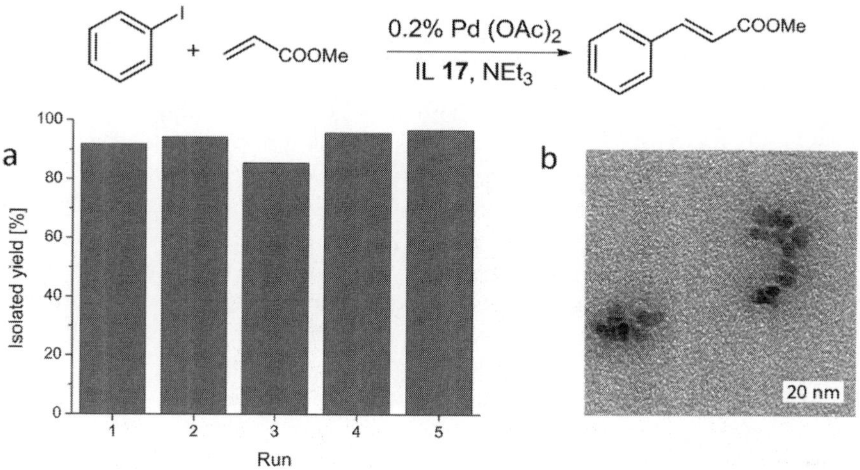

Figure 5. Heck reaction in IL 17. a) isolated yields of product for five consecutive runs with recycling of the catalyst-IL reaction medium. b) TEM image of Pd nanoparticles formed in IL 17 after the first run. Reproduced with permission from reference (40). Copyright 2014 RSC.

Cellulose Dissolution

Dissolution of cellulosic biomass is crucial for the optimal utilization of this valuable and most abundant renewable resource. Although some processes exist where cellulose materials are forced open without dissolution, these employ harsh conditions or destroy the native structure, causing derivatisation which is not always desirable. For most applications, including spinning of clothing fibers, as well as depolymerisation and conversion into smaller chemicals, dissolution is generally a more appropriate choice.

However, cellulose does not dissolve in common solvents due to an extended network of inter- and intramolecular hydrogen bonds between the glucose units in its chains. Before some ionic liquids were found to be very efficient solvents for such a process, the dissolution of cellulose could only be achieved in multi-component systems, such as metal complexes in water or highly concentrated solutions of lithium chloride in dimethylacetamide. The mechanism of dissolution in ionic liquids is generally thought to work via the disruption of the hydrogen bond network by the cation and formation of new hydrogen bonds between the cellulose strands and the basic anion (*45*).

Emim acetate is currently considered the best solvent for cellulose, which shows a solubility of ca. 20 wt% at 100 °C in this compound (*46*, *47*). Remarkably, it has been recently reported that Emim can even dissolve wood (*48*, *49*). Due to the structural similarity of hydrophilic IL **23** to Emim OAc, the solubility of microcrystalline cellulose (MCC) in **23** was assessed. Hydrophilic

ILs are notoriously hygroscopic. As water is an anti-solvent for cellulose, ILs are therefore often dried under vacuum for extended periods before cellulose dissolution experiments. This of course is cost- and time-consuming. In our experiments, we did not try to reduce the water content (about 6 % at the time of measurement) in the IL to a minimum. Dissolution experiments were performed in an open round bottom flask equipped with a mechanical stirrer. During the dissolution at 100 °C water slowly evaporated from the ionic liquid, which did not appear to interfere with the process. Dissolution was confirmed using a light microscope with a hot stage. In our experiments we observed dissolution of MCC in **23** up to 16.9 wt%. After dissolution, the cellulose was precipitated with water and analyzed by powder x-ray diffraction (XRD). The XRD patterns are shown in Figure 6. The native cellulose type I crystallinity fingerprint of MCC is lost after dissolution with the appearance of signals characteristic of more amorphous patterns, confirming that actual dissolution of cellulose took place. Once again this demonstrates that the renewable IL is a good substitute for commercially available ILs for this application and performs better than many other ILs that have been used for this purpose.

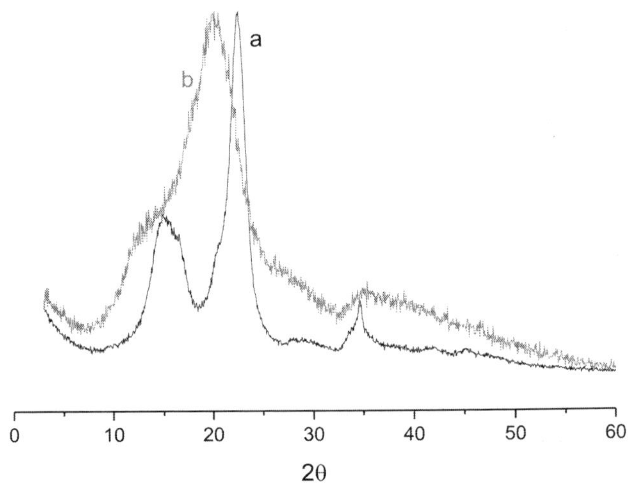

*Figure 6. XRD patterns of MCC a) before and b) after dissolution in IL **23**. Reproduced with permission from reference (40). Copyright 2014 RSC.*

Functionalization of Cellulose

Cellulose is a renewable material characterized by many desirable properties including its mechanical strength and stiffness, the many functionalities available for derivatisation, biocompatibility and not least its natural abundance and low cost.

This has led to its application for the preparation of many different materials, ranging from the classic and established paper based systems to the more sophisticated gels (*50*), supports for catalysts (*51*, *52*) and enzymes (*53*, *54*), membranes (*55*) and electronics (*56–59*).

A typical synthetic transformation applied during the preparation of cellulose-based materials is cross linking. Cellulose-based fibers for instance, are often cross-linked with different reagents to align them and decrease wrinkles. On the other hand, functionalized cellulose materials have been used for the removal of toxic contaminants, such as heavy metals (*60–62*) and aromatics (*63*) from wastewater.

In principle, hydroxyl groups on position 2, 3, and 6 of glucose, the constituting monomer of cellulose chains, are available for functionalization, with the primary O-6 position being the most reactive. In this regard, the bifunctional imidazolium compounds described above could be used to cross-link different chains of cellulose, producing new materials with ionic functionalization, which might be useful for instance for the selective removal of anions from solution. Employing different ImZw for the synthesis of the cross-linker, the morphology of the material could be modulated, and properties such as hydrophobicity could be easily tuned, according to the choice of ImZw.

To evaluate this hypothesis we synthesized the bifunctional compound **24** by esterification of **1** with ethanol using trifluoromethanesulfonic acid (TfOH) as the catalyst (Figure 7).

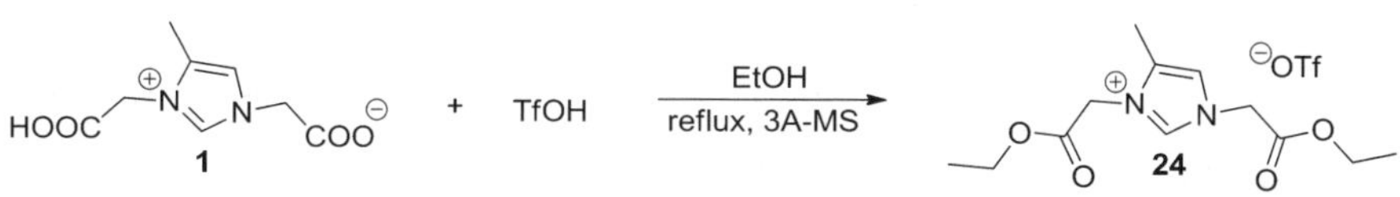

Figure 7. Synthesis of esterified ionic liquid used in the functionalization of cellulose

MCC was dissolved in IL **23** and reacted with varying amounts of **24** in the presence of TfOH under vacuum at 100 °C. Different concentrations of added **24** gave rise to composites with different morphologies: After extensive washing of the products, lower concentrations of **24** gave fibrous materials, while the addition of higher concentrations (50 and 75 %) of **24** resulted in compacted hard materials. Preliminary analysis of the obtained materials revealed the incorporation of the imidazolium moiety. For instance, infrared spectra were characterized by the presence of new peaks at 1755, 1536 and 1626 cm^{-1} indicating the formation of ester bonds and the presence of the imidazolium ring. Solid state CP/MAS NMR showed peaks in the region of 170 ppm, indicative of ester bonds. Moreover, elemental analysis confirmed the increased presence of nitrogen in accordance with increasing amounts of imidazoliums used (Table 6). As a control, MCC was dissolved in the IL **23** in absence of compound **24**, then precipitated and washed in the same way as the functionalized material. Elemental analysis for the blank confirmed no incorporation of nitrogen into the material. A thorough evaluation and characterization of these materials is currently ongoing in order to elucidate their structural features and potential application in detail.

Table 6. Elemental analysis of cellulose materials as a function of the amount of 24

entry	*Amount of 24 added [%][a]*	*N content [%]*	*C/N ratio*
1	0	0.27	162.24
2	20	3.43	13.12
3	50	6.16	7.75
4	75	6.05	7.85

[a] mol% calculated in respect to anhydroglucose units. N content and C/N ratio determined by elemental analysis after extensive washing in hot acetonitrile.

Conclusion

We have developed a sustainable alternative family of imidazolium zwitterions obtained from amino acids and carbohydrate fragments. Various transformations can be applied to the carboxylic functionalities common to all of these molecules, generating bifunctional cross-linkers and monomers. Moreover, ImZw can be converted to ionic liquids in a streamlined, low cost and sustainable decarboxylation process. The resulting compounds can be used with the same efficiency as their petroleum-derived counterparts in different applications. In addition to the results presented here, N-heterocyclic carbene catalysts, PILs, and carbon materials obtained on the basis of this new family of ImZw are currently under evaluation.

References

1. Cai, C. *Prepr. - Am. Chem. Soc., Div. Energy Fuels* **2014**, *59*, 352.
2. Mazzotta, M. G.; Gupta, D.; Saha, B.; Patra, A. K.; Bhaumik, A.; Abu-Omar, M. M. *ChemSusChem* **2014**, *7*, 2342–2350.
3. Zakrzewska, M. E.; Bogel-Lukasik, E.; Bogel-Lukasik, R. *Chem. Rev.* **2011**, *111*, 397–417.
4. Wilkes, J. S.; Wasserscheid, P.; Welton, T. In *Ionic Liquids in Synthesis*, 2nd ed.; Wasserscheid, P., Welton, T., Eds.; Wiley-VCH: Weinheim, 2007, pp 1–6
5. Urbanik, E.; Zabielska-Matejuk, A.; Skrzypczak, A.; Pernak, J. *Material und Organismen* **1997**, *31*, 247–263.
6. Demberelnyamba, D.; Kim, K. S.; Choi, S.; Park, S. Y.; Lee, H.; Kim, C. J.; Yoo, I. D. *Bioorg. Med. Chem.* **2004**, *12*, 853–857.
7. Ao, M.; Xu, G.; Zhu, Y.; Bai, Y. *J. Colloid Interface Sci.* **2008**, *326*, 490–495.
8. Arduengo, A. J. *Acc. Chem. Res.* **1999**, *32*, 913–921.
9. Ketz, B. E.; Ottenwaelder, X. G.; Waymouth, R. M. *Chem. Commun.* **2005**, *45*, 5693–5695.
10. Fechler, N.; Fellinger, T. P.; Antonietti, M. *Adv. Mater.* **2013**, *25*, 75–79.
11. Radziszewski, B. *Ber. Dtsch. Chem. Ges.* **1882**, *15*, 2706–2708.

12. Debus, H. *Justus Liebigs Ann. Chem.* **1858**, *107*, 199–208.
13. Esposito, D.; Kirchhecker, S.; Antonietti, M. *Chem. Eur. J.* **2013**, *19*, 15097–15100.
14. Esposito, D.; Antonietti, M. *ChemSusChem* **2013**, *6*, 989–992.
15. Painter, R. M.; Pearson, D. M.; Waymouth, R. M. *Angew. Chem., Int. Ed.* **2010**, *49*, 9456–9469.
16. Rasrendra, C. B.; Fachri, B. A.; Makertihartha, I. G.; Adisasmito, S.; Heeres, H. J. *ChemSusChem* **2011**, *4*, 768–777.
17. Mohan, D.; Pittman, J.; Charles, U.; Steele, P. H. *Energy Fuels* **2006**, *20*, 848–889.
18. Dapsens, P. Y.; Mondelli, C.; Kusema, B. T.; Verel, R.; Pérez-Ramírez, J. *Green Chem.* **2014**, *16*, 1176.
19. Venderbosch, R. H.; Prins, W. *Biofuels, Bioprod. Biorefin.* **2010**, *4*, 178–208.
20. Scott, D. S. WO Patent 88/00935, 1988.
21. Peterson, D. G.; Reineccius, G. A. *Flavour Fragrance J.* **2003**, *18*, 215–220.
22. Zgherea, G.; Stoian, C.; Peretz, S. *J. Liq. Chromatogr. Relat. Technol.* **2011**, *34*, 1268–1282.
23. Zimmermann, J.; Ondruschka, B.; Stark, A. *Org. Proc. Res. Dev.* **2010**, *14*, 1102–1109.
24. Chiappe, C. In *Eco-friendly Synthesis of Fine Chemicals*; Ballini, R., Ed.; *RSC Green Chemistry Series*; Royal Society of Chemistry: Cambridge, UK, 2009; Vol. 3, pp 155−185.
25. Cole, A. C.; Jensen, J. L.; Ntai, I.; Tran, K. L. T.; Weaver, K. J.; Forbes, D. C.; Davis, J. H., Jr. *J. Am. Chem. Soc.* **2002**, *124*, 5962–5963.
26. Galgano, P. D.; El Seoud, O. A. *J. Colloid Interface Sci.* **2011**, *361*, 186–194.
27. El Seoud, O. A.; Pires, P. A.; Abdel-Moghny, T.; Bastos, E. L. *J. Colloid Interface Sci.* **2007**, *313*, 296–304.
28. Xiong, Y.-B.; Wang, H.; Wang, Y.-J.; Wang, R.-M. *Polym. Adv. Technol.* **2012**, *23*, 835–840.
29. Fernandez-Castano, M., Master Thesis, Max Planck Institute for Colloids and Interfaces, Potsdam, Germany, 2013.
30. Clare, B.; Sirwardana, A.; Macfarlane, D. R. *Top. Curr. Chem.* **2010**, *290*, 1–40.
31. Seddon, K. R.; Stark, A.; Torres, M.-J. *Pure Appl. Chem.* **2000**, *72*, 2275–2287.
32. Villagran, C.; Deetlefs, M.; Pitner, W. R.; Hardacre, C. *Anal. Chem.* **2004**, *76*, 2118–2123.
33. Fukaya, Y.; Iizuka, Y.; Sekikawa, K.; Ohno, H. *Green Chem.* **2007**, *9*, 1155–1157.
34. Fukumoto, K.; Yoshizawa, M.; Ohno, H. *J. Am. Chem. Soc.* **2005**, *127*, 2398–2399.
35. Handy, S. T. *Chem. Eur. J.* **2003**, *9*, 2938–2944.
36. Handy, S. T.; Okello, M.; Dickenson, G. *Org. Lett.* **2003**, *5*, 2513–2515.
37. Ohno, H.; Fukumoto, K. *Acc. Chem. Res.* **2007**, *40*, 1122–1129.
38. Tao, G.-h.; He, L.; Liu, W.-s.; Xu, L.; Xiong, W.; Wang, T.; Kou, Y. *Green Chem.* **2006**, *8*, 639–646.
39. Kruse, A.; Dinjus, E. *J. Supercrit. Fluids* **2007**, *39*, 362–380.

40. Kirchhecker, S.; Antonietti, M.; Esposito, D. *Green Chem.* **2014**, *16*, 3705–3709.
41. Xiang, Z. *J. Mol. Struct.* **2013**, *1049*, 149–156.
42. Beletskaya, I. P.; Cheprakov, A. V. *Chem. Rev.* **2000**, *100*, 3009–3066.
43. Phan, N. T. S.; Van Der Sluys, M.; Jones, C. W. *Adv. Synth. Catal.* **2006**, *348*, 609–679.
44. Prechtl, M. H.; Scholten, J. D.; Dupont, J. *Molecules* **2010**, *15*, 3441–3461.
45. Zhang, J.; Zhang, H.; Wu, J.; Zhang, J.; He, J.; Xiang, J. *Phys. Chem. Chem. Phys.* **2010**, *12*, 1941–1947.
46. Wang, H.; Gurau, G.; Rogers, R. D. *Chem. Soc. Rev.* **2012**, *41*, 1519–1537.
47. Zhao, H.; Baker, G. A.; Song, Z.; Olubajo, O.; Crittle, T.; Peters, D. *Green Chem.* **2008**, *10*, 696–705.
48. Sun, N.; Rahman, M.; Qin, Y.; Maxim, M. L.; Rodríguez, H.; Rogers, R. D. *Green Chem.* **2009**, *11*, 646–655.
49. Brandt, A.; Gräsvik, J.; Hallett, J. P.; Welton, T. *Green Chem.* **2013**, *15*, 550–583.
50. Heath, L.; Thielemans, W. *Green Chem.* **2010**, *12*, 1448–1453.
51. Cai, J.; Kimura, S.; Wada, M.; Kuga, S. Nanoporous Cellulose as Metal Nanoparticles Support. *Biomacromolecules* **2009**, *10*, 87–94.
52. Gelesky, M. A.; Scheeren, C. W.; Foppa, L.; Pavan, F. A.; Dias, S. L. P.; Dupont, J. *Biomacromolecules* **2009**, *10*, 1888–1893.
53. Akkaya, A. *Fibers Polym.* **2013**, *14*, 22–27.
54. Mahmoud, K. A.; Lam, E.; Hrapovic, S.; Luong, J. H. *ACS Appl. Mater. Interfaces* **2013**, *5*, 4978–4985.
55. Ma, H.; Burger, C.; Hsiao, B. S.; Chu, B. *J. Mater. Chem.* **2011**, *21*, 7507–7510.
56. Song, H.; Luo, Z.; Zhao, H.; Luo, S.; Wu, X.; Gao, J.; Wang, Z. *RSC Adv.* **2013**, *3*, 11665–11675.
57. Buchtová, N.; Guyomard-Lack, A.; Le Bideau, J. *Green Chem.* **2014**, *16*, 1149–1152.
58. Thiemann, S.; Sachnov, S. J.; Pettersson, F.; Bollström, R.; Österbacka, R.; Wasserscheid, P.; Zaumseil, J. *Adv. Funct. Mater.* **2014**, *24*, 625–634.
59. Chiappone, A.; Nair, J. R.; Gerbaldi, C.; Jabbour, L.; Bongiovanni, R.; Zeno, E.; Beneventi, D.; Penazzi, N. *J. Power Sources* **2011**, *196*, 10280–10288.
60. O'Connell, D. W.; Birkinshaw, C.; O'Dwyer, T. F. *J. Chem. Technol. Biotechnol.* **2006**, *81*, 1820–1828.
61. Anirudhan, T. S.; Nima, J.; Sandeep, S.; Ratheesh, V. R. N. *Chem. Eng. J.* **2012**, *209*, 362–371.
62. O'Connell, D. W.; Birkinshaw, C.; O'Dwyer, T. F. *Bioresour. Technol.* **2008**, *99*, 6709–6724.
63. Vismara, E.; Melone, L.; Gastaldi, G.; Cosentino, C.; Torri, G. *J. Hazard. Mater.* **2009**, *170*, 798–808.

Chapter 5

Opportunities for the Replacement/ Minimization of Selective Hazardous Solvents: Applications, Concerns and Approaches To Identify Alternatives

Buchi Reddy Vaddula,[1,3] Swathi Yalla,[2] and Michael A. Gonzalez*,[3]

[1]Oak Ridge Institute for Science and Education (ORISE), Oak Ridge, Tennessee 37831
[2]Guest Researcher to the US EPA
[3]Sustainable Technology Division, National Risk Management Research Laboratory, U.S. Environmental Protection Agency, 26 West Martin Luther King Drive, MS 483, Cincinnati, Ohio 45268
***E-mail: gonzalez.michael@epa.gov**

The use of hazardous organic solvents in large-scale processes or at the individual user-scale has had tremendous impacts on our environment and human health. To address this, several governmental treaties and regulations have been enacted to minimize the effects of harmful chemical exposure to workers, consumers, and the environment. Additionally, the scientific community has been persistently engaged to provide solutions to these challenges by establishing guidelines, developing alternate methodologies and technologies, and promoting collaborations to reduce or eliminate solvent usage. This chapter details the use of solvents and the methodologies being developed and adopted to tackle these issues.

Introduction

Solvents play an important role peoples' lives, whether it's realized or not. These useful and functional chemicals are present in products such as paints, cosmetics, cleaners, and personal care products used daily. Also, solvents are incorporated in the production of furniture, electronics, automobiles and plastics, to name a few. While their role and functionality are essential to our everyday lives in terms of facilitated living and increased product choice and availability, they also provide a significant contribution to the global economics for the chemical and manufacturing sectors.

Over the past twenty years the tenants of Green Chemistry and Green Engineering have been developed, applied, improved, and incorporated into many other disciplines other than just chemistry and chemical engineering. From these advancements, researchers across the globe are identifying strategies, methods and tools that are being used to incorporate "greening" into their respective sectors and fields. The application of these advancements have also been integrated in to the replacement and elimination of solvents, not only for economic reasons but also for improving the environmental friendliness of the process or industry.

In this book chapter, the reader will learn of the role and functionality of solvents in society from daily use to industrial applications. As well as, the magnitude of solvents used globally and the reported emissions associated with their manufacture and use, along with the potential health and environmental effects attributed to solvent emission exposure. From this, the discussion will move to approaches that have been developed and implemented for the replacement of a solvent to a more environmentally-favorable one or its elimination entirely. Then, example case studies are provided to demonstrate the feasibility and economic benefits associated with replacing or eliminating a hazardous solvent. As with any replacement or elimination scenario, the reader will learn of methodologies that have been developed to evaluate the sustainability implications of this change, and the decision process that is used to ensure the choice is one that leads to a more sustainable situation, based on user preferences and constraints. With this information, it is anticipated the reader will learn of the life cycle attributes that must be considered when seeking out a process change that involves the replacement or elimination of a solvent.

Production of Hazardous Solvents

Any source of danger or risk that has the potential to harm, damage or exhibit adverse effects on the health of living beings, the environment, or property can be labeled a hazard. These hazards can originate from many sources, which include safety, physical, chemical, biological, ergonomic, and psychological. The identified hazards are also associated with the use and manufacture of chemicals, especially solvents. Since most solvents are organic chemicals and those with the ability to evaporate at or near room temperature, due to their high vapor pressure, have the potential to be further hazards, known as volatile organic compounds (VOCs). Under this category are solids that have the ability to sublime under ambient conditions. Furthermore, compounds identified as hazardous air pollutants (HAPs), are chemicals (consisting mainly of VOCs) that have the potential to cause serious environmental and health hazards.

While organic solvents are used predominately in many aspects of our day-to-day activities, they are a major cause of concern due to their physical and chemical properties, toxicity, widespread usage, consumer and worker exposure, and consumption in large quantities. To demonstrate their commercial significance, Table 1 identifies the top twenty chemicals produced in the United States (U.S.) for the year 2012 (*1*). Of this group, two chemicals (cumene and ethylene dichloride) are listed as hazardous air pollutants (HAPs) by the US EPA and six more chemicals (bolded in Table 1) are defined as volatile organic compounds (VOCs).

Table 1. The top twenty produced chemicals in U.S. in the year 2012 (*1*)

No.	*Chemical Name*	*CAS No.*	*2012 Production volume range (in million tons)*
1	Carbon dioxide	124-38-9	67.5-70.0
2	**Ethanol**	64-17-5	42.5-45.0
3	Sulfuric acid	7664-93-9	40.0-42.5
4	Tricalcium silicate	12168-85-3	40.0-42.5
5	**Ethene**	74-85-1	27.5-30.0
6	**Propane**	74-98-6	27.5-30.0
7	Calcium carbonate	471-34-1	27.5-30.0
8	Calcium oxide	1305-78-8	25.0-27.5
9	**Butane**	106-97-8	25.0-27.5
10	Iron oxide	1309-37-1	25.0-27.5
11	Hydrogen	1333-74-0	25.0-27.5
12	Ethane	74-84-0	20.0-22.5
13	Ammonia	7664-41-7	17.5-20.0
14	**Cumene**	98-82-8	17.5-20.0
15	**Isobutane**	75-28-5	17.5-20.0
16	**Propene**	115-07-1	15.0-17.5
17	Sulfur	7704-34-9	15.0-17.5
18	**Ethylene dichloride**	107-06-2	12.5-15.0
19	Sodium hydroxide	1310-73-2	12.5-15.0
20	Urea	57-13-6	12.5-15.0

Whereas, the global production of many of these identified VOCs is in the range of millions of tons per year. To further demonstrate the commercial significant of these chemicals, Table 2 provides the global production values of some of these HAPs in the recent years.

Table 2. Global production (in million tons) of some of the hazardous air pollutants (HAPs) in recent years.

No	*Chemical*	*Production (In million tons)*	*Year*
1	Acetaldehyde	1.33 (estimated)	2012
2	1,3-butadiene	11.9	2012
3	Chloroprene	0.44 (estimated)	By 2017
4	Formaldehyde	30	2013
5	Hydrogen fluoride	1	2013
6	Methanol	65	2013
7	Chlorine	56	2013
8	Phenol	10	2014
9	Methyl chloride	1.22	2004
10	Methyl-*tert*-butyl ether	15.2	2013
11	Phosgene	2.29	2006
12	Propylene oxide	7.76	2012
13	Styrene	32.1	2012
14	Vinyl acetate	5	2011
15	Acetone	6	2013
16	Ethylene glycol	18	2013
17	Xylenes	42.5	2010
18	Toluene	19.8	2010
19	Ethyl benzene	29.2	2010

From the list of chemicals in Table 2, 11 of them are identified as being utilized as commodity solvents.

Industry Wide Application of Organic Solvents

Organic solvents are employed for a variety of industrial purposes. The variety and magnitude of these tasks range from cleaning and maintenance of equipment to the manufacture of pharmaceutical drugs and to their use as a mobile-phase carrier. However, for many solvents, their benefits are also associated with their emission and release as volatile organic compounds (VOCs) across all stages of their life cycle: manufacture, use, reuse and disposal. These VOCs, if not contained, can pose serious health, safety, and environmental concerns. A brief description of selected industrial sectors that use solvents is provided in sections below. To demonstrate the quantity and variability of applications of solvents in use, Table 3, provides the total release of hazardous air pollutants (HAPs) from these identified various industrial sectors in the United States for the year 2012 (*2*).

Paints and Associated Products

Paints and associated products such as stains, varnishes, inks, lacquers, and others are used in applications ranging from providing fire resistance, noise and water insulation, avoidance of metal corrosion, reflective coatings, and for aesthetic purposes, to name a few. Most of these paints and associated products are comprised of more than 30% solvent, by volume. The main role of a solvent(s) in paints and associated products is to provide dissolution of various components during their preparation, such as pigments, and resins (*3*). In order for the product to maintain a uniform consistency and facilitated ease of use, the solvent choice plays an important role in the wetting and dispersion of pigments, dissolution of solid/viscous binder components, and controlling viscosity and stability during storage. Generally, the solvents commonly used for these applications are based on aliphatic and aromatic hydrocarbons, glycol ethers, ketones, alcohols and esters of acetic acid. Additional solvents used include toluene, xylene, perchloroethylene, methylene chloride, trichloroethylene, or mixture of these solvents (*4*).

Table 4, listed below, provides a list of the top 20 HAPs, as defined by U.S. EPA on their Air Toxics Web Site (*5*), released by the paints and coating industry during the product manufacturing phase.

While pure organic solvents are an important constituent in almost all paints, water-based coatings, which also contain organic solvents, also have as equally an important function and role. Water-based paints may contain up to 15% organic solvents by volume. This addition of an organic solvent component is important and allows the products to maintain high durability and provide an improved luster. In these water-based paint products, the choice of solvent imparts an effect on the flow properties needed for application, glossiness of the paint coating, needed for durability and aesthetic value. Additionally, the solvent composition aids in controlling the drying behavior of the paint coating, and governs its (the solvent's) release into the air from the applied liquid film coating as the paint dries. The choice of solvent flexibility also helps in tuning the properties of the coating, which may directly result in improving its environmental benignity, performance, and the properties of materials used.

Additional advantages of using solvents in paints and associated products include:

- Reduction in energy consumption and minimization of costs due to ready evaporation of solvents at room temperature (10 times faster when compared to other low-VOC solvent alternatives).
- Varied application conditions (temperature and humidity)
- Improved flow and leveling characteristics of the product

Table 3. TRI On-site and Off-site Reported Disposed of or Otherwise Released (in pounds), for Hazardous Air Pollutants, By Industry, U.S., 2012 (*2*)

No.	*Industry*	*Total On- and Off-site Disposal or Other Releases (lbs)*
1	Metal Mining	732,103,727
2	Chemicals	223,413,369
3	Electric Utilities	172,835,697
4	Paper	137,074,027
5	Primary Metals	113,476,535
6	Hazardous Waste/Solvent Recovery	90,814,540
7	Plastics and Rubber	30,029,250
8	Food/Beverages/Tobacco	29,524,424
9	Petroleum	23,502,556
10	Transportation Equipment	21,519,004
11	Fabricated Metals	18,317,008
12	Stone/Clay/Glass	11,565,363
13	Defense related	8,114,476
14	Wood Products	7,225,160
15	Printing and Publishing	6,025,285
16	Machinery	4,397,795
17	Cement	3,680,269
18	Coal Mining	3,647,163
19	Furniture	3,112,724
20	Electrical Equipment	2,831,892
21	Miscellaneous Manufacturing	2,269,810
22	Petroleum Bulk Terminals	1,543,964
23	Chemical Wholesalers	1,301,084
24	Textiles	1,270,660
25	Computers/Electronics Products	1,100,114
26	Leather	742,136
27	Apparel	21,147
	Total	**1,651,459,177**

Table 4. Total Release (in pounds), for facilities in NAICS 325510 – Paints and Coating, for Hazardous Air Pollutants, U.S., 2012 (top 20 HAPs)

No.	*Chemicals*	*Total release (lbs)*
1	Methanol	927,290
2	Xylene (mixed isomers)	581,008
3	Toluene	479,028
4	Zinc compounds	392,621
5	Styrene	172,500
6	Ethylbenzene	113,317
7	Methyl isobutyl ketone	105,761
8	Methyl methacrylate	83,460
9	Diisocyanates	78,062
10	Chromium compounds	71,708
11	Ethylene glycol	64,477
12	Dichloromethane	41,630
13	Phenol	28,127
14	n-Hexane	18,927
15	Vinyl acetate	17,747
16	Phthalic anhydride	15,816
17	Naphthalene	9,719
18	Manganese compounds	9,318
19	Hydrogen fluoride	8,624
20	Formaldehyde	8,471
21	Lead compounds	8,450

Upon application of a paint coating, the rapid evaporation of the solvent is an important function that governs the drying rate of the paint and results in a film of coating. The fugitive evaporated solvent, if precautions are not taken, can cause human health hazards to the exposed worker and pollute the surrounding atmosphere and environment (*6*).

The exposure of organic solvents, even in low concentrations, may cause temporary symptoms such as dizziness, headache and euphoria. However, exposure at higher concentrations may lead to respiratory congestion, anesthesia, and circulatory disturbances which may ultimately result in death. Chronic exposure may damage the central nervous system leading to emotional and cognitive disorders (*6*).

To counter these immediate near-field effects, efficient air exhaust and proper ventilation at the workplace are required to protect workers from any potential toxicological and human health effects of solvents. Additional examples of protective measures include solvent recovery plants and incineration of solvents to control the release of solvents and to deter their harmful effects on human health and environment.

Other alternative methods for the prevention of these hazardous effects are the development of low-solvent and low-VOC solvent coatings. However, solvent-free coatings, such as completely water-based paints and solid based paints, would be an ideal choice (*7*). To further expand on this potential, the impact of the hazardous effects of these solvents can be reduced by assessing alternative environmentally-benign solvents that can be used to replace harmful ones (*8*). This concept will be discussed in more detail in a later section of this chapter.

Pharmaceuticals and Fine Chemicals

Organic solvents also play an important role in the pharmaceutical and fine and bulk chemical industry. They are commonly used in all steps of preparation of an active pharmaceutical ingredient (API) or in the production of fine and bulk chemicals with uses ranging from reaction media, extraction/separation and purification, to cleaning of reactor and separation equipment. This importance of organic solvents to the chemical (pharmaceutical and fine and bulk chemical) industry is demonstrated by this industry being one of the top consumers and producers of organic solvents. Organic solvents are also widely used in pharmaceutical ingredient formulation, although any residual solvent remaining in the final dosage form is strictly regulated by the FDA (*9*).

With this reliance on their consumption, as well as production, the chemical industry is one of the major contributors to the total release of VOCs. Table 5 provides the total release of hazardous air pollutants from various processes in chemical industry sector. In the overall total industry releases, the chemical industry is the second largest contributor in U.S., with the metal mining industry being largest generator of HAPs.

The mass of solvents used in an average batch chemical process operation accounts for approximately 80% of the total process mass, with its use posing and contributing to a major concern in terms of health and safety (*10*). This effect implies the solvent plays a greater role in determining the human toxicity impact of a given process due to their larger mass proportion than the reacting components and resulting product. However, more often than not, the retrosynthetic design of a chemical route doesn't always take into account the rôle and environmental impact of the organic solvents used. And more so, they are merely considered only as a reaction medium (*11*). To avoid this from continuing to occur, it is imperative researchers and process operators follow and apply the principles of green chemistry and engineering in the movement towards increased sustainability of chemical operations and processes (*12*).

Table 5. TRI On-site and Off-site Reported Disposed of or Otherwise Released (in pounds), for facilities in NAICS 325 - Chemicals, for Hazardous Air Pollutants, U.S., 2012 (top 20 of 155 chemicals)

No.	*Chemical*	*Total On- and Off-site Disposal or Other Releases (lbs)*
1	Manganese compounds	45,066,591
2	Methanol	30,332,921
3	Acetonitrile	17,335,230
4	Formaldehyde	15,617,437
5	Acrylonitrile	10,898,898
6	Chromium compounds(except chromite ore mined in the transvaal region)	8,038,393
7	Hydrochloric acid (1995 and after "acid aerosols" only)	7,637,275
8	Carbonyl sulfide	7,376,257
9	n-Hexane	6,869,557
10	Acrylamide	5,189,100
11	Toluene	4,807,712
12	Hydrogen fluoride	4,728,825
13	Manganese	4,460,053
14	Ethylene glycol	4,097,955
15	Acrylic acid	3,678,094
16	Styrene	3,535,398
17	Phenol	2,999,481
18	Nickel compounds	2,527,499
19	Xylene (mixed isomers)	2,468,856
20	Acetaldehyde	2,298,948

In view of the hazards associated with organic solvents, efforts are ongoing or have been put into place to replace toxic organic solvents with relatively more benign ones or to decrease the amount of solvents used and their exposure potential in a process. As an example, several neoteric solvents are being designed to counter the problems associated with the use of hazardous organic solvents (*9*). This concept will be discussed more in-depth in Section 2.4 of this chapter.

Printing Operations

Printing operations include screen-printing, flexography, lithography, letterpress, and gravure on substrates such as paper, plastic, metal and ceramic. While the printing results may be similar, the operations for each of these processes are not same. However, the use of organic solvents is a common feature in printing, finishing, imaging, and plate making operations. These solvents play a significant role in determining the type of inks used in finishing operations, fountain solutions, blanket & roller washes, dampener roller washes, deglazers, offset inks, flexographic inks, gravure inks, screen process, general cleaners, and degreasers. For example, fountain solutions primarily contain isopropyl alcohol (IPA) as a solvent, which serves as a component that moistens the non-image area of the printing plate to avoid ink from depositing.

Those printing operations that involve the use of solvents like toluene, certain glycol ethers, ethylene glycol, xylene (mixed isomers), methyl isobutyl ketone (MIBK), *N*-Methyl-2-pyrrolidone (NMP), 1,2,4-trimethylbenzene, methanol, *n*-butyl alcohol, and *n*-hexane for various operations in press rooms contribute to in-plant VOC generation in the air and challenges with a printers effluent. Printing and finishing operations can contribute, to a large extent, to VOC emissions and is significant, as each of these processes requires inks with varied properties. For instance, the nature (physical and chemical properties) and concentration of solvents in inks vary depending on the viscosity requirement of the process' associated inks. Thus, this is a complex process that needs solvents with specific intensive and extensive properties to function properly.

Additionally, the printing and publishing industry is the fifteenth largest contributor of VOC in the U.S., for the year 2012. To address this, the reduction of VOC emissions in printing operations is being tackled with extensive research efforts. It has been identified cleaning operations in the printing industry contributes to the majority of the overall VOC emissions associated with this industry. In order to confront this issue, several companies have devised cleaning solutions that contain relatively less hazardous and less toxic solvents (*13*). Table 6 below provides the total release of hazardous air pollutants from various processes in the printing and publishing industry.

The impetus of solvents to produce VOCs in various industries and applications over centuries is due to their volatility and solvency properties. Though these solvents pose a significant risk to the environment and workers health, their continued use is due to the paucity of efficient replacements. The health hazards associated to exposure of VOCs include headaches, respiratory illness, asthma, skin complications, cancer, and compromised pulmonary function. This emphasizes the urgent need to develop suitable benign alternatives to solvents that produce VOCs.

The number of options for improving the quality of air for workers and their safety through VOC exposure reduction has been steadily increasing. Several volatile solvents are now being replaced with neoteric solvents in applications such as cleaning formulations, inks, and lubricants, which otherwise contain high levels of VOCs. However, currently, it may not be possible to substitute all VOC generating solvents. But, the advantages offered by solvents from renewable sources require further research for development. The development of alternate technologies, such as the UV dried inks, is a significant advancement in reducing VOC emissions and product quality.

This environmental responsibility also requires updating information and improving the working knowledge on the availability of new techniques and tools for performing an operation more efficiently with improved safety and environmental-friendliness. The correct selection of solvents and equipment for a desired operation is not only important in terms of productivity, and cost but also meeting environmental and safety regulations.

Solvent Cleaning, Conditioning and Degreasing

Grease, oil, wax, fats, and soil are examples of contaminants removed from various plastics, metals or glass items using organic solvents by a physical process called “solvent degreasing” or “solvent cleaning”. This process is primarily used to remove oils, grease, wax, and oil-borne soils from materials that have undergone a process or use such as welding, soldering, molding, stamping, and die-casting. Different methods are used to carry out the degreasing process and the choice of solvent is determined by the type of equipment used for the contaminant removal process. There are also specific processes such as cold cleaners, vapor degreasers, and conveyorized degreasers. The organic solvents generally used in solvent cleaning and degreasing are ketones, alcohols, chlorinated hydrocarbons, and petroleum distillates. The type of solvent (family) used for the degreasing process are chosen based on the solubility of the substance to be stripped, toxicity, and other physical properties of the solvent. For components containing grease, oil, and waxes, cleaning is more effectively carried out with chlorinated solvents rather than hydrocarbon or fluorocarbon solvents.

Examples of industries that use solvent degreasing are metal-working such as aviation, automobile, defense machinery, electronics, shipping, and refrigeration. The types of products cleaned can range from tiny transistor parts to aircraft components. Solvent cleaning applications are also carried out in industrial sectors such as paints, plastics, printing, chemicals, rubber, paper, and electric power. Examples of specific applications of solvents that are the used are petroleum distillate and Stoddard solvents in the maintenance of cold cleaners to various types of solvents when a higher degree of cleaning and quality is necessary. For vapor degreasers processes the use halogenated solvents like 1,1,1-trichloroethane, trichloroethylene, and perchloroethylene are favored due to their high vapor pressure.

Parts cleaning are an important industrial process, which prepares a surface prior to finishing or protection of sensitive components. For example, in the electroplating industry, part cleanliness is crucial since any remaining droplets of oil from the previous process may prevent in effective and desired adhesion of the coating material.

Table 6. On-site and Off-site Reported Disposed of or Otherwise Released (in pounds), for facilities in NAICS 323/51 - Printing and Publishing, for Hazardous Air Pollutants, U.S., 2012

No.	*Chemical*	*Total On- and Off-site Disposal or Other Releases (lbs)*
1	Toluene	5,217,574
2	Certain glycol ethers	588,466
3	Ethylene glycol	119,329
4	Xylene (mixed isomers)	39,813
5	Methyl isobutyl ketone	39,044
6	Methanol	13,730
7	Lead	3,384
8	n-Hexane	2,555
9	Chromium	673
10	Diethanolamine	500
11	Ethylbenzene	366
12	Nickel compounds	145
13	Phenol	92
14	Lead compounds	42
15	Mercury compounds	11
16	Antimony	0
17	Manganese	-
	Total	**6,025,725**

Adhesives

The term adhesive is defined as a substance that binds an adherent (i.e. surface of materials) with a certain degree of strength depending on the type of adhesive material used. Other terms for an adhesive include glue, paste, cement, or mucilage. The word adhesive also contain prefixes to describe the nature of the adhesion and type of adhesive, the application conditions, and the material type to be bonded (*14*).

Traditionally, solvent-based adhesives are developed from resins such as acrylic, polychloroprene, polyurethane, and synthetic and natural rubbers. The volatile organic compounds (VOCs) that are generated from the use of these solvent-based adhesives in significant quantities are considered to have harmful environmental and health implications. The top 20 hazardous air pollutants released by adhesive manufacturing in the year 2012 is provided in the Table 7.

Table 7. Total Release (in pounds), for facilities in NAICS 325520 – Adhesive manufacturing, for Hazardous Air Pollutants, U.S., 2012

No.	*Chemical*	*Total Release (lbs)*
1	Toluene	198,469
2	n-Hexane	105,640
3	Methyl methacrylate	87,385
4	Dichloromethane	70,781
5	Diisocyanates	49,900
6	Vinyl acetate	43,504
7	Methanol	30,892
8	Xylene (mixed isomers)	16,896
9	Vinylidene chloride	15,552
10	Glycol ethers	12,061
11	Ethylene glycol	10,835
12	Lead compounds	9,773
13	Chromium compounds	7,216
14	Tetrachloroethylene	6,355
15	Manganese compounds	5,925
16	Styrene	5,174
17	Methyl isobutyl ketone	4,911
18	Ethylbenzene	4,192
19	Trichloroethylene	3,918
20	Ethyl acrylate	2,715

As in the paint industry, the characteristics and role of the solvent is crucial to having a successful end product. For adhesives, the drying and bonding time of a solvent-based adhesive is varied and dependent on the method of application and the process of assembly. The heat resistance and bonding strength of solvent-based adhesives are typically intermediate and provide a good bond line of flexibility. The organic solvents used in solvent-based adhesives render them less sensitive to any contamination remaining on the bonding surfaces.

As with many applications involving solvents, the faster the evaporation rate of the solvent upon application, the greater the adhesion and viscosity of the resulting adhesive film. The materials to be adhered together can be bonded as soon as the adhesive is applied to their surfaces or upon evaporation of a portion of the carrier solvent. However, the adhesion must occur before the adhesive completely dries out. Allowing for proper set time before initiating bonding increases the adhesive bond initial strength and quick processing of the final assembly.

Although a variety of solvents are used in the preparation of adhesives, the commonly used solvents are acetone, heptane, methyl ethyl ketone (MEK), naphtha, methylene chloride, *n*-butyl acetate, mineral spirits, and toluene. Common types of solvent-based adhesives are non-pressure sensitive, pressure sensitive, and contact adhesives.

Pressure sensitive adhesives form a bond between the two surfaces when significant pressure is applied (e.g. labels and tapes). Whereas, non-pressure sensitive adhesives do not require the direct application of pressure (e.g. sealants and mastics) and contact adhesives are used when a high shear-resistance is required with strong bonding (e.g. laminates and footwear).

Styrene-butadiene rubber (SBR), styrenic block copolymers, acrylic polymers, and natural rubber are generally used in solvent-based pressure sensitive adhesives. And, butyl rubber or SBR adhesives are generally used in non-pressure sensitive applications. For contact adhesives polychloroprene is commonly employed.

Ketone based organic solvents are used in the preparation of solvent-based adhesives. MEK and acetone are the primary ingredients in solvent mixtures for nitrile, neoprene, and urethane-based adhesives. The main solvent in pressure sensitive chlorinated rubber and resin-type adhesives is acetone. The other ketone, methyl isobutyl ketone (MIBK), is a component in the solvent blend for acrylic, nitrile rubber, polyvinyl chloride (PVC), and PVC-PVA copolymer adhesives. The nature of ketone used in the adhesive is determined by the required rate of evaporation (*15*).

Plastics and Rubber Materials

Many of the diverse products produced by the industrial sector require the use of process solvents. Solvents may be used for finishing, cleaning, preparation etc. For example, solvents are used in various processes within the polymer production industry, including the manufacture of plastics and synthetic rubber. In the U.S., the plastics and rubber industry employs over 1,000,000 people and operates over 16,000 plants as per 1997 estimates. The revenue associated with this sector's product sales total over $160 billion, of which the tire industry contributes over $14 billion itself (*16*).

The total release of HAPs from the plastic and rubber industry is greater than 38 million pounds per year. Of all VOC releases by U.S. industries, this sector is estimated to be the 10th largest HAPs contributor and 7th largest contributor of VOCs (*17*). The emission contribution from these industries is identified as small and is in the range of 10-50 mg per kg of rubber processed. But, due to the large number of products produced per year their contribution is quite significant. This industry emits VOCs such as toluene, methanol, methyl ethyl ketone, 1,1,1-trichloroethane, carbon disulfide, dichloromethane, hexane, methyl isobutyl ketone, xylene and 1,1,1-trichloroethane in very large quantities (*18*). Table 8 below illustrates the total release of hazardous air pollutants from various processes within the plastics and rubber industry for the year 2012.

Table 8. TRI On-site and Off-site Reported Disposed of or Otherwise Released (in pounds), for facilities in NAICS 326 - Plastics and Rubber, for Hazardous Air Pollutants, U.S., 2012 (top 20 of 64 chemicals)

No.	*Chemical*	*Total On- and Off-site Disposal or Other Releases (lbs)*
1	Styrene	14,686,440
2	Carbon disulfide	6,969,353
3	Toluene	2,152,116
4	Methanol	915,340
5	Xylene (mixed isomers)	807,533
6	Trichloroethylene	675,420
7	Di(2-ethylhexyl) phthalate	663,307
8	Methyl methacrylate	445,419
9	Ethylene glycol	365,060
10	Phenol	360,970
11	*n*-Hexane	336,743
12	Dichloromethane	335,918
13	Certain glycol ethers	244,338
14	Methyl isobutyl ketone	210,550
15	Antimony compounds	159,862
16	Hydrochloric acid	155,207
17	Vinyl acetate	121,165
18	*N,N*-Dimethylformamide	75,782
19	Ethylbenzene	66,751
20	Formaldehyde	48,749

In the production of tires, many different layers and pieces of rubber are joined together by applying dearomatized aliphatic hydrocarbon solvents to ensure their durability and safety. Additionally, anti-ozonants are used to increase the protection of rubber from ozone attack. These anti-ozonants include unsaturated rubbers such as nitrile-butadiene, styrene-butadiene, polybutadiene, and natural rubbers. Methyl isobutyl ketone (MIBK) is used in the preparation of anti-ozonants from this respective raw materials.

In plastics industry, solvents are used for plastic welding, which is a process for joining softened surfaces of plastic materials. The solvents, upon application, temporarily dissolve the polymer surface, when applied to the surface at room temperature. As the surface of the polymer is dissolved, this results in freely moving polymer chains, which can entangle with chains from the other dissolved polymer surface to be welded. Upon welding the polymer surfaces become intact. This is the result of the solvent permeating through the polymer, followed by evaporating and leaving the polymer chains rigid. For example, in household plumbing, the common application of this technique is used in connecting ABS and PVC pipes. The adhesion of plastics (polystyrene, polycarbonate, etc) is also achieved by using a solvent welding process.

The chlorinated solvent dichloromethane, apart from its presence in paint strippers, is also used in the chemical welding of some plastics. Some examples include sealing electric meter cases, water buckets, etc. It can also be used to solvent weld polymethylmethacrylate and polycarbonate and is the primary solvent present in weld plumbing.

In the manufacturing of plastics, the polymerization of isoprene is facilitated with alkyllithium catalysts. These homogeneous catalysts are used due to their solubility in hydrocarbon solvents. The resulting non-polar hydrocarbon rubbers such as polyisoprene, polybutadiene, butyl, styrene-butadiene, polybutadiene, ethylene-propylene, ethylene-propylene-diene, and natural rubbers and their vulcanizates dissolve/swell considerably in hydrocarbon solvents (*19*). These are just two examples of why hydrocarbon solvents are preferred by this industry.

Role and Function of Solvents

Physical and Chemical Properties Influencing the Choice of Solvents

The overall influence of a solvent on chemical reactivity is known as the solvent effect. This effect influences the choice of solvent for a particular reaction and results in enabling control of the reaction rate, solubility of the reagents, and stability of the intermediates. The thermodynamic and kinetic control of a chemical reaction is more pronounced upon choosing an appropriate solvent. The dissolution of a solute in a solvent depends upon forming favorable interactions with the solvent. The change in free energy of both the solvent and solute determines the extent of dissolution.

Several factors influence the free energy of solvation, such as the formation of a cavity in the solvent. As the ordered structure of the solvent increases, solvent-solvent interactions become decreased with the formation of this cavity, leading to unfavorable enthalpy and entropy values for the solution. On the other hand, it is becoming preferred for the bulk solute to break apart into solute particles. This breaking apart is unfavorable in terms of enthalpy, but favorable in terms of entropy. Also, the cavity formed in the solvent should now become occupied by the solute. The resulting mixture has more disorder and with these favorable solute-solvent interactions a greater entropy is achieved when compared to the entropy values of the independent solute and solvent. The similarity of these interactions between solute-solvent and solvent-solvent often results in dissolution. Hence the term "like dissolves like" signifies, nonpolar solvents dissolving nonpolar solutes, and polar solvents dissolving polar solutes.

The equilibrium constant of a reaction can also be influenced by varying the solvent and its chemical properties, which allow for creating stability of the product(s) or the reactant(s). This preferential stabilization of the product(s) or reactant(s) determine the direction of equilibrium shift. Various non-covalent interactions of solvents, such as Vander Waals forces, hydrophobic effects, H-bonding etc., with either the product or reactant results in stabilization and the respective resulting equilibrium constant.

The change of solvent can also influence the ionization equilibrium of an acid or a base. The dielectric constant of a solvent and its preferential solvation affects the ionization equilibrium through stabilization of certain species in acid-base solution. Additionally, the effect of the solvent is due to its inherent acidity or basicity. Thus, the acidity or basicity of a reaction mixture can be influenced by altering the solvent. For example, the stabilization of an ionic species is far greater in water when compared to dimethyl sulfoxide (DMSO) or acetonitrile. This is due to water's relatively highly polarity and as a result the dissociation equilibrium of an acid is more pronounced in water than DMSO and acetonitrile.

Solvents also have the ability to influence a chemical's predominant isomer form. For example, 1,3-dicarbonyl compounds can exist in either of the two tautomeric forms: cis and diketo. The cis form is more predominant in solvents with a low polarity whereas the diketo form is preferred in solvents with a higher polarity. This preference for conformation is the result of the intramolecular hydrogen bonding being more pronounced in low polarity solvents, which leads to a greater cis-enol form. But in higher polarity solvents, the competitive H-bonding of 1,3-dicarbonyl compounds with the polar solvents is favored over the intramolecular H-bonding. This results in the diketo form being predominant.

Reaction solvents can also influence the order of a chemical reactivity and their respective reaction rates. However, during the design of a chemical reaction or determining reactivity of a reaction, the role of solvents is often considered passive. As scientists learn more about chemical reactivity of solvents, this knowledge is creating new strategies for developing chemical reaction and increasing the greenness of the overall reaction.

Chemical reaction rates are influenced by solvent effects and can be explained by transition state theory. The differential in solvation effect between the transition state and starting materials attributed by the reaction solvent affects the rate of a reaction. The solvent molecules play a role by stabilizing the transition state of the reaction via orientation. However, the solvent molecules can only stabilize the transition state and the starting materials to a certain extent. If the transition state is stabilized to a greater extent than the starting materials, the reaction moves forward with a higher rate. Similarly, if the solvent molecules orient to stabilize the starting materials, then the reaction progress would be slow or nonexistent. The reoriented solvent molecules which aid either the starting materials or the transition state needs to relax to the ground state. These effects are generally observed in reactions with solvents that are rapidly relaxing and weakly dipolar.

The above approach does not apply for reactions that proceed rapidly, as the stabilization of the transition state with strongly dipolar and slowly relaxing solvents do not play a significant role. In these fast proceeding chemical reactions, the rate of the reaction is largely affected by the dynamic contributions of the solvent such as viscosity, density, and friction.

Hughes and Ingold (*20*) studied the effect of solvation on nucleophilic substitution and elimination reactions by considering the electrostatic interactions between solvents and charged molecules. These reactions were classified into different types based on the charge of initial and transition states. However, it was assumed the extent of solvation will be affected by changing the magnitude of charge, changing delocalization, loss of charge, and dispersal of charge.

From this it was understood the rate of a reaction either increases or decreases by changing solvent polarity and the charge developed in the activated complex with the reactant molecules. If the charge in the activated complex is greater than the charge of the starting materials, increasing or decreasing the polarity of solvent will increase or decrease the rate of a chemical reaction, respectively, or vice versa.

The nature of the solvent used in these reactions significantly affects the rate and the mechanism of a reaction. Some solvent conditions prefer one mechanism route to another. As an example, those solvents that stabilize a resulting intermediate carbocation are more suitable for S_N1 reactions. This stability is generated by a polar solvent solvating the intermediate carbocation and decreasing its energy when compared to the starting material thus accelerating the S_N1 reactions.

For S_N2 reactions, poor solvation of the nucleophile increases the rate of a reaction. While stabilizing the intermediate carbocation is important for accelerating S_N1 reactions. Destabilizing the starting nucleophile is important for accelerating S_N2 reactions. The strength of the nucleophile determines the mechanism of a reaction when both S_N1 and S_N2 reactions are viable. The use of protic solvents for S_N2 reactions are not a good choice as the nucleophilicity of the nucleophile is linked to its basicity. The excellent basicity of stronger nucleophiles leads to acid-base type of reaction with protic solvents, thus hindering the strength of the nucleophile and decelerating the rate of reaction.

Apart from the substitution reactions discussed above, transition metal catalyzed reactions and free radical reactions are also dramatically influenced by solvation from the solvent used. Therefore, it is of great importance to consider the appropriate solvent for the role and results desired. But, there also needs to be emphasis on the tangential effects of selecting a solvent and the implications that arise from that selection.

Cause for Concern

Health Concerns

Organic solvents, in spite of their wide area of applications and numerous benefits, may be the cause of most hazards in view of worker and consumer health. These solvents, if not properly and effectively contained, can enter the human body by any of three routes – oral ingestion, nasal inhalation, and skin absorption. The resulting health issues associated with solvent exposure depend on several factors, such as the nature of solvent, duration of exposure, and the route of entry. Upon entering the body, the symptoms may be short-term or long-term depending on the length and the nature (acute or chronic) of exposure. The short term effects of solvent exposure include drowsiness, nausea, headache, and dermatitis. Extensive effects from high-level solvent exposure include sedation, fainting, suffocation and death. Low-levels of chronic exposure may have effects on the function of the heart, liver, kidney, lungs, stomach, and central nervous system and fertility complications in men and women.

Almost all organic solvents are poisonous if swallowed or inhaled in sufficient quantities, and cause dermatitis upon sufficient skin contact. At high concentrations most solvents can cause necrosis (dizziness, nausea, fatigue, loss of coordination and coma). Long- term exposure to high concentrations can lead to brain damage.

Some of the most dangerous solvents include aromatic hydrocarbons. For example, benzene is considered to be carcinogenic as it can cause acute myeloid leukemia (AML) in humans (*21*). As a result, benzene has been banned as an ingredient in home products. The major hazards associated with aromatic hydrocarbons are induced by inhalation, although they may also be absorbed through the skin. It has been determined their effect on the body is as strong as those experienced with narcotics. Benzene has been used in applications that include lacquer thinners, paint and varnish removers, aerosol spray cans, fluorescent dye solvents etc.

Another class of solvents that also demonstrate extensive hazard are the halogenated hydrocarbons. For example, halogenated hydrocarbons like chloroform were used as anesthetics in the past but were found to be extremely toxic. When skin is exposed to a halogenated hydrocarbon, dermatitis may form. This is the reaction of the fatty layer of the skin being dissolved by the chemical. Additional health effects from exposure are hepatic and kidney damage.

It has been recently established that exposure of 1,1,2-trichloroethene in the workplace increases the risk of developing Parkinson's disease by 6 times (*22*). This is further emphasized by the European countries having categorized trichloroethene as a serious carcinogen. While the use of trichloroethylene is extensive, such as it being used as a degreasing agent in the metal finishing industry, it has been banned in many applications worldwide.

The petroleum distillates, which are commonly used and are referred to as aliphatic hydrocarbons, are considered to be less hazardous when compared to other solvents. While the narcotic effect of these solvents is mild, they can cause significant bodily damage like pulmonary edema from ingestion. Aspiration into the lungs may cause lung irritation and eventual death. These solvents are widely used in rubber cement thinners, clean-up solvents, silk screen poster inks, and paint thinners.

The ester-based solvents are commonly used in resin, lacquer, and plastics. These solvents have been known to cause anesthetic effects and have also been found to be nose, throat, and eye irritants.

The toxicity of glycol ethers and their corresponding acetates was found to be much higher than previously expected (*23*). Kidney damage and anemia are known effects caused by exposure to the methyl and butyl derivatives of cellosolve. Mixed glycol ethers have been reported to cause increased rate of miscarriages, birth defects, sterility at low levels and testicular atrophy. However, glycol ethers are used in various applications which include color photography, paints, lacquer thinners, and aerosol sprays.

Ketones are known to cause upper respiratory tract irritation, eye irritation and necrosis. The exposure of ketones is indicated by odor warning properties. Ketones upon prolonged exposure to skin cause dry, cracked, and scaly skin due to defatting of the dermal layer. Ketones have found several applications such as in cleaning solutions, waxes, lacquer thinners, plastics, oils and vinyl screen inks. The Table 9 briefly provides the health hazards associated with organic solvents.

Table 9. Organic solvents and commonly associated toxicities

Toxicity	*Solvent type*[a]
Reproductive toxicity	Glycol ethers, N-methyl-2-pyrrolidone
Carcinogenicity	Halogenated solvents
Neurotoxicity	Most organic solvents
Hepato and kidney toxicity	Most organic solvents
Defatting of skin, contact dermatitis	Most organic solvents

[a] The levels of exposure may vary with each solvent.

Environmental Concerns

The majority of organic solvents used in the chemical and manufacturing industries are hazardous and possess significant toxic properties. Furthermore, solvents are costly and they constitute a major portion of industrial waste. So, in addition to paying for the solvent, the utilities associated with heating and moving the solvents throughout a process and facility and their handling, the process also bears cost associated with their treatment and disposal. This resulting waste is a serious environmental concern especially for the halogenated solvents, which are more costly to dispose. Chlorinated solvents, such as methylene chloride, chloroform, and tetrachloroethene have half-lives of 2.5 - 3.5 months and transport long distances via air routes, thus contributing to their persistence and multi-region presence. They also have high persistence values in water and soil. However, atmospheric degradation of chlorinated solvents is quite easy. Trichloroethylene, a three chlorine atom containing solvent, is more reactive than the other chlorinated solvents. Their resulting ozone depletion at the ground level is partly contributed by VOCs, and their persistence in the environment causes potential contamination of the ground water.

Also, solvent spills and leaks associated with solvents are very likely to cause serious damage to the human health and property, apart from water, air, and land pollution. It is estimated that 0.1 billion liters of drinking water can be contaminated by only one liter of an organic solvent. The depletion of ozone levels by possible atmospheric emission of VOCs can cause irreparable damage to animal life and plant life, apart from causing respiratory diseases and affecting crop production. Moreover, it is an expensive affair to dispose these hazardous wastes.

Continuous monitoring of VOC emissions by state and federal government agencies and regulations to control their releases aid in minimizing serious damage caused to the environment. The use of VOCs in organizations requires permits and compliance with the regulations in place to prevent pollution, control use, and safer treatment of hazardous waste.

Any violation of the permitted regulations and environmental policies can result in lawful action against the organization which may result in cancellation of the permits, prohibition of the organization, and prosecution of the responsible personnel.

Safety Concerns

The use of organic solvents also poses a serious safety risk during manufacturing, production, and use-phase operations. In general, organic solvents are flammable and their use requires specialized equipment, handling, storage, and disposal. Organic solvents commonly possess lower flash points and should be handled with care. The higher the density of the VOC vapors, the greater the seriousness of risk they possess. And, with the vapors being heavier than air they may not be adequately vented or exhausted. Therefore, this type of solvent is more persistent in the environment when used and are not easily removed. With their higher levels of fugacity, these VOC's can cause concerns with their potential ignition if they are to come into contact with open flames, electrical sources, or hot surfaces. Such sources of ignition include electric power points, welding, sparks, and flashes of smoke. Accumulation of solvent vapors in confined places increases the propensity of serious health and property risks. So, this is the reason users of solvents are directed to use these products in well ventilated and large areas. The use of hazardous organic solvents in industrial settings may increase property and liability risks from an insurance point of view and cause higher insurance premiums for the company. This is yet another strong justification for the use of solvent-less processes or those that utilize processes that employ green solvents.

Approaches To Assess Alternative Solvent Systems

Program for Assisting the Replacement of Industrial Solvents (PARIS)

The U.S. Environmental Protection Agency's (U.S. EPA) Solvent Substitution Software Tool, PARIS III (Program to Assist with the Replacement of Industrial Solvents), is an application that allows users to identify alternative solvents as either single solvent or solvent mixtures with either identical or similar specific chemical and physical properties to the original, but may also have lower environmental impacts when compared to the solvent under investigation for replacement. This software (*24*) was developed with the intended purpose to help users identify replacements for single solvent or solvent mixtures that are presently used in industrial processes, but pose significant danger to human health and the environment. Users can use PARIS III to identify potential replacement solvent systems to eliminate certain solvents used within their facilities and products, or assist those users who are propagating the use of greener solvents. Additionally, PARIS III can assist those users who are interested in introducing new environmentally benign solvents to the chemical marketplace. All these activities are performed with the goal of cutting down the harmful impacts to the environment from solvents used within industrial processes and products. This software application is designed for use by chemists, solvent designers, chemical and environmental engineers, consultants and technicians for either academic or commercial purposes.

PARIS III utilizes the coefficients of a solvent that helps quantify various chemical and physical properties (e.g., diffusivity, viscosity, activity coefficients, etc.), and the observation that universal behavior of solvents is governed by their properties as the theoretical basis for performing calculations to arrive at the identified potential replacement solvent system(s). To further emphasize the varied applications of PARIS III, the application uses specific solvent features such as environmental impact, equilibrium and dynamic properties, and safety and performance requirements.

The application of this mathematical representation for solvent design results in PARIS III producing a ranking of potential alternative solvents and solvent mixtures, based on the scores obtained via the mathematical algorithms. This is performed by assigning a higher ranking to those identified alternatives that have coefficients identical or as close to the original solvent system being evaluated for replacement. Upon entering the required information (e.g. CAS number and identified constraints) into PARIS III, the program generates within seconds a ranked solvent replacement list. The resulting ranking list can be further refined by minimizing or maximizing the constraints for those solvent properties that are of greatest importance to the user, which leads to identifying more alternative solvent systems or eliminating those that are not ideal for the situation under investigation. Additionally, PARIS III aids in advancing a pollution prevention approach and enhancing the economics of the system under evaluation, since the identified replacement systems can use existing capital equipment or can be adopted with minor modification to an existing chemical processes. Thus, facilitating the introduction and acceptance of greener and safer solvents across a multitude of applications.

Design for the Environment (DfE)

In an attempt to prevent pollution and the risks and hazards associated with hazardous chemicals, as well as incorporating sustainable molecular design into the chemical enterprise, the U.S. EPA has created the Design for the Environment (DfE) Program. This effort is designed to provide methodology, information, and tools to suggest alternatives to existing chemicals based on improving their environmental and human health profiles with respect to the chemical being replaced. Recently, the DfE methodology has been demonstrated on providing potential alternatives for those brominated flame-retardants that are used in electronics and commercial products (*25*). In addition to designing preferable replacement approaches, DfE also takes into consideration the overall life cycle impact of the product, service, or process on the environment and the human health. This life cycle approach is used to suggest an optimized product/process by considering all the impacts associated with that product or service and allow designers to utilize specially designed software tools to achieve sustainability-based goals. The objectives of the DfE program are:

- Assess alternatives for identification of safer chemicals by considering the entire life cycle of the chemical
- Promoting best and advisable practices in wide sectors (industries/ services) ranging from automobile manufacturing to cleaning products.
- Provide designing guidelines and promote products that are safer to the consumer and the environment.

The real challenge faced by modern day global manufacturers is to provide goods that are cost-effective and environmentally friendly while meeting the stringent consumer demands of product performance and engage in market competition from a global perspective. Businesses are now being encouraged to adopt and incorporate environmental and human health concerns when designing processes to meet these challenges, while contributing to the greening of their business and industry. While doing all this, they also have to achieve this in a fashion that allows them to remain and/or grow economically and in a sustainable manner. The incorporation of a DfE strategy into a business' research and operation can contribute to development of economically viable products, reduce environmental risks, meet or exceed EPA regulations, increase worker safety and create opportunities to expand their business and market share.

The partnerships created by this program provide a platform for industries, environmental organizations, academic groups, and others to study the impacts of alternative chemicals on human health and environment. This partnership assessment helps an industry identify and select safer chemicals for their products and processes, while reducing the hazards associated with dangerous chemicals, and minimizing the environmental risks. This approach ensures development of data to ensure the alternatives being analyzed are safer through informed substitution.

This alternative assessment approach is also applied to chemicals under the Toxic Substances Control Act (TSCA) chemicals when necessary as a risk management measure. Apart from regulatory requirements, this approach inspires industry to adopt safer alternatives whose safety profile has been established by conducting full-scale studies. In addition, this approach can also demonstrate alternatives that perform better or highlight their application even if there are limitations on the chemicals.

The DfE evaluation process is based on the complete analysis of the environmental and health hazards of a chemical as defined by its methodology. This assessment helps to safely move from a problematic chemical to a chemical of choice based on these studies rather than moving to an alternative that is poorly understood and could possess greater hazardous properties than the chemical being replaced.

The DfE program has been applied to perform an alternative assessment approach to evaluate traditionally used flame-retardants in furniture and printed circuit boards (*25*, *26*), and nonylphenol ethoxylate surfactants (*27*). This approach is now being applied to brominated flame retardants like decabromodiphenylether (decaBDE) and hexabromocyclododecane (HBCD) in polystyrene foam, and finding alternatives to bisphenol A and certain phthalates (*28–31*).

Solvent-less/Mechano-Chemical Mixing

A solvent-less or dry media approach is the most preferable and sought-after choice for performing organic chemical transformations. The primary goal of this approach is to replace an existing hazardous solvent with a "no solvent" environment, and carry out reactions under neat conditions. Apart from averting the risks associated with hazardous solvents, several advantages are also garnered using this approach; such as facilitated reaction work-up, achieving higher reaction rates, overall environmental-friendliness, and opportunities for cost savings and efficiency. In addition, this approach can lead to an overall advantage of improved process sustainability. However, drawbacks to this approach do exist and need to be addressed. These include a resulting inhomogeneity or a biphasic reaction mixture, reactants with high viscosity, and the absence of a mechanistic role played by solvents. However, with the advent of novel mixing and reactor strategies, it is now being demonstrated these identified drawbacks are being resolved.

While this approach was considered as being non-traditionally, this is no longer the case given the emphasis that has been placed on green chemistry approaches over the past 20 years. Through these research contributions, solvent-less approaches are being reported as dramatically accelerating several reactions (*32*, *33*). Furthermore, under solvent-less reaction conditions, there are many techniques that can be applied to attain the activation energy necessary to allow the reaction to progress. These techniques include:

- Mechanochemical mixing in a ball mill or planetary mixer,
- Heat-melt
- Ultrasound/microwave irradiation (*34*)

This additional energy provided to overcome a reaction's activation energy is the result of energy generated from pinpoint heat sources. This allows the reaction to reach higher temperatures at the reaction points on the molecule, while the bulk temperature remains considerably lower.

This overall solvent-less approach is further ameliorated when used in conjunction with microwave or ultrasound as the additional energy source for speeding up the reaction. These reactions performed under solvent-less conditions can also be heated to high temperatures in an oven. But, this approach is generally not favored. Alternatively, by using a heat-melt technique, the solid reactants are mixed well and melted at elevated temperatures. The resulting molten reactants are now in a liquid state allowing for facilitated interactions with one another resulting in the desired product. An additional advantage is gained upon cooling, as the product can easily precipitate into its solid form.

The mechanism for the reaction of liquid reactants under solvent-less conditions is similar to the mechanism identified for conventional batch reactions, except for solvent-less conditions that have reactants which are involved in highly concentrated reactions. This is the result of these types of reactions not having no dilution effect. The mechano-chemical mixing of solid reagents results in frictional heating of the reaction mixture and formation of surface defects. This mixing process provides increased mass transfer of the reactants and prevents degradation by avoiding hot spots from exothermic reactions.

Studies have also been carried out investigating the mechanistic pathways of solvent-free reactions involving solid reagents. When these solid reactants are mixed mechano-chemically (e.g. by ball milling, grinding, etc), one of the reagent molecules in the mixture falls into the crevices of the other solid reagent. The desired reaction between these solids leads to the product beginning to form at the interface of the two crystalline reactant materials as a result of the mixing process and further distorts their structures. As larger quantities of product continue to form, the product and reactant mixture phase further disintegrates leading to exposure of more reactant surface area for further propagation of the reaction.

Most of the recently reported literature in solvent-less chemistry focuses on the principles of green chemistry that are directed to material chemistry, organic chemistry, and biomass transformation. This approach can also be applied to the extraction of natural products. However, as of now, there is little literature available to inform on this direction. As an example, plant materials can be used to produce a solvent-free liquor that can be utilized in the synthesis of nanoparticles. This was demonstrated by the synthesis of silver nanoparticles using the extract of *capsicum annum L* (*35*).

Although advances in solvent-less chemistry are being made in the areas of chemistry, the majority of reports have been reported in the field of synthetic organic chemistry. These solvent-less organic transformations can be further classified into either thermal or photochemical reactions. Several types of reactions have been reported using thermal solvent-less synthesis conditions. These include C-C bond formation via condensation reactions (*36*), Wittig reactions (*37*), Reformatsky and cycloaddition reactions (*38, 39*). Other transformations include substitutions (*40*), eliminations (*41*), additions (*42*), oxidations (*43*), isomerisations (*44*), reductions (*45*), phenol couplings (*46*), oxidative couplings (*47*), pinacol couplings (*48*), and polymerisations (*49*). Photochemical transformations studied using solvent-less conditions include addition (*50*), polymerization and dimerization (*51*), isomerization (*52*), decarbonylation (*53*), and cyclization (*54*). Additonally, several reactions have been developed with solvent-less procedures, and many more reactions are theoretically feasible.

To extend this approach of solvent-less conditions for carrying out chemical reactions, the use of supercritical CO_2 (SC-CO_2) is also being explored. This approach is applied in those cases where the reactants "explode" when neat components come into contact with one another. Thus, it is preferable to use a solvent. It is always advisable to carry out reactions under solvent-less conditions to avoid or reduce the volumes of solvents used. This has resulted in the number of reactions involving solvent-less conditions reported continuing to grow.

Solvent-less catalytic transformation reactions have also been reported with several catalysts being explored and demonstrated for this purpose. As an example, copper (I) iodide and copper metal balls when used separately under ball milling conditions for the same organic reaction afforded similar yields. However, when similar reactions were performed in the absence of solvent while using a copper catalyst and ligand for a Sonogashira coupling, this resulted in greater conversions (*55*). Also, demonstrated using a solvent-less approach are polymer-supported and solid catalysts, oxidation catalysts, enantioselective catalysts, and biocatalysts. However, photochemical reactions are relatively less explored under solvent-less conditions. But, the reports presented do offer encouraging results and conditions.

To realize the wide application of solvent-free reactions, this area has been studied through a variety of lenses. This includes recent advancements in multi-component reactions and catalyst-free reactions which have been well-explored in the absence of reaction solvents.

As demonstrated, the solvent-less synthesis arena is growing exponentially and has the huge potential to become one of the major alternative approaches in the greening of synthetic organic chemistry along with that of materials and inorganic chemistry. Although large-scale solvent-less reactions are not common, multi-ton scale industrial reactions are being carried out, especially, in gas-phase. For example, in the development of a potential anti-tubercular drug, the use of solvent-less conditions is applied in many of the synthesis steps, for this multi-step reaction scheme. The application of these solvent-less conditions increases the overall yield by three-fold and reduces overall solvent usage in the scheme by one-third. Solvent-less synthesis strategies can also allow for achieving suitable and best alternative conditions for the development of efficient and more sustainable chemistry methodologies. Since the solvent-less approach does not require separation and removal of reaction solvents, this could significantly reduce the total energy demand for the process. A plethora of reactions studied shows the promise that solvent-less reactions can hold for more complex organic transformations.

Neoteric Solvents

It is generally thought that by merely replacing a solvent by one that is more benign results in the process become greener. This is not always the case and careful consideration of the overall effects of the change must be taken into account. The environmental impact of a process is a portmanteau of several factors such as conversion, E-factor, selectivity, energy usage, reaction time, and life cycle assessment (LCA) of the materials and processes used. Various solvent alternatives have been proposed or demonstrated as a greener alternative to VOCs, as previous discussed. The anticipated environmental-friendliness of these solvents was based on their source of production, physical properties, or environmental impact. However, the complete safety and environmental profile of all of these solvents has not been determined. However, SC-CO_2, water and solvents from renewable sources have been identified to be better when compared to ionic liquids and fluorous solvents (*56*).

Supercritical Carbon Dioxide (SC-CO_2)

Carbon dioxide (CO_2) when held at or above its critical point (critical temperature and critical pressure) exists in a fluid form. This form is referred to as supercritical carbon dioxide (SC-CO_2). In this phase, there is no distinction between the gaseous and liquid state. From an initial point of view, SC-CO_2 has been identified as an efficient greener alternative because it is naturally abundant, non-flammable, and proven to be non-toxic, all of which makes it suitable for a variety of process applications. As for its role as a reaction solvent and in chemical extraction (i.e. separations), it is appealing as a commercially and industrially important solvent alternative. From a more in-depth viewpoint, its use requires significant fiscal expense in the construction of high-pressure vessels and pumps, increased cost of operations and energy usage, and increased vigilance for workers due to high pressure and temperature workplace conditions.

From a chemical property standpoint, SC-CO_2, is a non-hydrogen bond donor and is considered as a non-polar solvent. Although it is considered as a non-polar solvent, it is really a quadrupolar molecule. Also, SC-CO_2 is an excellent solvent for use with small molecular weight non-polar compounds and has been reported as a superior solvent for Diels-Alder reactions. One report states that a Diels-Alder reaction can accelerate up to ten times faster in SC-CO_2 when compared to acetonitrile or ether (*57*).

SC-CO_2 can also exhibit solvent effects as previously demonstrated for other solvents. An example for this is that of acetylacetone, which can exist in two tautomeric forms, keto, and enol. The enol form is favored in non-polar solvents and the keto form in polar solvents. Acetylacetone favors the enol form in *n*-hexane, whereas it is density dependent in SC-CO_2. As the density of SC-CO_2 increases, the keto form becomes more pronounced, which may attribute to a changing of the CO_2 quadrupole moment as a result of increased pressure.

SC-CO_2 has been used successfully on the commercial scale for the extraction of caffeine from coffee beans (*58*) and to remove metals and pesticides from crops without damaging the desired plant material. On a reduced scale, this method has been used in the dry cleaning industry as an environment-friendly alternative to traditional organic solvents (*59*), as an extraction solvent in laboratories (*60*), an extractant of organic pollutants in soils and other samples (*61*), and the synthesis of nanoparticles (*62*).

The extraction of coffee (decaffeination) using SC-CO_2 is commercially important and this extraction technique replaced an earlier method, which used dichloromethane, a chlorinated solvent identified as being harmful to the environment and unsafe to human health. To further extend the concept of going solvent-less and applying a holistic approach to chemical process design is the development of coffee beans that do not contain caffeine thus eliminating the need for a chemical process for the extraction of caffeine entirely.

Adding to the functionality of SC-CO_2 is the ability for it to be miscible with oxygen, hydrogen, and carbon monoxide gases. Additionally, SC-CO_2 is non-reactive to oxygen gas, which makes this property useful for oxidation reactions, as it serves as a diluent and allow for reactions well within the safety range for oxidation reactions using oxygen. To facilitate this, a number of heterogeneous and homogeneous oxidation catalysis have been developed and tested in this solvent medium. Also suitable for use with SC-CO_2 are heterogeneous catalysts. This is due to the viscosity range of SC-CO_2 and the resulting solutes possess the ability to have increased diffusion into the heterogeneous catalyst allowing for taking advantage of those catalysts with high surface areas. Homogeneous catalysts, however, need appropriate ligands that facilitate dissolution of the metal complex into SC-CO_2 to be able to ensure the catalyst is available for participation in the desired chemical reaction.

The catalytic hydrogenation of SC-CO_2 leads to the production of carbon monoxide, methanol, and formic acid, thus used as a chemical feedstock. This catalytic hydrogenation takes place in the presence of triethylamine and generates formic acid. In the same manner, methyl formate can be produced with the addition of methanol through a two-step process via the formation of formic acid.

The application of SC-CO_2 has also received considerable attention as a polymer foaming agent (*63*). In this application, SC-CO_2 is either used solely or as a co-solvent with other to mitigate, or avoid the use hazardous solvents. These polymerization reactions are of significant interest and is a huge success in the manufacture of polytetrafluoroethene (PTFE) or commonly known as Teflon. SC-CO_2 is also an effective solvent for polymerization reactions due to its high affinity for fluorinated compounds.

From these examples, it is of little doubt that SC-CO_2 can serve as an efficient alternative for use in various industries across several applications, and serve as a suitable replacement for commonly used traditional solvents.

Ionic Liquids

Room temperature ionic liquids (RTIL's), which are otherwise simply called Ionic liquids (ILs), are salts that exist in the form of a liquid at room temperature. In some contexts, the term is used to define organic salts with melting points under 100°C or some arbitrary temperature. The concept of "greenness" associated with ILs is due to their very low vapor pressure, ease of recycling, stability at high temperatures, containment, and reduced effort needed with product recovery. The ILs can also be tailored according based on the need of the process simply by changing the anion or cation. Since they can be designed as a needed requirement, they are also called "designer solvents". The IL properties such as miscibility and moisture stability can be tuned by altering the anion or by substitutions on the cation. Similarly, modification of other properties like viscosity, melting point, density, polarity, water miscibility, acidity/basicity, and ability of coordination can also be achieved. Due to their ionic character, ILs have been studied for a wide variety of applications such as solvents in organic transformations, extractions, electrochemistry, and separations. Chemical reaction rates experienced when performed in an IL are greatly increased and variety of catalysts (including biocatalysts) can be employed in the presence of ILs. Additionally, biphasic solutions of SC-CO_2 and ILs have also been studied for organic transformations. ILs have also been explored as a suitable replacement for aromatic (non-polar) solvents in a variety of chemical processes. They can also allow for significant control on reactivity and stereoselectivity. To demonstrate the continued functionality of ionic liquids, they can also a serve as reagent supports (as in solid phase synthesis) by tethering a functional groups covalently to react with the dissolved reactants. These types of ILs are called task-specific ionic liquids (RTILs).

ILs have been investigated/found useful for applications in other industries such as pharmaceutical, cellulose processing, algae processing, dispersants, gas handling, nuclear fuel processing, solar thermal energy, food and bio-products, waste recycling and batteries. For the above reasons, ILs hold a strong potential as a substitute for VOCs.

The great potential of ILs can be efficiently tapped as long as they are well contained. Their use in a large-scale industry can have a significant economic impact. Not only can they be costly to produce, they have been found to be toxic especially in water (*64*). Their aquatic toxicity is more severe than many traditional organic solvents. Moreover, the preparation and purification of ILs themselves require organic solvents. But, new studies are being reported for their synthesis under solvent-less conditions (*65*).

The life cycle assessments of these compounds have to be performed to fully evaluate their greenness. A study was carried out combining *in silico* and laboratory methods to determine the economic, environmental, and energy impacts attributed to their preparation (*66*). Many factors were considered such as availability of raw materials, chemical recycling and disposal, and energy required for reactor vessel and coolant operation. The study revealed the choice of counter ions and solvent used in their synthesis play a greater role in the cost-efficiency, toxicity, and energy demand of the process. By choosing a simpler and cheaper cation, the impact of the overall process is reduced by 50%. The study also identified care should be taken when deciding the ILs anion, as their environmental persistence and toxicity are dependent on this variable. However, the sustainability of ionic liquid synthesis can be further improved by using biorenewable feedstocks. However, LCA studies need to confirm this and it is important the study includes the impact of the growing and harvesting phase of the biorenewable feedstock. Ionic liquids synthesized via this biorenewable feedstock strategy have been demonstrated for use a Heck synthesis reaction (*67*). The applications and utilities associated with ionic liquids have thus successfully showcased their potential to be considered as greener alternatives for VOCs.

Water

Water is not only a cheaper solvent; it can be abundant depending on geographic region and is environmentally benign. But, it also shows a completely different reactivity when compared to traditional and novel organic solvents. The use of water as a reaction solvent for performing organic transformations is varied and has garnered significance in many applications such as the modification of biomolecules chemically (*68*), and the preparation of bioactive molecules (*69*), to name just a few.

Aqueous organic transformations could be the answer to the organic chemistry future. The types of organic reactions are broad including reaction of carbocation and carbanion equivalents (*70*), transition-metal catalysts (*71*), pericyclic reactions (*72*), reactions of carbenes (*73*) and radicals (*74*), and oxidations-reductions (*75, 76*). Several organic solvent-based cleaners are replaced with water-based cleaners for cleaning operations (*77*).

The upside of using water as a green solvent are its nontoxic (liquid + vapor) feature, it is non-flammable, can be abundantly available depending on geographic region, can offer better availability than organic solvents in less developed countries), and it is relatively inexpensive (also depending on geographic region).

While there is a suitable upside for using water as a solvent there are also significant downsides. These include water becomes corrosive at high temperatures and pressures, due to its high specific heat value it can behave as a heat sink, has a high boiling point and moderate vapor pressure which lead to the potential for high energy costs associated with its heating, cooling and pumping, it expands when frozen, has poor solubility of organic materials, the potential for azeotrope formation, and it's reaction range is smaller compared to organic solvents. Additionally, water is a vital resource and should be conserved & protected. So, before considering the use of water as a reaction medium, suitable sustainability assessments should be performed to ensure water is a viable alternative, and its choice is not going to create a burden shift from one stage to the life cycle to another.

Biomass Derived Solvents

Glycerol Based Solvents

Glycerol is classified as an alcohol and contains multiple hydroxy groups (polyol). It is a highly viscous, colorless, and odorless liquid and is considered greener than many other solvents due to its high boiling point, low vapor pressure, non-toxicity, and biodegradability. It is synthetically prepared on the industrial scale as a product in the formation of epichlorohydrin from propylene. Also, large stockpiles of glycerol are produced as a by-product in the synthesis of soap and biodiesel. The transesterification reactions of plant oils in the synthesis of biodiesel and soaps produces glycerol as by-product. The generation of large amounts of excess glycerol has resulted in a considerable price drop over the past few years. To offset its production as a byproduct, it is sometimes present in soap as a softening agent. Research studies are now being carried out to develop new applications for glycerol and its derivatives (*78*).

As a result of this abundance of glycerol, studies have been performed for its use as a reaction solvent in various reactions such as palladium catalyzed carbon-carbon coupling (*79*), nucleophilic substitution, and reduction reactions (*80*). The results from these studies have shown glycerol to produce comparable yields when compared to traditional organic solvents. A pitfall of using a glycerol obtained as a crude by-product from plant oils is the need for pretreatment and refining steps prior to use in organic transformations or other purposes. When crude glycerol was employed in a Heck carbon-carbon coupling reactions, it produced lower yields when compared to pure "clean" glycerol (*81*). However, an aza-Michael addition reaction was demonstrated using an "on-glycerol" methodology to show very high yields with both crude and pure glycerols (*82*).

Derivatives of glycerol, like glycerol ethers, are also useful solvents. Like glycerol, they are also dipolar hydrogen bond donor solvents. But, their properties can change based on the type and number of substitutions on the glycerol base molecule. The use of glycerol ethers as good reaction solvents are due to their hydrogen bonding nature as demonstrated in the epoxidation of cyclooctene (*83*). The advantages associated with glycerol and glycerol ethers are due to their environmental friendliness, beneficial effect on reaction rates, low toxicity, low solubility with commonly used organic solvents, low price, and applications with SC-CO_2 makes it a viable choice and an alternative green solvent.

2-Methyltetrahydrofuran

2-Methyltetrahydrofuran (2-MeTHF) is viewed as an environmental-friendly solvent and is derived from renewable sources. Because of its origin it is being promoted as a greener alternative to traditional organic solvents, especially THF. 2-MeTHF is produced from levulinic acid or furfural, both of which are biomass derived. Furfural, is a five membered heterocyclic aldehyde produced from agricultural waste, and levulinic acid is produced from sucrose.

Owing to it is versatility, 2-MeTHF is stable in both bases and acids. It is a strong Lewis base, non-polar solvent, with variable miscibility in water, has a higher boiling point, improved performance and derived from a renewable resource. 2-MeTHF is being recognized as a suitable replacement for tetrahydrofuran (THF), a common and widely used solvent. THF is obtained by the intramolecular cyclization of 1,4-butanediol in the presence of acid. 2-MeTHF has showed promising results in various types of organic transformations (*84*), due to its stability in similar conditions as THF and the benefits described above. 2-MeTHF has been used in biocatalysis (*85, 86*), organocatalysis (*87*), and organometallic transformations (*88*), and biomass processing (*84*). It has also been identified as a suitable replacement for dichloromethane (*89*).

It has also been successfully demonstrated as a solvent in hydride reductions, Grignard reactions, and lithiations due to its stability in bases (*88*). However, it can form peroxides with air like THF, which can be dodged by adding stabilizers. Preliminary toxicology results are also demonstrating promise as a safe alternative (*90*). Thus, lending to its potential for use in pharmaceutical processes. Many industries have recognized its potential and are starting to assess its implementation as a replacement solvent in their processes.

Limonene

Limonene, a cyclic terpene, exists in two enantiomeric forms of which D-limonene is more abundant in natural sources. It is classified as a non-polar and non-hydrogen bond donor solvent. It is readily available from the rind of citrus fruits and the name limonene is derived from the word lemon. D-limonene possesses a strong citrus odor and is used as an effective cleaning agent. Industrially, it is available as a by-product in juice production from citrus fruits.

Limonene while classified as volatile organic compound is biodegradable and rapidly oxidizes to carbon dioxide and water rendering it harmless to ozone layer. It is used in the extraction of oils replacing alkanes as a solvent (*91*) and it has been used as a cleaning solvent for many purposes such as stripping paint, removal of oil from machine parts, alternative to turpentine, constituent in room fresheners, paints, and glues. D-Limonene is also used in place of xylene in histology labs as clearing agents, a chain transfer agent and as a solvent in ring-opening metathesis polymerization reactions (*92*).

Specific Examples of Solvent Replacement Systems

Case Study 1: Cleaning Solvent Alternatives for Printed Circuit Boards Used in Aerospace Applications

The commercial and military aerospace industries use countless electronic devices that contain printed wiring assemblies and circuit boards in their products. In order to maintain the needed high-level quality assurance standards, these electronic devices require cleaning of the residual flux on the boards, which is used to facilitate soldering of the devices electronic components. Alcohol-based solvents have been demonstrated to be very effective at removing this residual flux material due to the polar –OH functionality on the molecule. Along with alcohols, multi-halogenated solvents such as, 1,1,2-trichloro-1,2,2-trifluoroethane or 1,1,1-trichloroethane have also been used in these cleaning formulations. However, these halogenated solvents were found to cause depletion of the ozone layer resulting in the banning of their usage in industrial applications (*93*, *94*). Since the cleaning of these printed circuit boards was still required, it was imperative a suitable alternative be identified. After a series of evaluations and tests, *n*-propyl bromide, was marketed in formulations along with alcohol to meet this industrial cleaning solvent demand.

The importance of solvent effectiveness is dependent on the type of flux use in the manufacture of these printed circuit boards. This is demonstrated by the several types of flux materials that are available and used in the manufacture of these circuit boards. For low solids flux, also known as no-clean flux, no cleaning solvent is required for their removal. This type of flux also works very well with *n*-propyl bromide-based formulations. Water soluble fluxes are also used on circuit boards and their cleaning can be facilitated by using deionized water. And, finally traditional rosin based fluxes can be cleaned with a water-based saponifier cleaning agent. This demonstrates the ability of varying solvent systems to perform a task once relegated to only hazardous VOC type solvents.

Other traditional solvent formulations are also available for flux removal from circuit boards. However, they are hazardous in terms of environmental and human health viewpoints. These solvents include hydrofluoroethers (HFEs), chlorofluorocarbons (HCFC-225), and hydrofluorocarbon (HFC-4310). HCFC-225 is known to cause ozone depletion and will begin to be banned in 2015. From this information, the best alternatives appear to be the use of no-flux resins or water soluble resins.

However, airtight/airless degreasers for industrial cleaning can be used with *n*-propyl bromide/alcohol based formulation or the other alternatives described above. With the increased cost of these alternatives, the use of airtight degreasers is justified since the solvents can be efficiently reused resulting in reduced process emissions. However, low solids flux or water-based cleaning fluxes are the better alternatives.

To demonstrate the applicability of a new alternative solvent system, an aerospace subcontractor that once used 1,1,1-trichloroethane for their cleaning printed circuit boards operations adopted the *n*-propyl bromide solvent system as an alternative. This California-based subcontractor with activities in the manufacture and repair of aircraft components for use in both commercial and military aircrafts, like the C-17 and the C-130. They also manufacture and repair pumps and manufacture air locking devices and braking systems. This demonstrates the variety and capacity of their business, which is important when transitioning successful case studies into other business lines in the company.

As stated, 1,1,1-trichloroethane was used previously by this company for cleaning of remaining rosin flux on printed circuit as required by military specifications. Because of this specification they had were relegated to the use of a vapor degreaser operation for cleaning, in order to achieve mandatory specifications. In addition, this company had another commercial operation that used a water soluble flux for soldering on printed circuit boards for another application. In this process, the cleaning of the residual water soluble flux was carried out in mechanical shaker using only plain deionized water.

To address the costs and hazards associated with using a halogenated-based solvent in a higher temperature environment, the company investigated the use and eventually adopted using a water-based cleaner for the removal of residual rosin flux for both military and commercial destined printed circuit boards. In this example, the company purchased a large dishwasher ($24,950) for this cleaning operation and an evaporator ($6,950) for disposing the used cleaner. And, since this water-based cleaner does not contain any organic solvents, the disposal regulations are not as stringent as those for organic-based solvent allowing for easier disposal. The total cost for this process modification was $37,378, including taxes, and amounted to only $3,925 per year amortized over a 10 year usage period, including capital costs.

In the case of using 1,1,1-trichloroethane and an *n*-propyl bromide solvent as cleaners, their annual consumption and operation cost was approximately $25,000. While the cost for using deionized water in the commercial PCB cleaning operation was just $4 per year. Additionally, the disposal cost for 1,1,1-trichloroethane and *n*-propyl bromide was $400/year.

As for the water-based cleaner, it costs approximately $12 /gallon and 10 parts by volume is required for the cleaning. With a dishwasher tank capacity of 5 gallons, this would total to a cost of $7,800 for the cleaner per year, basied on five batches of operation a day for 260 days. The cleaning phase is then followed by a rinsing phase with deionized water which has a cost of approximately $20 / year. And finally, the water-based cleaning formulation disposal needs, after its removal by evaporation, comes to a cost of approximately $1200 per year, based on the same operation schedule identified above.

The total labor costs associated with the 1,1,1-trichloroethane and DI water cleaning was $11,375. Whereas, the labor costs for the water-based cleaner operation was $4,550. The electricity expenses for yearly operation of a large dishwasher and evaporator was valued at $3,190 and the assumed electricity expenses for the 1,1,1-trichloroethane versus the deionized water systems was determined to be negligible.

After this cost analysis (see Table 10 below), it was calculated that switching to a water-based cleaner, including new equipment purchase cost, was 44% lower than the previously used 1,1,1-trichloroethane and deionized water system.

Table 10. Running costs comparison of the adopted water-based cleaner vs 1,1,1-trichloroethane and deionized water systems

Cost item	*Water-based cleaner system ($)*	*TCA and DI water system ($)*
Employment	4,550	11,375
Energy	3,190	-
Discarding waste	1,200	400
Equipment	3,925	-
Cleaner	7,800	25,000
Deionized water	20	4
Total	**20,685**	**36,779**

Case Study 2: Vapor Degreasing Solvent Substitution at Durex Inc.

The vapor degreaser employed at Durex Inc. for removal of oils from metal parts is of the open-top variety. This piece of equipment is equipped with a free board chiller and a closed cover. To further describe this operation, this facility employs over 100 personnel and their processes include metal fabrication, stamping, and assembly.

In this operation, the degreasing process previously used was based on a 1,1,1-trichloroethane solvent. To move away from the use of this solvent, studies were carried out to assess and prioritize alternatives. Through this study, the company identified a commercially available option, Techtride® NPB DG, which is a non-HAP solvent. This formulation is comprised of 1,2-epoxybutane, n-propyl bromide, and 1,3-dioxolane. The performance of this solvent is very much comparable to 1,1,1-trichloroethane and the main advantage for its use and choice is the ability to use their existing equipment.

This cleaning agent (a product of Parts Cleaning Technologies; www.partscleaningtech.com) was identified and chosen by Durex for use in their vapor degreasing unit. In addition to the quantities detailed above, the new cleaning solvent has offered several advantages in terms of safety, economics, and environment. Also, there were very low (nil) atmospheric emissions associated with its use, solvent purchase costs were reduced and safety levels were increased as a result of having less harmful exposure to employees. This whole process was further ameliorated by recycling the spent oil at the manufacturer's location. As a result, the company has minimized the quantity of wastes to be shipped for disposal. Which further eliminated the need of reporting the chemical to Release and Pollution Prevention Reports (RPPR).

Case Study 3: Alternative Solvent Assessment for Vapor Degreasing Operations

Bihler of America, Inc. is a New Jersey based company which employs over 150 personnel and uses trichloroethylene and 1,1,1-trichloroethane for their cleaning operations. When the company switched their primary product line from computer disc slide doors to electrical supplies, unfinished surgical needles, and metal fabrication, they also used this opportunity to change their cleaning processes to reduce/avoid the generation of hazardous air pollutants. The company has continuously made efforts to improve their cleaning processes by trying various formulations such as Abzol (active ingredient: *n*-propyl bromide), 3M's engineered hydrofluoroether (HFE) fluid, and pressurized hot water and aqueous detergent washes.

During the process of evaluating the alternative, there were some employee exposures and maintenance issues associated with Abzol, which would turn acidic if not properly monitored. The hydrofluoroethers option, manufactured by 3M consisted of isopropyl alcohol, ethyl nonafluoroisobutyl ether, methyl nonafluoroisobutyl ether, 1,2-trans-dichloroethylene, ethyl nonafluorobutyl ether, and methyl nonafluorobutyl ethyl ether) was deemed expensive, but the benefits of this blend is it is not an identified HAP and not an ozone depleting chemical. Essentially, it contains minimal VOCs. The use of detergent and hot pressurized water was also a viable option. But, the detergent emitted out of the plant is considered a pollutant and needed to be addressed.

Since, Bihler employed many pressurized hot water cleaners and a degreaser that used a HFE 72DA solvent blend many options involving their use were considered. This on-site open-top vapor degreaser allows for reduce emission and employee exposure, and equipped with a refrigeration unit for passive control.

As with many other vapor degreasers of this type, the basket is lowered into and out of the unit for cleaning. The refrigeration area condenses the VOCs from the degreased parts back into the tank. The resulting spent solvent is sent back for processing at the solvent supplier's site. The HFE-72DA solvent used in the vapor degreaser unit is expensive, but is bundled with several benefits making it a worthy alternative option for the company. The benefits associated with its use are handling, storage, environment, regulations, fees etc.

However, the water used in the water-based cleaners are designed to be recycled for reuse. After completion of the cleaning operation, the dirty, oily water is filtered and separated for regeneration. Any resulting loss of water during the regeneration process is replenished. The separated oil is sent to waste treatment sites.

Apart from exploring the suitable solvent substitutes, the company also performed an evaluation of the products to be cleaned. This resulted in optimized cleaning procedures of the parts to the satisfaction and requirements of their clients resulting in saving labor, energy, and time. Also, installation of computer controlled oil application on the parts also reduced up-front cleaning efforts.

In addition, the company also adopted energy saving options for their plants by installing automatic sensors for lighting, air compressors with variable frequency drives, among others made them pro-environment that helps maintain safety. After all, these efforts the company eliminated the need for reporting 1,1,1 trichloroethane and trichloroethylene to Release and Pollution Prevention Reports (RPPR) due to their elimination.

Green and Sustainability Aspects of Alternatives Assessment

Non-governmental organizations, academics, industry, and governments are all working together to identify more sustainable and safer chemical alternatives. These Chemical Alternatives Assessment (CAA) methodologies, guidance and tools can guide the user through the selection criteria to arrive at potentially safer alternatives. In addition, these tools can also be used in combination with other methodologies or life cycle assessment or risk assessment to allow for more informed decisions to be made regarding chemical substitutions.

The intent of chemical alternatives assessment is to not only to gain a better understanding of new chemicals entering the commercial marketplace, but to also to apply this thinking into understanding how an existing chemical behaves in a living system and the environment. It is with this newly gained information that researchers can begin to grasp the interconnectedness between a molecule's design and its propensity for creating impact. Through these better informed decisions about material and chemical substances, assessments into creating situations to avoid undesired or unintended economic, environmental, and social impacts can be made. As illustrated earlier in this chapter, the U.S. EPA's DfE Program is a prominent example of a chemical alternatives assessment and how this approach can be used to encourage more sustainable decision-making. While groups globally are working individually to develop their alternative assessment methods, there are also collaborative efforts between some groups to accelerate the development of these tools and methodologies, as well as bring a collective standardization of terms, methodologies and assumptions. One such example is Furniture Flame Retardancy Partnership headed by the DfE program. This collaborative initiative among chemical manufacturers, the furniture industry, and EPA and environmental groups bring together each groups expertise to provide information and examples that has reinforced the transition for flame retardants used in furniture applications to safer alternatives.

To gain a better understanding of the procedure associated with the thought process that is involved when conducting a chemical alternative assessment. We will not walk you through the keys steps.

Key Steps To Conducting an Alternatives Assessment

Step 1: *Judging the Possibility for Assessment of Alternatives*

As we all know, there are reasons why a specific chemical is used in a unique application. As demonstrated earlier in this chapter, solvents are an excellent example of how a unique chemical intensive or extensive property can be exploited for facilitating a chemical reaction. When determining the need for an alternative, not only does the specific role of the chemical need to be a priority, but the alternatives identified should have a larger safety and operation window in terms of health and the environment, be economically favorable and commercially available, should be able to appease stake holders requirements, and provide a result that will be a lasting change. It is recognized this is asking a lot, but the above listed benefits must be taken into consideration when judging the possibility of an alternative replacement.

Step 2: *Gathering Information on Possible Alternatives*

Given the criteria identified above, acquiring data to support a justification for an alternative replacement is the next step. The DfE methodology is designed to gather needed information on the problematic chemical and the proposed alternatives before embarking on performing the assessment of these alternatives. The type of information needed includes data regarding the manufacturing process of the each chemical, including the raw materials involved, and by-products and side products from the production process, application and use-phase data for each identified chemical and chemical, physical and toxicological data for each. This approach also collects and utilizes literature information for suitable alternatives that have been identified for the same or similar chemicals.

Step 3: *Organizing Interested Parties*

In order for this process to be effective and have any significant impact, it is imperative the input from actual users and purchasers of chemical to be replaced be received and considered. These stakeholders are the entities that will be employing the proposed alternatives into their processes, products and technologies. Their input is invaluable and these groups can also identify criteria and constraints not identified in the first two steps. To do this, DfE considers the input from diverse groups to establish the project scope, identify and propose alternatives, and assist the manufacturer and acceptance of identified alternatives. To further expand the scope of the assessment, interested parties from the entire supply chain of the chemical to be replaced are invited. The participation of these key stakeholders are for the entire duration of the project. This is also necessary since their involvement will assist with facile adoption of the alternative product, improving credibility of the Program and methodology, and incorporation into mainstream usage in other manufacturing sectors. These stakeholder groups include manufacturers of the chemical and associated products, academic and governmental institutions, consumers and retailers, related researchers from relevant areas and waste handlers.

Step 4: *Identification of Possible Alternatives*

With the information collected from the previous three stages now in hand, the DfE Program along with its stakeholders, begins focusing on identifying those alternatives that have reduced or minimum impact on the production process and use-phase and provide a ranking of those alternatives identified. The stage may also involve demonstrating the applicability of identified alternatives by the manufacturers and users.

Step 5: *Assessment of the Hazard*

When conducting an assessment it is important to identify and set system boundaries and constraints. This is not any different from that of assessing hazard and done by recognizing and setting hazard endpoints. This assessment of hazard levels, for example low, medium, and high, for each potential alternative is performed for various hazards associated with each chemical such as toxicity, irritations, persistence etc.

The hazard assessment is performed by acquiring data from sources such as literature, classified information provided to EPA by businesses and chemical manufacturers, information available within the EPA from its own research and from external societies, organizations and governments.

If the data is not available for a specified endpoint, DfE conducts a hazard evaluation based on expert judgment and data from estimation techniques, such as structure activity relationship (SAR) methods. This hazard evaluation is also based on the alternatives assessment criteria found within the DfE methodology. This method ensures all or as many of the possible indicators are taken into consideration for assessing the alternatives and is based on utilizing the information which has been made available. To assist with the differentiation of the toxicities among the proposed alternatives, specific key endpoints which have been identified are used.

Step 6: *Application of Cost and Life Cycle Aspects*

The assessment report generated by the DfE Program is then provided to the interested parties. Within this report includes all necessary results and supplemental information that were taken into account during the decision making process. The information may include all the product manufacturing procedures, life cycle assessment results, and use-phase methods that are noteworthy. Additionally, this report may include economics, costs and technological information and practices. While the primary goal of alternative assessment is to identify alternatives that provide safer outcomes than the predecessor chemical, the economics associated with the use of the alternative is also of great interest. Aside from a regulatory driver that propagates change of a chemical or process technique, economics of the process or chemical is the next driver for change.

Step 7: *Application of Research Results in the Decision Making Process*

Chemical alternatives assessments conducted through the DfE Program combine information from five sources, and assigns an order of preference when evaluating the data. These sources include: (1) publicly available empirical data; (2) confidential empirical data received by EPA under Toxic Substances Control Act (TSCA) regulations; (3) analogs and structure activity relationship based estimations; (4) expert judgment of EPA staff; and, (5) confidential empirical data on experimental studies supplied by the chemical manufacturers for the alternatives assessment. As discussed, this information assists in preparing each assessment report, which includes a detailed discussion of the relevant products, materials, and technologies for the functional use category; narrative and tabular summaries of the assessment. Also provided in these reports are discussions of general exposure scenarios and life-cycle considerations attributed to each proposed alternative and a summary of the environmental, health, technical, economic, and social factors taken into account during the decision-making process. The chemical alternatives assessment report may also contain information on alternative technologies that might influence manufacturing practices.

In the context of DfE, chemical alternatives assessments are guided by these seven principles (given above) to ensure their utility and value. The first principle advocates that the alternatives should be readily available or likely to become commercially available; and the second principle speaks about the technological feasibility of the alternatives. With these two principles, it is assumed the identified alternative would satisfy all the physical, chemical, technical and functional requirements as the chemical it is substituting. The DfE Program also recognizes, in some cases that a chemical substitution may not be a best alternative as exemplified in the specific case studies in section 3.0 above. For example, an alternative material with inherent flame retardancy (without the need for a flame retardant chemical) or a barrier within the material may present preferable environmental and human health profiles than that for an alternate chemical flame retardant. The third principle suggests that the cost and performance of the identified alternative should deliver comparable or better value as the chemical it is replacing. Fourth, to demonstrate the benefits of substitution toxicologically, all alternatives should have an ameliorated environmental and health profiles, as validated by the chemical alternatives assessment. The fifth principle asserts that relevant societal and economic factors should be considered by the proposed alternatives; and, the sixth principle, all interested parties should participate and be engaged in all stages of the chemical alternatives assessment process. Finally, a lasting change should be imparted by the alternatives and should serve as examples to propagate change in other applications. The integration of these principles into the six key steps of chemical alternatives assessment methodology is outlined above.

Chemical alternatives assessment also focuses on finding alternative chemicals that are safer and highly-functional that can be used to replace hazardous chemicals to reduce or eliminate environmental problems rather than assessing the intensity of the problem associated with the use of such hazardous chemicals. In addition to identification of inherently safer functional alternatives, chemical alternatives assessment considers related trade-offs and burden shifting to minimize the potential of unintended and intended impacts. The goals of chemical alternatives assessment are to promote pollution and risk prevention, technological innovation, and to reduce the unquantifiable risks by suggesting feasible options. For instance, chemical alternatives assessment would study the availability of solvent alternatives or safer solvents, or suggest strategies for an alternative processing to eliminate the initial risk altogether rather than trying to quantify certain solvents risks.

The procurement/attaining of complete test data to perform a chemical alternatives assessment may be more challenging than performing the assessment, since the needed data may not be available for public use or simply, may not exist. The High Production Volume Challenge Program at the U.S. EPA is compiling better test data on the most commonly used chemicals in industry, however, robust test data is still incomplete on chemical hazards. An EPA research program has been conducted in computational toxicology to study the human health risk assessment by applying computational approaches. This program aims to assess the human health by including transcriptomics, chemoinformatics, and high-throughput *in vitro* assays. These studies help in dose-response assessment and hazard identification by providing alternative approaches (*95*).

This DfE methodology can also apply to EPA identified priority chemicals, and assist with characterizing the hazard posed by a chemical on a full range of environmental and human health endpoints, including expert predictive models and primary data sources. Examples of chemical alternatives assessments accomplished by DfE include studies on flame retardants in printed circuit boards and furniture and the current chemical alternatives assessments for alternatives of bisphenol-A for use in thermal paper and the flame retardant chemicals hexabromocyclododecane and decabromodiphenylether used in extruded and expandable polystyrene foam for insulation, and nonylphenol ethoxylate based surfactants.

The outcome of an effective chemical alternatives assessment provides industry and other interested parties with valuable information to facilitate their decision when choosing chemicals that are safer and allows for minimizing or eliminating the potential for unintended adverse impacts, like those that result from transitioning to a poorly understood substitute which may be potentially more hazardous. Chemical alternatives assessments do not themselves specify a favored alternative; however, the assessment outcome assists with making an alternative chemical selection. In other cases, chemical alternatives assessments are being used in policymaking concomitantly with other environmental and economic assessments.

Conclusion

As demonstrated, there have been significant advancements in the development of methodologies, strategies, and tools specifically designed to reduce our dependency on solvents for chemical and manufacturing processes. We are only at the beginning, but as with any newly introduced concept or procedure it will take time and effort, in the absence of policy or regulation, to be fully integrated. Through technological innovation, continued development and business integration, the adoption of solvent-less strategies will become increasingly prevalent in the industrial and manufacturing sectors.

The strategies and information provided in this chapter are to provide the reader with the current state of the literature and research being conducted to decrease our reliance on solvents and to minimize and/or reduce the human health and environmental impacts attributed to emission during their use and manufacture. Additionally, the sustainability implications associated with these changes need to be evaluated and quantified to ensure the alternative choice is one that leads to a more sustainable situation, based on user preferences and constraints.

Acknowledgments

B.R.V. was supported in part by an appointment to the Research Participation Program for the U.S. Environmental Protection Agency, Office of Research and Development, administered by the Oak Ridge Institute for Science and Education through an interagency agreement between the U.S. Department of Energy and the EPA.

References

1. *Fact Sheet on Top Uses of Chemicals: A Snapshot of the Data*; 2012 Chemical Data Reporting Results; U.S. Environmental Protection Agency: 2012.
2. TRI On-site and Off-site Reported Disposed of or Otherwise Released (in pounds), for Hazardous Air Pollutants, By Industry. In *Toxic Release Inventory Explorer*; U.S. Environmental Protection Agency: 2012.
3. Streitberger, H.-J. In *Paints, Coatings and Solvents*; Stoye, D., Freitag, W., Eds.; 2nd ed.; Wiley-VCH: Weinheim, 1998; pp 101−104.
4. Dick, F. D. *Occup. Environ. Med.* **2006**, *63*, 221–226.

5. Toxics Release Inventory (TRI) database search for the NAICS code #325510; The Right-To-Know network (rtknet.org). http://www.rtknet.org/db/tri/search (accessed Aug 28, 2014).
6. Hassan, A. A. E. H.; Elnagar, S. A. E. M.; El Tayeb, I. M.; Bolbol, S. A. E. H. *Open J. Saf. Sci. Technol.* **2013**, *3*, 87–95.
7. Freitag, W.; Stoye, D. *Paints, coatings and solvents*, 2nd ed.; Wiley-VCH: Weinheim, New York: 1998.
8. Dursun, D.; Sengul, F. *Resour., Conserv. Recycl.* **2006**, *47*, 316–331.
9. Grodowska, K.; Parczewski, A. *Acta Polym. Pharm.* **2010**, *67*, 3–12.
10. Tucker, J. L. *Org. Process Res. Dev.* **2006**, *10*, 315–319.
11. Constable, D. J. C.; Jimenez-Gonzalez, C.; Henderson, R. K. *Org. Process Res. Dev.* **2006**, *11*, 133–137.
12. Jiménez-González, C.; Constable, D. J. C. Green Chemistry and Green Engineering in the Movement Toward Sustainability. In *Green Chemistry and Engineering: A Practical Design Approach*; John Wiley & Sons: Hoboken, NJ, 2011; pp 1−106.
13. Beers, N.; Koppes, M.; Rupert, L. *Pigm. Resin Technol.* **1998**, *27*, 289–297.
14. Pocius, A. V. *Adhesion and Adhesives Technology*, 3rd ed.; Carl Hanser Verlag: Munich, 2012; pp 1−15.
15. Cheremisinoff, N. P. *Industrial Solvents Handbook*, 2nd ed.; Marcel Dekker: New York, 2003; pp 77−81.
16. *Profile of the Rubber and Plastics Industry*, 2nd ed.; EPA Office of Compliance Sector Notebook Project; U.S. Environmental Protection Agency: 2005.
17. TRI On-site and Off-site Reported Disposed of or Otherwise Released (in pounds), for All chemicals, By Industry. In *Toxic Release Inventory Explorer*; U.S. Environmental Protection Agency: 2012.
18. Wypych, G. Rubber and Plastics. In *Handbook of Solvents*, Wypych, G., Ed. ChemTec Publishing: Toronto, 2001; p 1025.
19. Franta, I. Synthetic Rubbers. In *Elastomers and Rubber Compounding Materials*; Elsevier: Prague, 1989; Vol. 1, pp 76−299.
20. Hughes, E. D.; Ingold, C. K. *J. Chem. Soc.* **1935**, 244–255.
21. Smith, M. T. *Environ. Health Perspect.* **1996**, *104*, 1219.
22. Goldman, S. M.; Quinlan, P. J.; Ross, G. W.; Marras, C.; Meng, C.; Bhudhikanok, G. S.; Comyns, K.; Korell, M.; Chade, A. R.; Kasten, M.; Priestley, B.; Chou, K. L.; Fernandez, H. H.; Cambi, F.; Langston, J. W.; Tanner, C. M. *Ann. Neurol.* **2012**, *71*, 776–784.
23. Glycol Ethers; Technology Transfer Network - Air Toxics Web Site; U.S. Environmental Protection Agency. http://www.epa.gov/ttn/atw/hlthef/glycolet.html (accessed Oct 6, 2014).
24. PARIS III - Solvent Substitution Software Tool; U.S. Environmental Protection Agency. http://www.epa.gov/nrmrl/std/parisIII/parisIII.html (accessed Oct 6, 2014).
25. *Partnership to Evaluate Flame Retardants in Printed Circuit Boards*; Design for the Environment: An U.S. EPA partnership program; U.S. Environmental Protection Agency: 2008.

26. *Alternatives Assessment on Flame Retardants Used in Flexible Polyurethane Foam* Design for the Environment: An U.S. EPA partnership program; U.S. Environmental Protection Agency: 2014.
27. *DfE Alternatives Assessment for Nonylphenol Ethoxylates*; Design for the Environment: An U.S. EPA partnership program; U.S. Environmental Protection Agency: 2012.
28. *Alternatives assessment for bisphenol A (BPA) in thermal paper*; Design for the Environment: An U.S. EPA partnership program; U.S. Environmental Protection Agency: 2014.
29. *An Alternatives Assessment for the Flame Retardant Decabromodiphenyl Ether (DecaBDE)*; Design for the Environment: An U.S. EPA partnership program; U.S. Environmental Protection Agency: 2014.
30. *Flame Retardant Alternatives for Hexabromocyclododecane (HBCD)*; Design for the Environment: An U.S. EPA partnership program; U.S. Environmental Protection Agency: 2014.
31. Alternatives to Certain Phthalates Partnership; Design for the Environment: An U.S. EPA partnership program. http://www.epa.gov/dfe/pubs/projects/phthalates/index.html (accessed Aug 28, 2014).
32. Orita, A.; Jiang, L.; Nakano, T.; Ma, N.; Otera, J. *Chem. Commun. (Cambridge, U. K.)* **2002**, 1362–1363.
33. Konnert, L.; Lamaty, F.; Martinez, J.; Colacino, E. *J. Org. Chem.* **2014**, *79*, 4008–4017.
34. Khumraksa, B.; Phakhodee, W.; Pattarawarapan, M. *RSC Adv.* **2014**, *4*, 20454–20458.
35. Li, S.; Shen, Y.; Xie, A.; Yu, X.; Qiu, L.; Zhang, L.; Zhang, Q. *Green Chem.* **2007**, *9*, 852–858.
36. Rodríguez, B.; Bruckmann, A.; Rantanen, T.; Bolm, C. *Adv. Synth. Catal.* **2007**, *349*, 2213–2233.
37. Balema, V. P.; Wiench, J. W.; Pruski, M.; Pecharsky, V. K. *J. Am. Chem. Soc.* **2002**, *124*, 6244–6245.
38. Chen, X. a.; Zhang, C.; Wu, H.; Yu, X.; Su, W.; Cheng, J. *Synthesis* **2007**, *2007*, 3233–3239.
39. Diaz-Ortiz, A.; de la Hoz, A.; Langa, F. *Green Chem.* **2000**, *2*, 165–172.
40. Ji, S.-J.; Wang, S.-Y.; Zhang, Y.; Loh, T.-P. *Tetrahedron* **2004**, *60*, 2051–2055.
41. Toda, F.; Okuda, K. *J. Chem. Soc., Chem. Commun.* **1991**, 1212–1214.
42. Cave, G. W. V.; Raston, C. L. *Chem. Commun. (Cambridge, U. K.)* **2000**, 2199–2200.
43. Kesavan, L.; Tiruvalam, R.; Rahim, M. H. A.; bin Saiman, M. I.; Enache, D. I.; Jenkins, R. L.; Dimitratos, N.; Lopez-Sanchez, J. A.; Taylor, S. H.; Knight, D. W.; Kiely, C. J.; Hutchings, G. J. *Science* **2011**, *331*, 195–199.
44. Tyagi, B.; Mishra, M. K.; Jasra, R. V. *Catal. Commun.* **2006**, *7*, 52–57.
45. Vass, A.; Dudás, J.; Tóth, J.; Varma, R. S. *Tetrahedron Lett.* **2001**, *42*, 5347–5349.
46. Toda, F.; Tanaka, K.; Iwata, S. *J. Org. Chem.* **1989**, *54*, 3007–3009.
47. Zhang, X.; Wang, L. *Green Chem.* **2012**, *14*, 2141–2145.
48. Tanaka, K.; Kishigami, S.; Toda, F. *J. Org. Chem.* **1990**, *55*, 2981–2983.

49. Fortenberry, D. I.; Pojman, J. A. *J. Polym. Sci., Part A: Polym. Chem.* **2000**, *38*, 1129–1135.
50. Tang, C. P.; Chang, H. C.; Popovitz-Biro, R.; Frolow, F.; Lahav, M.; Leiserowitz, L.; McMullan, R. K. *J. Am. Chem. Soc.* **1985**, *107*, 4058–4070.
51. Sonoda, Y. *Molecules* **2010**, *16*, 119–148.
52. Zimmerman, H. E.; Zuraw, M. J. *J. Am. Chem. Soc.* **1989**, *111*, 2358–2361.
53. Orski, S. V.; Poloukhtine, A. A.; Arumugam, S.; Mao, L.; Popik, V. V.; Locklin, J. *J. Am. Chem. Soc.* **2010**, *132*, 11024–11026.
54. Caronna, T.; Liantonio, R.; Logothetis, T. A.; Metrangolo, P.; Pilati, T.; Resnati, G. *J. Am. Chem. Soc.* **2004**, *126*, 4500–4501.
55. Fulmer, D. A.; Shearouse, W. C.; Medonza, S. T.; Mack, J. *Green Chem.* **2009**, *11*, 1821–1825.
56. Clark, J. H.; Tavener, S. J. *Org. Process Res. Dev.* **2006**, *11*, 149–155.
57. Cott, D. J.; Ziegler, K. J.; Owens, V. P.; Glennon, J. D.; Graham, A. E.; Holmes, J. D. *Green Chem.* **2005**, *7*, 105–110.
58. Raventós, M.; Duarte, S.; Alarcón, R. *Food Sci. Technol. Int.* **2002**, *8*, 269–284.
59. Zhang, X.; Han, B. *Clean: Soil, Air, Water* **2007**, *35*, 223–229.
60. Reverchon, E.; Porta, G. D.; Senatore, F. *J. Agric. Food Chem.* **1995**, *43*, 1654–1658.
61. Hawthorne, S. B.; Miller, D. J. *J. Chromatogr. Sci.* **1986**, *24*, 258–264.
62. Ohde, H.; Hunt, F.; Wai, C. M. *Chem. Mater.* **2001**, *13*, 4130–4135.
63. Han, X.; Zeng, C.; Lee, L. J.; Koelling, K. W.; Tomasko, D. L. *Polym. Eng. Sci.* **2003**, *43*, 1261–1275.
64. Thuy Pham, T. P.; Cho, C.-W.; Yun, Y.-S. *Water Res.* **2010**, *44*, 352–372.
65. Varma, R. S.; Namboodiri, V. V. *Chem. Commun. (Cambridge, U. K.)* **2001**, 643–644.
66. Kralisch, D.; Stark, A.; Korsten, S.; Kreisel, G.; Ondruschka, B. *Green Chem.* **2005**, *7*, 301–309.
67. Handy, S. T.; Okello, M. *Tetrahedron Lett.* **2003**, *44*, 8399–8402.
68. Wang, Q.; Chan, T. R.; Hilgraf, R.; Fokin, V. V.; Sharpless, K. B.; Finn, M. G. *J. Am. Chem. Soc.* **2003**, *125*, 3192–3193.
69. Polshettiwar, V.; Varma, R. S. *Chem. Soc. Rev.* **2008**, *37*, 1546–1557.
70. Richard, J. P.; Amyes, T. L.; Toteva, M. M. *Acc. Chem. Res.* **2001**, *34*, 981–988.
71. Li, C. J. *Chem. Rev. (Washington, DC, U. S.)* **1993**, *93*, 2023–2035.
72. Klijn, J. E.; Engberts, J. B. F. N. Pericyclic Reactions in Water. Towards Green Chemistry. In *Green Solvents: Reactions in Water*; Li, C.-J., Ed.; Wiley-VCH: Weinheim, 2010; Vol. 5, pp 229–271.
73. Amyes, T. L.; Diver, S. T.; Richard, J. P.; Rivas, F. M.; Toth, K. *J. Am. Chem. Soc.* **2004**, *126*, 4366–4374.
74. Perchyonok, V. T. *Radical reactions in aqueous media*; The Royal Society of Chemistry: Cambridge, U.K., 2010.
75. Brink, G.-J. T.; Arends, I. W. C. E.; Sheldon, R. A. *Science* **2000**, *287*, 1636–1639.
76. Wu, X.; Li, X.; Hems, W.; King, F.; Xiao, J. *Org. Biomol. Chem.* **2004**, *2*, 1818–1821.

77. Wolf, K.; Morris, M. *Switching to water-based cleaners in repair and maintenance parts cleaning*; Pollution Prevention Center, Institute for Research and Technical Assistance, USA: 1999.
78. Pagliaro, M.; Ciriminna, R.; Kimura, H.; Rossi, M.; Della Pina, C. *Angew. Chem., Int. Ed.* **2007**, *46*, 4434–4440.
79. Wolfson, A.; Dlugy, C. *Chem. Pap.* **2007**, *61*, 228–232.
80. Wolfson, A.; Dlugy, C.; Tavor, D.; Blumenfeld, J.; Shotland, Y. *Tetrahedron: Asymmetry* **2006**, *17*, 2043–2045.
81. Wolfson, A.; Litvak, G.; Dlugy, C.; Shotland, Y.; Tavor, D. *Ind. Crops Prod.* **2009**, *30*, 78–81.
82. Gu, Y.; Jerome, F. *Green Chem.* **2010**, *12*, 1127–1138.
83. Garcia-Marin, H.; van der Toorn, J. C.; Mayoral, J. A.; Garcia, J. I.; Arends, I. W. C. E. *Green Chem.* **2009**, *11*, 1605–1609.
84. Pace, V.; Hoyos, P.; Castoldi, L.; Domínguez de María, P.; Alcántara, A. R. *ChemSusChem* **2012**, *5*, 1369–1379.
85. Shanmuganathan, S.; Natalia, D.; van den Wittenboer, A.; Kohlmann, C.; Greiner, L.; Dominguez de Maria, P. *Green Chem.* **2010**, *12*, 2240–2245.
86. Simeo, Y.; Sinisterra, J. V.; Alcantara, A. R. *Green Chem.* **2009**, *11*, 855–862.
87. Shanmuganathan, S.; Greiner, L.; Domínguez de María, P. *Tetrahedron Lett.* **2010**, *51*, 6670–6672.
88. Aycock, D. F. *Org. Process Res. Dev.* **2006**, *11*, 156–159.
89. MacMillan, D. S.; Murray, J.; Sneddon, H. F.; Jamieson, C.; Watson, A. J. B. *Green Chem.* **2013**, *15*, 596–600.
90. Antonucci, V.; Coleman, J.; Ferry, J. B.; Johnson, N.; Mathe, M.; Scott, J. P.; Xu, J. *Org. Process Res. Dev.* **2011**, *15*, 939–941.
91. Mamidipally, P. K.; Liu, S. X. *Eur. J. Lipid Sci. Technol.* **2004**, *106*, 122–125.
92. Mathers, R. T.; McMahon, K. C.; Damodaran, K.; Retarides, C. J.; Kelley, D. J. *Macromolecules* **2006**, *39*, 8982–8986.
93. *Chemicals in the Environment: Freon 113*; Office of Pollution Prevention and Toxics, U.S. Environmental Protection Agency: 1994.
94. *Chemical Summary for Methylchloroform*; Office of Pollution Prevention and Toxics, U.S. Environmental Protection Agency: 1994.
95. Wang, N. C. Y.; Jay Zhao, Q.; Wesselkamper, S. C.; Lambert, J. C.; Petersen, D.; Hess-Wilson, J. K. *Regul. Toxicol. Pharmacol.* **2012**, *63*, 10–19.

Advances in Materials and Chemical Reactions

Chapter 6

Greener Oxidation of Alcohols, Glycosides and Sulfides

Mo Hunsen*

Department of Chemistry, Kenyon College, 310 Tomsich Hall, 200 North College Road, Gambier, Ohio 43022, United States
***E-mail: hunsenm@kenyon.edu**

Catalytic reactions significantly contribute towards the development of sustainable chemistry approaches that do not harm the environment. We have been actively developing novel catalytic reactions including enzyme catalyzed polymerization and green oxidation reactions. While chromium reagents have proven to be powerful and versatile oxidizing agents, their carcinogenicity makes them less attractive for any large scale or industrial applications. We have developed greener reactions for preparation of aldehydes, ketones, and carboxylic acids directly from alcohols. Our method is also useful for selective oxidative ring-opening of glycosides and for oxidation of sulfides to sulfones. In these reactions, we use pyridinium chlorochromate (PCC) as a catalyst together with a co-oxidant that is used to regenerate the catalyst. As a co-oxidant, we show that the electrolytically recyclable periodic acid works very well. Hence, using PCC as a catalyst, instead of a stoichiometric amount, we are able to reduce the amount of chromium waste generated by up to a 100 fold while still maintaining the power and versatility of chromium oxidation reactions.

Introduction

Oxidation reaction is an important tool in the transformation of organic compounds (*1*). There are a number of methods available in the literature. For example, Jones' reagent (*2*) or chromium trioxide in sulfuric acid is a classic condition for preparation of ketones from secondary alcohols and carboxylic acids from primary alcohols. Jones condition is not suitable for preparation of aldehydes due to further oxidation of the aldehydes into carboxylic acids,. A milder reagent, pyridinium chlorochromate (PCC), was later developed by the Corey group (*3*). PCC was successfully used to prepare aldehydes from primary alcohols and ketones from secondary alcohols. However, the search for greener processes for oxidation of organic compounds is still an active area of investigation. We have been developing greener alternatives to existing oxidation methods (*4–6*).

We describe here a method for catalytic oxidation reactions of various organic compounds using periodic acid as the co-oxidant and pyridinium chlorochromate (PCC) as the catalyst. We have also carried out some of the oxidation reactions using other chromium reagents including chromium trioxide (CrO_3), pyridinium fluorochromate (PFC) and the results are comparable. Due to its reasonable price and long shelf life, our preferred catalyst is PCC and this work focuses on using PCC with periodic acid to oxidize alcohols, sulfides, and carbohydrates.

Results and Discussion

1. Catalytic Oxidation of Alcohols to Aldehydes & Ketones

Chromium (VI) is a known carcinogen (*7–10*) and this limits its utility in large scale applications. We sought to maintain the power of chromium oxidation reactions while reducing the amount of toxic waste that is generated. To do that, we have identified that periodic acid (H_5IO_6) is a suitable co-oxidant maintaining chromium at a higher oxidation state. As such using 2 mol% of PCC with respect to the substrates, alcohols, and one equivalent of periodic acid we have successfully oxidized primary alcohols to aldehydes and secondary alcohols to ketones (Schemes 1 and 2) (*6*). As shown below, our approach is mild and it enabled the oxidation of a variety of aliphatic and aromatic alcohols including those with electron withdrawing groups, electron donating groups, and benzylic as well as homobenzylic alcohols. Furthermore, it is worth noting that sensitive functional groups such as the cyclopropyl and propargyl groups were not affected by the reaction condition. However, in cases where an electron withdrawing group is attached to the aromatic ring of benzyl alcohols, we have observed that some further oxidation into carboxylic acids has occurred resulting in moderate yields (Scheme 2).

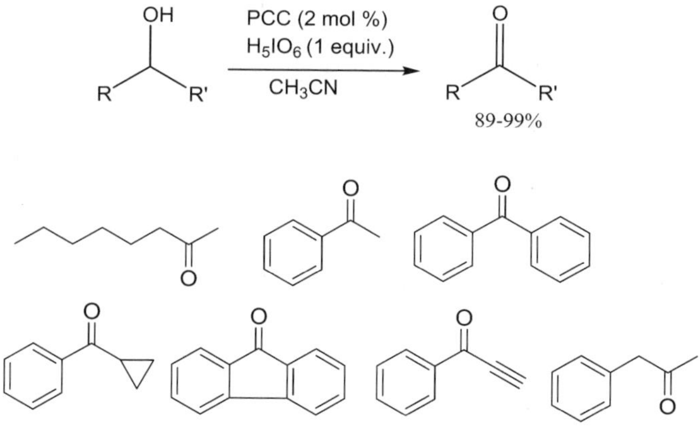

Scheme 1. PCC catalyzed oxidation of alcohols to ketones.

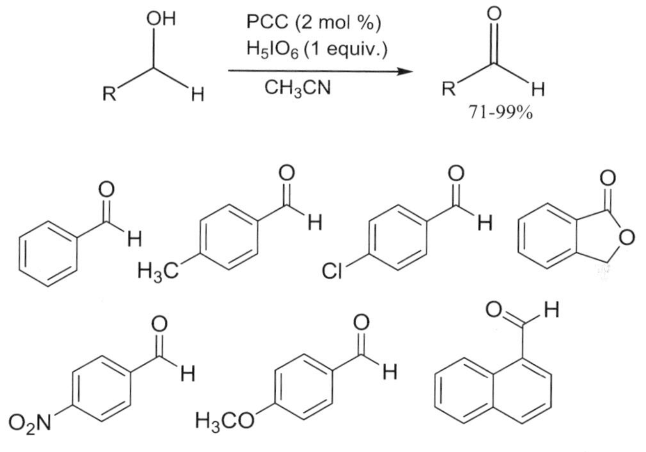

Scheme 2. PCC catalyzed oxidation of alcohols to aldehydes.

We hypothesize that a chlorochromatoperiodate complex forms in the reaction mixture (Scheme 3) resulting in PCC staying at a higher oxidation state throughout the reaction. This hypothesis is consistent with our observation that no green coloration of the reaction mixture was seen during the reaction suggesting the absence of Cr (III) species (Scheme 3).

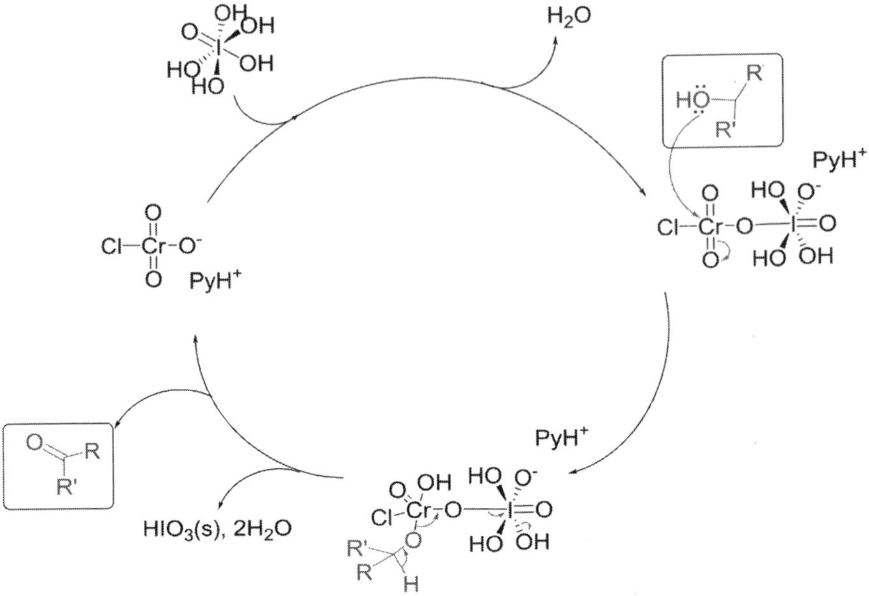

Scheme 3. Proposed mechanism for the PCC catalyzed oxidation of alcohols to aldehydes and ketones.

While our approach is useful for preparation of aldehydes, our attempt to prepare an ortho dialdehyde surprisingly gave an important lactone, phthalide. Phthalides and their analogs have shown interesting biological activities including antifungal and cytotoxic activities (*11*, *12*). These lactones also have potential utility in polymer chemistry as they make interesting substrates for ring opening polymerization of lactones to polyesters (*13*, *14*). One possible explanation could be the initial formation of a monoaldehyde that reacts with the intramolecular hydroxymethyl group to form a hemiacetal that is then oxidized to the observed lactone (Scheme 4).

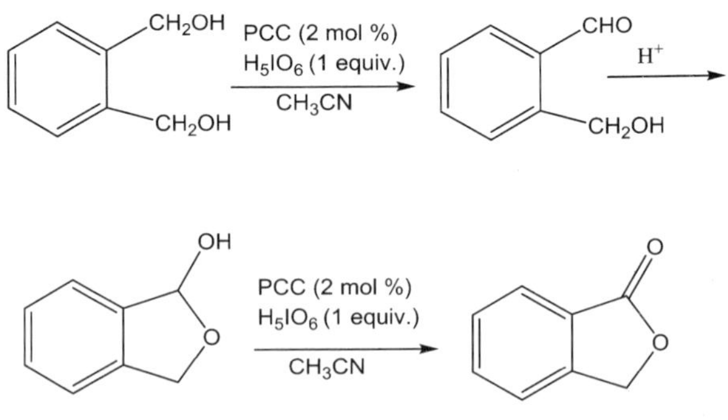

Scheme 4. PCC catalyzed oxidation of an ortho diol to a lactone.

2. Catalytic Oxidation of Alcohols Directly to Carboxylic Acids

Direct oxidation of alcohols to carboxylic acids is a very useful transformation in organic chemistry. Intrigued by some further oxidation of aldehydes into carboxylic acids we have observed during our preparation of aldehydes from primary alcohols, we pursued direct oxidation of alcohols to carboxylic acids.

We attempted the direct oxidation by using two equivalents of periodic acid with a catalytic amount of PCC (2 mol% with respect to substrates) and we were delighted to see that primary alcohols could be oxidized quantitatively into carboxylic acids (*5*). As shown in Scheme 5 below, our approach is rather versatile and it enabled oxidation of a variety of primary alcohols. Our method worked smoothly for direct oxidation of electron poor as well as electron rich benzylic and homobenzylic alcohols (Scheme 5). Aliphatic alcohols and diols were also effortlessly oxidized to the corresponding carboxylic acids in high yields.

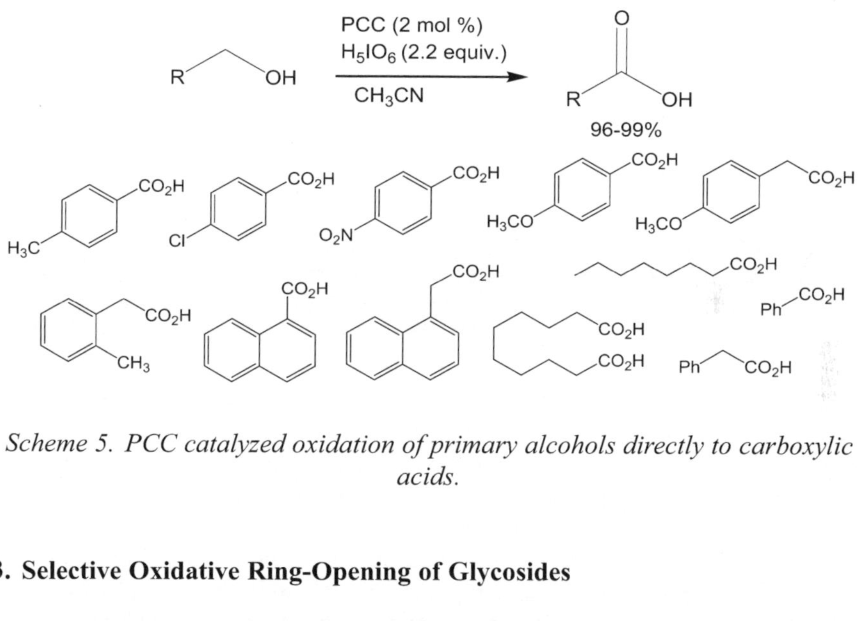

Scheme 5. PCC catalyzed oxidation of primary alcohols directly to carboxylic acids.

3. Selective Oxidative Ring-Opening of Glycosides

With the goal of developing a method for oxidatively ring-opening glycosides, we have further investigated our catalytic oxidation approach. Previously, excess chromium trioxide has been shown to oxidatively open glycoside rings (*15*). However, the carcinogenicity of chromium reagents prohibits any large or wide scale applications of this very useful transformation. As discussed above, our strategy involves maintaining the power of chromium reagents but using these reagents catalytically. Hence, the problem then becomes identifying co-oxidants that will regenerate the catalyst under the ring cleavage condition. Previously, we have reported the preparation of a novel polyketoester directly from cellulose in a one-pot acetylation and oxidation approach (Scheme 6) (*16*).

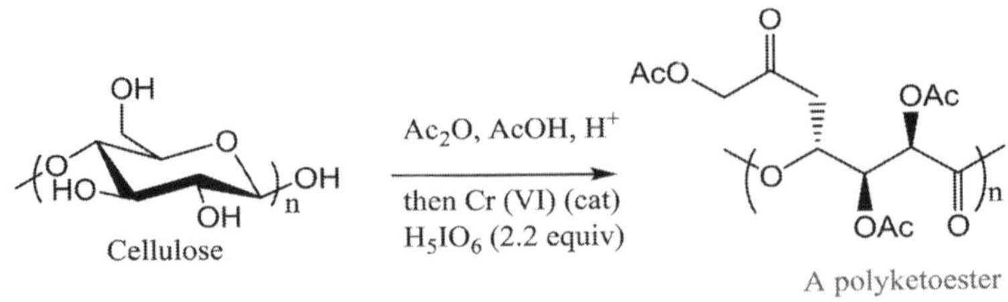

Scheme 6. Preparation of a polyketoester directly from cellulose in a one-pot acetylation and oxidation protocol.

In the above reaction, we have discovered that periodic acid is still a viable co-oxidant for the purpose of regenerating the catalyst efficiently under the reaction conditions shown above (Scheme 6). It is worth mentioning that the byproduct iodic acid could be re-oxidized to periodic acid by electrolysis. Eventhough electrochemical manufacturing of chemicals costs energy, there are promising improvements that could make it more attractive in the Chemical Industry (*17*). In addition, by using less than two equivalents of periodic acid, we have shown that preparation of a copolymer with polyglycoside and polyketoester segments is possible through a limited ring opening of glycosides. Here the oxidation of glycosides stops when all periodic acid that regenerates the catalyst is consumed. Further investigation is in progress to determine the scope and limitation of this oxidation reaction especially when other than 1-4 linkages are present.

Most interestingly, this catalytic oxidation reaction selectively oxidized β-glycosides while the α-glycosides were not amenable to the ring opening observed for the β-glycosides (Scheme 7). While 1-O-methyl β-D-glucose tetraacetate is oxidized to the corresponding ketoester, the corresponding α-glycoside did not generate a ketoester. One possible cause for this difference is the difference in the stereoelectronic environment around the anomeric carbon. Compared to the previous harsh reaction condition used by others (*15*), where excess chromium trioxide was used to carry out the oxidation of glycosides, our approach is milder and more attractive. We anticipate that our method might be useful in selective skeletal modification of complex carbohydrates which are more likely to contain both α- and β-linkages as well as different types of glycosidic linkages (e.g. 1-4 versus 1-2). Further investigation in this direction is in progress and will be reported in due course.

Furthermore, other groups have shown that the above type of ketoesters could be converted to iminocyclitols (Figure 1) - that are important glycosidase inhibitors and relevant in the treatment of diabetes, cancer, and immune diseases – via reductive amination reactions (*18–24*). Since our approach generates significantly lower amount of chromium waste compared to previous methods, we anticipate that it will make the scale up for preparation of iminocyclitols from ketoesters more attractive.

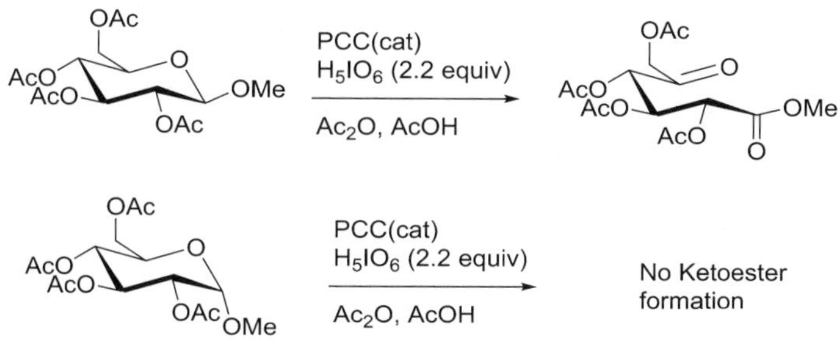

Scheme 7. PCC catalyzed oxidative ring-opening of glycosides with periodic acid.

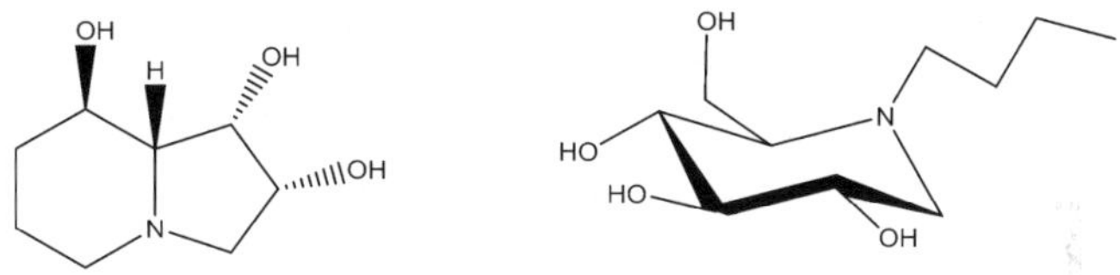

Figure 1. Iminocyclitols - important class of glycosidase inhibitors

4. Oxidation of Sulfides to Sulfoxides and Sulfones

Oxidation of sulfides to sulfoxides and sulfones is another area of interest in developing green organic chemistry methodologies. There are some interesting approaches for these transformations in the literature but many of these methods have limited substrate scope (*25–30*). In this regard, we have investigated the strengths and limitations of our approach by using a variety of sulfides and varying the amount of the co-oxidant. When one equivalent of periodic acid was used with a catalytic amount of PCC (2 mol% with respect to substrates), the results for the formation of sulfoxides are mixed (Scheme 8 and Table 1). For some of the sulfides, a mixture of sulfoxides and sulfones were obtained and we have indicated the mole % ratio in Table 1. We have observed some degree of further oxidation to sulfones both for sulfides with electron donating groups (entries 3, 4, and 7) and those with electron withdrawing groups (entries 2, 5, and 6). The percent yield for entry 7 was relatively low due to the water solubility of the product and we recommend that alternative work up approaches be pursued in such a case.

Since the oxidation of sulfides to sulfoxides is in general not difficult to carry out and there are various methods to do that, we were not disappointed with the results but we were rather curious about the possibility of oxidizing sulfides directly to sulfones with our catalytic approach. Indeed, we were pleased to observe that our method delivers sulfones almost quantitatively directly from sulfides (Scheme 9 and Table 2). Sulfides with electron withdrawing substituents (entries 2, 5, and 6) as well as those with electron donating substituents (entries 3 and 4) were oxidized smoothly to their corresponding sulfones. Interestingly, a hydroxymethyl group remained intact during the oxidation of sulfides (entry 7 in Tables 1 and 2), but some of the product was lost during workup due to its water solubility and we recommend that alternative work up approaches be pursued in such a case.

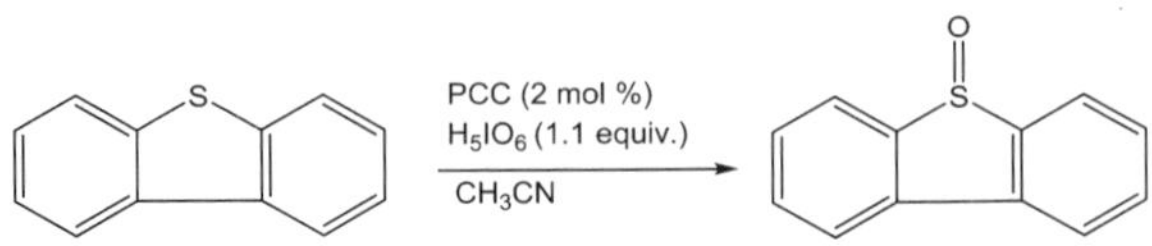

Scheme 8. Oxidation of sulfides to sulfoxides.

Table 1. Oxidation of sulfides to sulfides.

Entry	Substrate	Product	Yield % ($RSOR':RSO_2R'$) (mol % ratio)
1	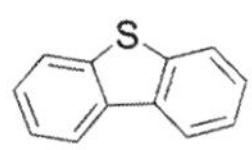	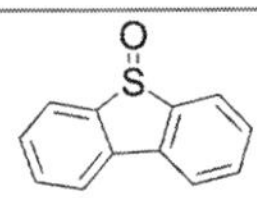	99 (100:0)
2	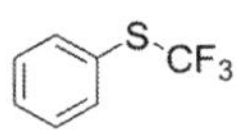	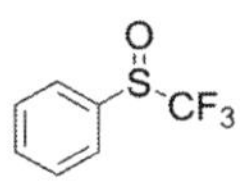	66 (80:20)
3	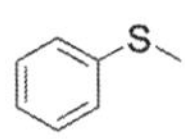	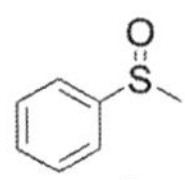	98 (98:2)
4	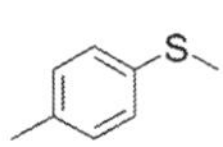	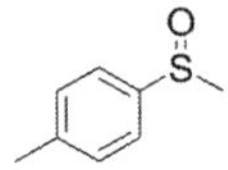	99 (65:35)
5	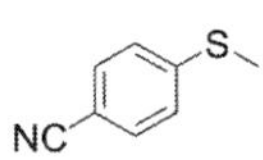	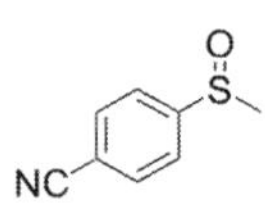	99 (86:14)
6	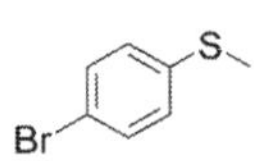	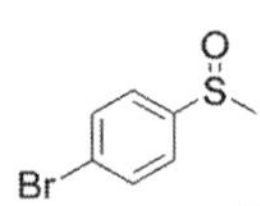	99 (82:18)
7	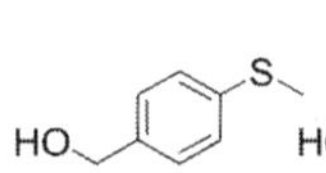	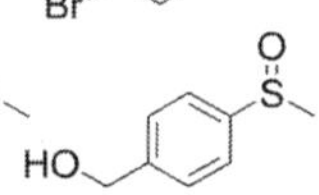	51 (75:25)

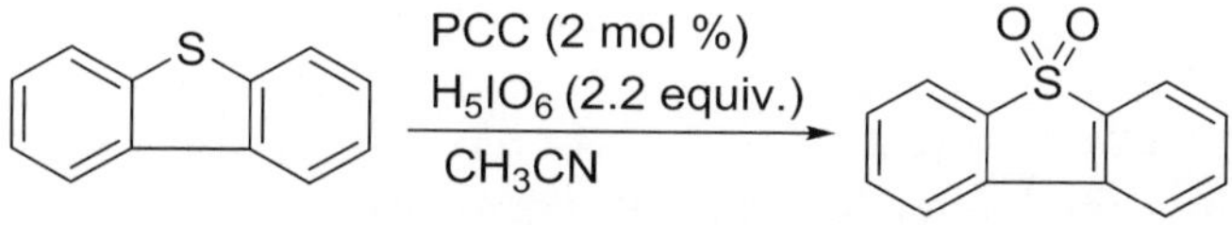

Scheme 9. Oxidation of sulfides to sulfones.

Table 2. Oxidation of sulfides to sulfones

Entry	Substrate	Product	Yield %
1	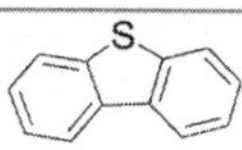	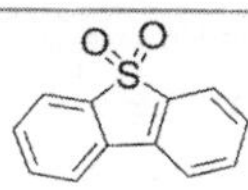	95
2	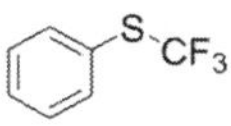	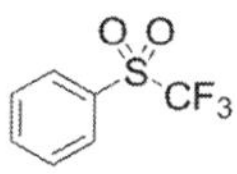	76
3	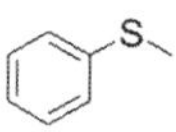	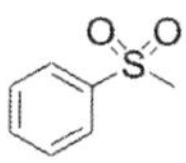	99
4	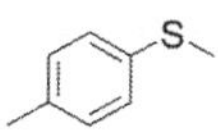	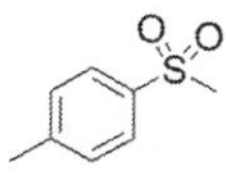	99
5	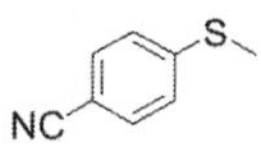	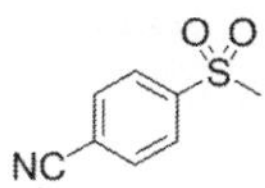	99
6	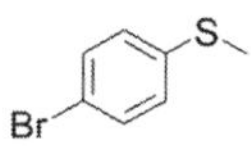	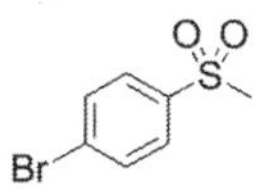	98
7	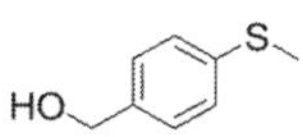	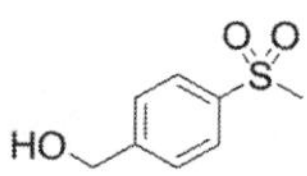	48

Conclusions

In summary, we have developed a versatile method for oxidation of a variety of organic compounds including alcohols, sulfides, and carbohydrates. Our approach of using a catalytic amount of PCC with periodic acid as the co-oxidant is useful in the preparation of aldehydes, ketones, carboxylic, acids, sulfones, and ketoesters of glycosides.

Experimental

Oxidation of 1-O-Methyl 2,3,4,6-tetra-O-acetyl-β-D-glucose Is Representative for Oxidative Ring Opening of Glycosides

To a mixture of 50 mL acetic acid, 5 ml acetic anhydride, and 0.90 g (2.5 mmol) of 1-O-methyl 2,3,4,6-tetra-O-acetyl-β-D-glucose was added (in ice bath) 2.28 g (10 mmol) of H_5IO_6 and 0.022 g PCC (2 mol%). The reaction mixture was stirred in ice-bath for 1 h and at r.t. overnight. The iodic acid precipitate was recovered by filtration under vacuum and the solvents were evaporated under reduced pressure using a rotary evaporator. The residue was dissolved in 100 mL ethyl acetate, and washed with brine and saturated aq. $NaHSO_3$. respectively. It was then dried over anhydrous sodium sulfate and the solvent was removed under reduced pressure to give methyl 2,3,4,6-tetra-O-acetyl-D-xylo-hex-5-ulosonate (0.84 g, 90%). The ^{1}H- and ^{13}C- NMR spectra of the product was identical to those reported by Angyal (*15*) and Pistia (*20*).

Synthesis of Methyl Phenyl Sulfoxide (Table 1, entry 3) Is Representative for Preparation of Sulfoxides

To 30 mL of acetonitrile was added 1.21 g (5.3 mmol) of H_5IO_6 and stirred vigorously at r.t. for 15 min. Methyl phenyl sulfide (0.621 g, 5 mmol) was then added (in ice-water bath) followed by addition of 22 mg (2 mol %) PCC in 10 mL acetonitrile (in two portions) and the reaction mixture was stirred at r.t. for 4 h. The reaction mixture was then diluted with 100 mL ethyl acetate and washed with 1:1 brine : water, saturated aq. $Na_2S_2O_3$ solution, and brine, respectively. It was then dried over anhydrous Na_2SO_4 and concentrated at reduced pressure to give the sulfoxide (0.69 g, 98 % Yield). The products are known compounds and they were characterized by comparing their NMR spectra with those reported by Aldrich and with authentic samples.

Synthesis of Dibenzothiophene Sulfone (Table 2, entry 1) Is Representative for Preparation of Sulfones

To 30 mL of acetonitrile was added 1.21 g (5.3 mmol) of H_5IO_6 and stirred vigorously at r.t. for 15 min. Dibenzothiophene (0.461 g, 2.5 mmol) was then added (in ice-water bath) followed by addition of 22 mg (2 mol %) PCC in 10 mL acetonitrile (in two portions) and the reaction mixture was stirred at r.t. for 4 h. The reaction mixture was then diluted with 100 mL ethyl acetate and washed with 1:1 brine : water, saturated aq. $Na_2S_2O_3$ solution, and brine, respectively. It was then dried over anhydrous Na_2SO_4 and concentrated under reduced pressure to give the sulfone (0.513 g, 95 % Yield). The products are known compounds and they were characterized by comparing their NMR spectra with those reported by Aldrich and with authentic samples.

Acknowledgments

MH thanks Kenyon College, The Henry Dreyfus Foundation, and HHMI for the generous funding.

References

1. Hudlicky, M. *Oxidations in Organic Chemistry*; ACS Monograph Series 186; American Chemical Society: Washington, DC, 1990; pp 1−46.
2. Bowden, K.; Heilbron, I. M.; Jones, E. R. H.; Weedon, B. C. L. *J. Chem. Soc.* **1946**, 39–45.
3. Corey, E. J.; Suggs, J. W. *Tetrahedron Lett.* **1975**, 2647–2460.
4. Hunsen, M. *J. Fluorine Chem.* **2005**, *126*, 1356–1360.
5. Hunsen, M. *Synthesis* **2005**, 2487–2490.
6. Hunsen, M. *Tetrahedron Lett.* **2005**, *46*, 1651–1653.
7. Nickens, K. P.; Patierno, S. R.; Ceryak, S. *Chem.-Biol. Interact.* **2010**, *188*, 276–288.
8. Costa, M. *Crit. Rev. Toxicol.* **1997**, *27*, 431–442.
9. Thompson, C. M.; Proctor, D. M.; Haws, L. C.; Hebert, C. D.; Grimes, S. D.; Shertzer, H. G.; Kopec, A. K.; Hixon, J. G.; Zacharewski, T. R.; Harris, M. A. *Toxicol. Sci.* **2011**, *123*, 58–70.
10. Wise, S. S.; Holmes, A. L.; Qin, Q.; Xie, H.; Katsifis, S. P.; Thompson, W. D.; Wise, J. P., Sr. *Chem. Res. Toxicol.* **2010**, *23*, 365–372.
11. Karmakar, R.; Pahari, P.; Mal, D. *Chem. Rev.* **2014**, *114*, 6213–6284.
12. Rukachaisirikul, V.; Rodglin, A.; Sukpondma, Y.; Phongpaichit, S.; Buatong, J.; Sakayaroj, J. *J. Nat. Prod.* **2012**, *75*, 853–858.
13. Hunsen, M.; Abul, A.; Xie, W.; Gross, R. *Biomacromolecules* **2008**, *9*, 518–522.
14. Hunsen, M.; Azim, A.; Mang, H.; Wallner, S. R.; Ronkvist, A.; Xie, W.; Gross, R. A. *Macromolecules* **2007**, *40*, 148–150.
15. Angyal, S. J.; James, K. *Aust. J. Chem.* **1970**, *23*, 1209–1221.
16. Hunsen, M. In *Materials, Chemicals and Energy from Forest Biomass*; Argyropoulos, D. S., Ed.; ACS Symposium Series 954; American Chemical Society: Washington, DC, 2007; pp 247−259.
17. Botte, G. G. *Interface* **2014**, *23*, 46.
18. Asano, N.; Nash, R. J.; Molyneux, R. J.; Fleet, G. W. J. *Tetrahedron: Asymmetry* **2000**, *11*, 1645–1680.
19. Watson, A. A.; Fleet, G. W.; Asano, N.; Molyneux, R. J.; Nash, R. J. *Phytochemistry* **2001**, *56*, 265–295.
20. Pistia, G.; Hollingsworth, R. I. *Carbohydr. Res.* **2000**, *328*, 467–472.
21. Heck, M.-P.; Vincent, S. P.; Murray, B. W.; Bellamy, F.; Wong, C.-H.; Mioskowski, C. *J. Am. Chem. Soc.* **2004**, *126*, 1971–1979.
22. Kajimoto, T.; Liu, K. K. C.; Pederson, R. L.; Zhong, Z.; Ichikawa, Y.; Porco, J. A., Jr.; Wong, C. H. *J. Am. Chem. Soc.* **1991**, *113*, 6187–6196.
23. Mitchell, M. L.; Lee, L. V.; Wong, C.-H. *Tetrahedron Lett.* **2002**, *43*, 5691–5693.

24. Moris-Varas, F.; Qian, X.-H.; Wong, C.-H. *J. Am. Chem. Soc.* **1996**, *118*, 76477652.
25. Engberts, J. B. F. N. In *The Chemistry of Sulphones and Sulphoxides*; Patai, S., Rappoport, Z., Stirling, C., Ed.; John Wiley and Sons: Chichester, UK, 1988; Vol. 107, pp 165–278.
26. Lakouraj, M. M.; Abdi, H.; Hasantabar, V. *J. Sulfur Chem.* **2011**, *32*, 435–441.
27. Priebe, W.; Grynkiewicz, G. *Tetrahedron Lett.* **1991**, *32*, 7353–7356.
28. Sato, K.; Hyodo, M.; Aoki, M.; Zheng, X.-Q.; Noyori, R. *Tetrahedron* **2001**, *57*, 2469–2476.
29. Simpkins, N. S. In *Sulphones in organic synthesis*; Tetrahedron organic chemistry series 10; Pergamon Press: Oxford, UK, 1993; Vol. 10, pp 5–50.
30. Rozen, S.; Bareket, Y. *J. Chem. Soc., Chem. Commun.* **1994**, 1959–1959.

Chapter 7

An EcoScale Comparison of Mechanochemistry and Solution Based Reactions

Kendra Leahy, Anthony M. Mack, and James Mack*

University of Cincinnati, Department of Chemistry, 301 Clifton Court, Cincinnati, Ohio 45221-0172
***E-mail: james.mack@uc.edu**

The development of environmentally benign conditions to conduct chemical reactions is needed to ensure the sustainability of chemistry. Mechanochemical reactions have been demonstrated to be an environmentally benign alternative to conducting chemical reactions. However, the question of how much greener is a mechanochemical reaction over a solution based reaction is still debated. By using the EcoScale, a direct comparison can be made between these two methods. Overall, mechanochemical reactions can be safer and less time consuming than conducting reactions in solution.

Introduction

Throughout the history of chemistry, solvents have been believed to be an essential component of a chemical reaction. Aristotle's proclamation "*corpora non agunt nisi fluida seu soluta*" (*transl.* "Compounds that are not fluid or dissolved, do not react") (*1*) and Karsten's suggestion (*2*)

> "*The reaction of two heterogeneous, solid, and under certain conditions reactive compounds can only occur if one of them can be transformed into a fluid induced by the interaction between the two compounds at a given temperature or due to pressure increased temperature, which then will induce the fluid state in the other compound.*"

represent examples that led chemists to believe chemical reactions cannot occur in the absence of a solvent or in the solid state. In addition, many notable Nobel

Prize winners earned this prestigious award specifically for their work in solution based research, including two of the first three Nobel Prizes awarded in chemistry. As more solvents became available to chemists, it was further noted that not only are solvents a necessity, but choosing the correct solvent for a particular reaction is of even greater importance. Sir Christopher Ingold stated

> "*Thus, in the unimolecular substitution of an alkyl halide, both the ions, which are formed in the initial heterolysis, have to be solvated*" (*3*). He re-affirmed this position sixteen years later and stated "*The energy of the initial heterolysis has to be set off against the energy of solvation of the ions (both ions, not one of them only, as has been mistakenly asserted). The energy difference approximates to the heat of ionisation in solution. Thus, in order that mechanism* S_N1 *shall operate with a practical degree of facility, a solvent of suitable polarity is essential (as has not always been fully appreciated.)*" (*4*)

These accounts and various others (*5–8*) have led chemists to believe a solvent is a necessary component of a chemical reaction. Because of the ingrained belief that solvents are necessary, little thought is given to conducting a chemical reaction in the absence of a solvent. This thought has perpetuated itself throughout chemistry, including the area of organic synthesis.

Conducting chemical reactions in the absence of a solvent has gained much prominence in the last decade (*9–13*). More specifically, conducting chemical reactions under mechanochemical conditions has been shown to be a viable solvent-free alternative for various chemical reactions.

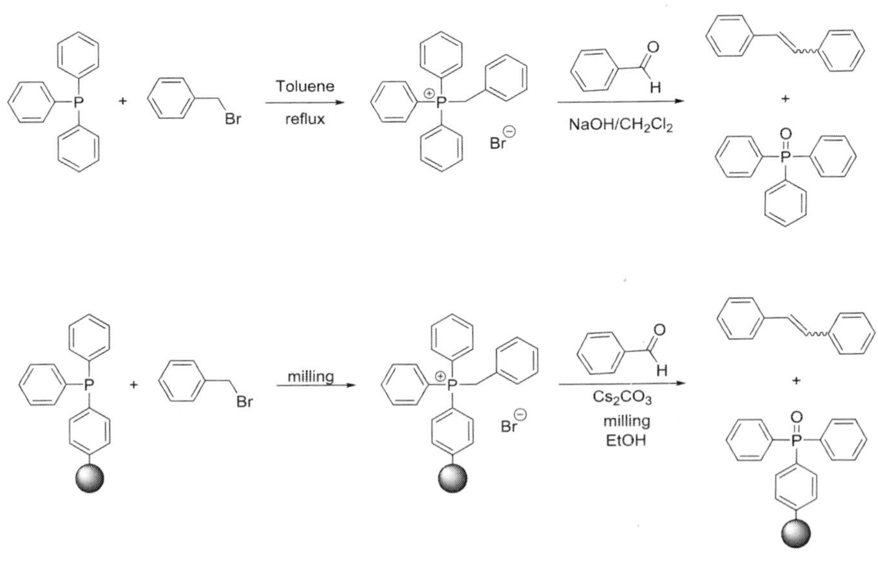

Figure 1. Comparison of soluion based Wittig (top) against mechanochemical Wittig (bottom)

Although various reports present the benefits of mechanochemical reactions, many have asked what specifically makes mechanochemical reactions safer than solution based reactions. In order to address this question, we wanted to test the Wittig reaction under mechanochemical conditions against the same reaction conducted in solution (Figure 1.). The goal of this work is to assess the greenness of the different reaction conditions; therefore the focus is placed on the amount of material used, regardless of the physical state of the reagents.

To assess the greenness of the reaction, we utilized Van Aken's EcoScale which was designed to determine the environmentally benign nature of different synthetic methodologies (*14*, *15*). In addition to measuring the EcoScale values, we wanted to determine the skill level needed to conduct a mechanochemical reaction. Therefore, the solution based reaction was conducted by a university professor, while the mechanochemical reaction was conducted by a 13 year old (with the supervision of a graduate student for safety reasons) who had no prior experience conducting chemical reactions. Anseleme, Warner and Anastas previously reported a procedure for the synthesis of stilbene via the Wittig reaction designed for an undergraduate laboratory (*16*). This report was the foundation of the procedure used for the solution based reaction. Shearouse and co-workers reported an environmentally benign Wittig reaction that utilized polymer supported reagents in tandem with mechanochemistry. This report was the foundation for the experimental procedure used for the mechanochemical reaction. Using a 1 mmol scale, the Wittig reaction in solution was compared to the Wittig reaction under mechanochemical conditions.

Comparison of the Reactions

Generation of the Phosphonium Salt

In both reaction protocols, the first step was the formation of the phosphonium salt using benzyl bromide in the presence of triphenylphosphine. To conduct the mechanochemical reaction, 0.139g of benzyl bromide and 0.495g of polymer supported triphenylphosphine were added to a 0.5 inch (diameter) by 2.0 (length) inch stainless steel reaction vial along with a stainless steel milling ball (3/16 inch diameter). The reaction vessel was sealed and milled in a Spex8000M Mixer/ Mill for two hours. By contrast, in solution 0.147g of benzyl bromide, 0.339g of triphenylphosphine, and 25mL of toluene were added to a 100 mL round-bottom flask along with a stir bar. The reaction was refluxed for 3 hours using a reflux condenser and a wax bath.

Formation of the Stilbene

At the conclusion of the initial two hours of the mechanochemical reaction, the reaction vessel was opened and 0.424g of cesium carbonate, 1mL of ethanol and 0.081g of benzaldehyde were added. The reaction vessel was sealed and the reaction was milled for an additional 2 hours. At the conclusion of the solution reaction, the flask was allowed to cool to room temperature over a 30 minute period. Once cooled, the phosphonium salt was filtered and allowed to dry for 15 minutes. The phosphonium salt was added to .091g of benzaldehyde, which had been pre-dissolved in 10 mL of methylene chloride. To the solution of benzaldehyde was added 10 mL of a 50% sodium hydroxide solution. The solution was allowed to reflux for an additional 30 minutes then allowed to cool to room temperature for 15 minutes.

Isolation of the Product

At the conclusion of the mechanochemical reaction, the reaction mixture was removed from the vial, suspended in ethyl acetate, and then filtered. The ethyl acetate was removed under reduced pressure and the product was isolated to give a 73% yield of stilbene in a 54:46 ratio of *cis* to *trans*.

At the conclusion of the solution based reaction, the reaction was cooled and washed with a brine solution and dried over magnesium sulfate. The methylene chloride was removed and the crude product was isolated. Column chromatography using cyclohexane and dichloromethane as the eluents was performed on the crude product mixture to give the stilbene in 55% yield in a 56:44 *cis* to *trans* ratio.

Calculating the EcoScale

Van Aken and co-workers developed a novel semi-quantitative post-synthesis analysis tool to compare various synthetic methods on safety, economical factors and ecological factors. The scale is defined by 6 factors: yield, price of the reaction components, safety, temperature/time, technical set-up and work-up conditions. The reaction conditions are scored from 0 to 100 where 0 is a completely failed reaction and 100 is defined as a reaction where compound A undergoes a reaction with inexpensive compound(s) B to give 100% yield of compound C at room temperature with minimal risk to the operator and minimal impact on the environment. Each reaction starts at 100 and is assessed penalty points based on a semi-quantitative scale. To calculate the final score of the reaction, the sum of the individual penalty points is subtracted from 100. A score of 75 or higher is given an excellent rating; a score ranging from 50-75 is given a rating of acceptable; and scores below 50 are deemed inadequate.

According to the EcoScale the penalty for yield is (100-% yield)/2. For the mechanochemical reaction the yield penalty is 13 points, while for the solution based reaction the penalty is 22 points. With respect to price of the components, the penalty is assessed based on the cost to make 10 mmol of product. Since the solution based reaction would cost less than $10 to synthesize 10 mmol of product, the reaction is deemed inexpensive and is not assessed any penalty points. On the contrary, the mechanochemical reaction costs between $10 and $50 to synthesize 10 mmol of product, making the reaction conditions expensive and thus 3 penalty points were assessed. Based on safety rating (i.e. toxicity, flammability and effect on the environment), the solution based reaction was assessed 25 penalty points, whereas the mechanochemical reaction was assessed 5 penalty points. Since the solution based reaction used standard glassware (i.e, round-bottom flask, stir bar, reflux condenser etc.) 0 penalty points were assessed for the reaction. On the other hand, because mechanochemistry requires special equipment (i.e., ball mill) it was assessed 2 penalty points. For temperature/time, the two reflux stages needed for the solution based reaction resulted in 5 penalty points. By comparison, the mechanochemical reaction was not awarded any penalty points since neither heating nor cooling was involved. In the work-up stage, simple extraction and column chromatography were used to isolate the product from the solution reaction; thus 13 penalty points were given. However, for the mechanochemical reaction, only simple filtration was used to isolate the product and therefore no penalty points were assessed.

Overall Assessment of the Two Methods

When comparing the EcoScale of the two reaction methodologies, it is clear the mechanochemical conditions are safer and more environmentally benign than traditional solution based conditions. According to the EcoScale rating, the mechanochemical Wittig reaction was more expensive due to the use of polymer-supported triphenylphosphine, which is safer but more expensive than non-supported triphenylphosphine. It also required special equipment. These two aspects led to an overall increase in 5 penalty points for the mechanochemical conditions over the solution conditions in these areas. However, the solution conditions gave a lower yield, were less safe, needed to be refluxed and the product purified by column chromatography; this led to a combined increase of 47 penalty points over mechanochemical conditions in these areas. All totaled, the solution reaction had an EcoScale rating of 35. Since this is below 50, it is deemed an inadequate reaction. On the contrary, the mechanochemical reaction had an EcoScale rating of 77. Since this is in the 75-100 range, the procedure is deemed excellent (Table 1 and Figures 2 and 3).

One area the EcoScale does not take into account, which should be mentioned here, is the time needed to conduct each experiment. The first step in both reactions was the generation of the phosphonium salt. Conducting the reaction in solution required the heating of the wax bath, setting up the reflux condenser and weighing the reactants, including the solvent. This took the university professor 45 minutes to setup. On the contrary, for the mechanochemical reaction the 13 year old needed to weigh the reactants and add them to the milling vial, which took a total of 21 minutes, less than half the amount of time. Once the solution based reaction was complete, the flask needed to be cooled to room temperature before filtration. Once filtered and dried, the next reaction needed to be set-up, requiring new glassware, different solvents and weighing out reagents, which took 43 minutes. By contrast, once the mechanochemical reaction was complete, the reagents were weighed and immediately added to the same reaction vial. It took 22 minutes to set-up that second step of the reaction, again almost half the time it took to set up the second step of the solution based reaction. At the conclusion of the second reaction conducted under mechanochemical conditions, the crude product was suspended in ethyl acetate and filtered, which took 30 minutes. While at the conclusion of the solution based reaction, the product was prepared for chromatography, which also took approximately 30 minutes. All totaled, from initial set-up to pure product the mechanochemical reaction took 5 hours and 11 minutes, while the solution based reaction took 7 hours and 37 minutes. Therefore not only is the mechanochemical reaction greener and safer, it took roughly 2/3 of the time to achieve the same outcome.

Table 1. EcoScale comparison of solution against mechanochemistry.

	Solution		*Mechanochemistry*	
Item numbers	*Item*	*Penalty*	*Item*	*Penalty*
Yield	55%	22	73%	13
Price of reaction components	Triphenylphosphine	0	Polymer supported triphenylphosphine	3
	Benzyl bromide	0	Benzyl bromide	0
	Benzaldehyde	0	Benzaldehyde	0
	Toluene	0	Ethanol	0
	Dichloromethane	0	Cesium carbonate	0
	Sodium hydroxide	0		
	Water	0		
	Cyclohexane	0		
Safety	Triphenylphosphine	0	Polymer supported triphenylphosphine	0
	Benzyl bromide	0	Benzyl bromide	0
	Toluene	5	Benzaldehyde	0
	Benzaldehyde	0	Ethyl acetate	5
	dichloromethane	0		
	Sodium hydroxide	10		
	Water	0		
	Cyclohexane	10		
Technical set-up	Common glassware stirring	0	Mechanochemistry	2
Temperature/ time	Reflux 3 hours	3	None	0

Continued on next page.

Table 1. (Continued). EcoScale comparison of solution against mechanochemistry.

	Solution		*Mechanochemistry*	
Item numbers	*Item*	*Penalty*	*Item*	*Penalty*
	Reflux 30 minutes	2		
Work-up	Liquid-liquid extraction	3	Simple filtration	0
	Silica gel chromatography	10		
Total Penalty Points		65	23	
EcoScale rating	100-65	**35**	100-23	**77**
Overall Assessment		Inadequate		Excellent

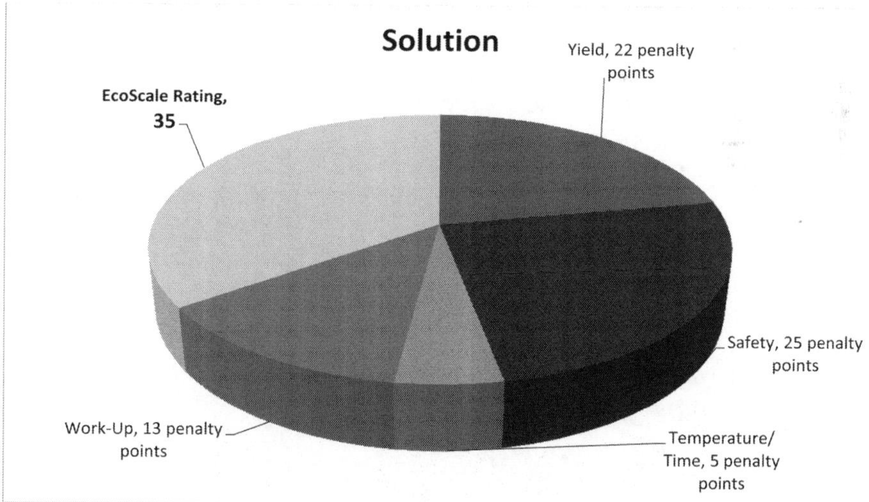

Figure 2. EcoScale rating of solution based Wittig reaction

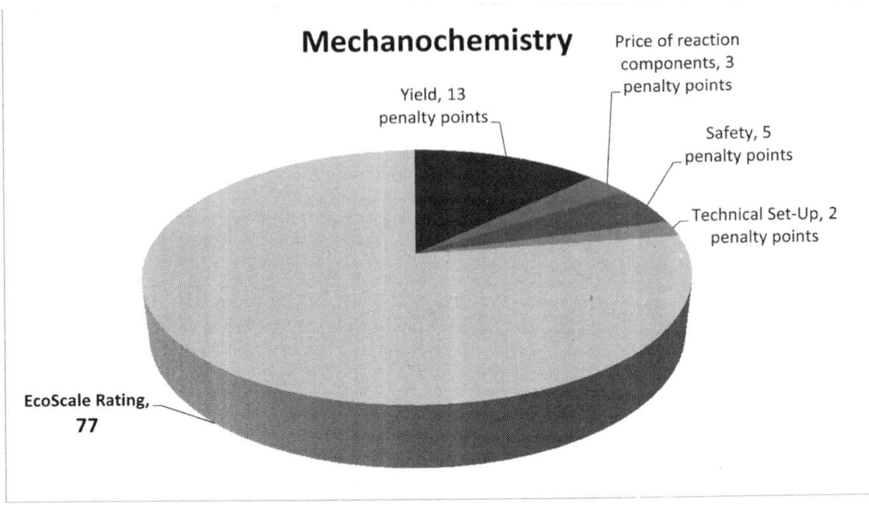

Figure 3. EcoScale rating of Wittig reaction conducted under mechanochemical conditions

Conclusions

The development of environmentally benign methodology is an important aspect of the future of chemistry. Mechanochemistry provides an effective environmentally benign alternative for conducting chemical reactions and isolating products. Mechanochemistry is safer, significantly reduces organic solvent use, can save time, and can provide similar results to what is reported in solution. From the work presented in this chapter, it was also demonstrated that a middle school student could conduct a mechanochemical reaction safely. It is expected that continued studies with mechanochemistry will lead to a significant advancement in the area of solvent-free synthesis.

References

1. Aristotle. *On the coming to being and passing away*; p 328b3.
2. Karsten, C. *Philosophy of Chemistry*; Druck und Verlag Van G. Reimer: Berlin, 1843.
3. Ingold, C. *Structure and Mechanism in Organic Chemistry*; Cornell University Press: Ithaca, NY, 1953; p 357.
4. Ingold, C. *Structure and Mechanism in Organic Chemistry*, 2nd ed.; Cornell University Press: Ithaca, NY, 1969; p 425.
5. Knorr, L.; Schmidt, J. *Justus Liebigs Ann. Chem.* **1896**, *293*, 70–120.
6. Claisen, L.; Falk, L.; Tingle, J. B.; Kerstiens, C. *Justus Liebigs Ann. Chem.* **1896**, *291*, 25–137.
7. Ohtaki, H. *Coord. Chem. Rev.* **1999**, *185–186*, 735–759.
8. Laidler, K. *Pure Appl. Chem.* **1990**, *62*, 2221–2226.

9. Michalchuk, A. A. L.; Tumanov, I. A.; Boldyreva, E. V. *CrystEngComm* **2013**, *15*, 6403–6412.
10. Takacs, L. *Acta Phys. Pol., A* **2012**, *121*, 711–714.
11. Boldyreva, E. *Chem. Soc. Rev.* **2013**, *42*, 7719–7738.
12. James, S.; Adams, C.; Bolm, C.; Braga, D.; Collier, P.; Friscic, T.; Grepioni, F.; Harris, K.; Hyett, G.; Jones, W.; Krebs, A.; Mack, J.; Maini, L.; Orpen, A.; Parkin, I.; Shearouse, W.; Steed, J.; Waddell, D. *Chem. Soc. Rev.* **2012**, *41*, 413–447.
13. Wang, G.-W. *Chem. Soc. Rev.* **2013**, *42*, 7668–7700.
14. Van Aken, K.; Strekowski, L.; Patiny, L. *Beilstein J. Org. Chem.* **2006**, *2*, 3.
15. Van Aken, K. Ecoscale. http://www.ecoscale.org/ (accessed November 5, 2014).
16. Warner, J. C.; Anastas, P. T.; Anselme, J.-P. *J. Chem. Educ.* **1985**, *62*, 346.

Chapter 8

Some Strategies in Designing Highly Efficient Photocatalysts for Degradation of Organic Pollutants in Water

Changlin Yu,[1,2,*] Wanqin Zhou,[1] Gao Li,[2] and Rongchao Jin[2,*]

[1]School of Metallurgy and Chemical Engineering, Jiangxi University of Science and Technology, 86 Hongqi Road, Ganzhou 341000, Jiangxi, China
[2]Department of Chemistry, Carnegie Mellon University, Pittsburgh, Pennsylvania 15213, United States
[*]E-mail: yuchanglinjx@163.com (C.Y.); rongchao@andrew.cmu.edu (R.J.)

Over the past decades, with the fast development of chemical industry, organic pollutants in water have become a major source of environmental pollution in waste fields. Photocatalysis, as a "green chemistry" technology, has attracted much attention in pollutants degradation due to its efficient degradation rate, high mineralization efficiency and low toxigenicity, ideally producing CO_2 and H_2O as the end-products. Photocatalytic degradation of organic pollutants includes fabrication of photocatalytic materials, research of reaction kinetics and working mechanism. This Chapter presents some new strategies in the design of highly efficient photocatalysts for degradation of organic pollutants, such as the design and fabrication of hetero-structured or microspherical photocatalysts, and sonochemical synthesis of photocatalysts with novel nano-structures. The formation of well-defined junctions between two semiconductors with matchied electronic band structure can effectively facilitate charge transfer and suppress the recombination of photo-generated electrons and holes, leading to extremely high activity and stability. Once the photocatalysis is applied on a large scale, operating costs should be emphasized, and recycling of photocatalysts becomes important in order to cut down the operating costs to an affordable level. Compared to the conventional nano-powder counterparts, microspherical photocatalysts appeared to be applicable because the microspherical photocatalysts can be readily recycled. As a mild and green preparation technique, sonochemical fabrication can meet the green requirement in the synthesis of photocatalyts with special morphology under low temperatures.

Progress in the Fundamentals of Environmental Photocatalysis

Photocatalysis is an attractive "green chemistry" technology, which is particularly important for both the increasingly serious environmental pollution and energy crisis since Fujishima's early finding that the hydrogen evolution reaction could occur when TiO_2 electrodes were irradiated with UV light (*1*). The concept of photocatalytis with a light-excited semiconductor as catalyst has attracted great attention since the 1970s (*2*).

The process of photocatalytic reactions in degradation of organic pollutants is illustrated in Scheme 1. As the first step in a heterogeneous photocatalytic process, generation of the electron (e^-)-hole (h^+) pairs, is initiated upon irradiation of light with photon energy greater than or equal to the band gap energy of the photocatalyst. The photo-generated electrons and holes can either recombine in the bulk to produce thermal energy, or migrate to the surface and react with the adsorbed species at the surface, enabling a series of reduction and oxidation reactions, which are applicable to the hydrogen generation or pollutant decomposition. Herein we focus on the latter process. Many evidences suggest that the mechanism in the environmental photocatalysis involves oxidation of pollutants by the extremely active radicals, such as the HO· radical (*3*), the $O_2^{\bullet -}$ radical (*4–6*), and so on. Hydroxyl radical (HO·) production can be expressed as follows: $H_2O + h^+ \rightarrow OH\cdot + H^+$. On the other hand, the photo-generated electrons on the surface of photocatalyst could be trapped by molecular oxygen in solution to form reactive oxygen species ($O_2^{\bullet -}$). Both HO· and $O_2^{\bullet -}$ radicals can effectively decompose most of the organic pollutants.

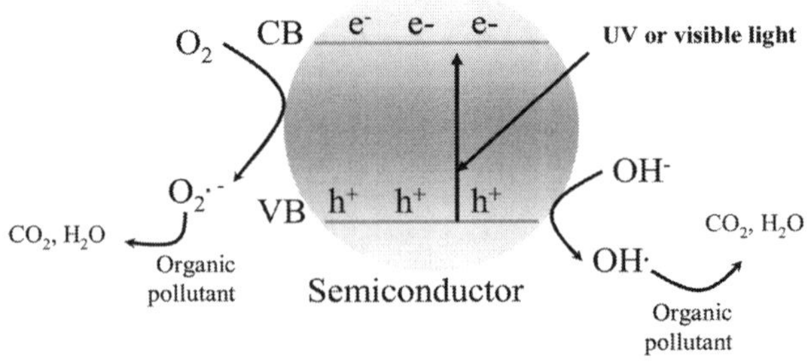

Scheme 1. Mechanism of semiconductor photocatalysis in degradation of organic pollutants.

Up to now, numerous organic substrates, including phenol, dyes, chlorinated compounds, aromatics, alkenes, alkanes, etc. have been tested as the pollutant models (*7*), and some pilot-scale tests have demonstrated that the photocatalytic reactions can be initiated by either UV or visible light (*8*, *9*). Moreover, some degradation pathways and reaction mechanisms have been illuminated, which helps to evaluate the toxicity of the intermediates (*10*).

Recent years have witnessed extensive work on the preparation of semiconductor photocatalytic materials (*11–17*). As for the large band gap semiconductors (such as TiO_2 and ZnO), much effort has been devoted to improving the kinetics of photocatalytic reactions by enhancing the separation efficiency of e^--h^+ pairs, or extending the wavelength of excitation light from UV to visible light. As an important photocatalyst, TiO_2 has some great advantages, such as its high oxidative power for organic pollutants, high photo-stability, low cost, and nontoxicity (*18–20*). However, the major drawback of TiO_2 is its large band gap of 3.2 eV, so wavelengths below 400 nm are necessary for excitation and only the small UV fraction in solar light, about 2-3%, can be utilized by TiO_2, which considerably limits the use of natural solar light or artificial visible light. To overcome this disadvantage, some approaches have been developed, for examples, coupling with a small bandgap semiconductor (*21*, *22*), addition of transition metals (*23*, *24*), and nonmetal doping (*25*, *26*); these methods are the common ones to extend the absorption by TiO_2 from UV light to the visible light regime. At the same time, alternative types of visible-light-driven photocatalysts have been rapidly developed in recent years, such as gold nanoclusters ($Au_{25}(SR)_{18}/TiO_2$) (*27*), $TiO_{2-x}N_x$ (*28*), Rh(Pd, Pt)/BiOX(Cl, Br, I) (*29*), *g*-C_3N_4 (*30*), plasmonic noble metal (e.g., gold, silver) nanoparticles (*31*), etc.

The optical property and catalytic performance of semiconductor photocatalysts strongly depend on their microstructures or morphologies. For example, the formation of interface or heterojunction between two semiconductors can largely enhance the separation of photo-generated electron–hole pairs, resulting in high activity and stability (*32*). It was found that the formation of phase junction between the surfaces of anatase and rutile nanoparticles could greatly enhance the photocatalytic activity for H_2 production (*33*). The formation of Ag_2O/Ag_2CO_3 over the surface of the Ag_2CO_3 brings up about 68 and 30 times increase in the activity in degrading methyl orange (MO) dye over the Ag_2CO_3 and Ag_2O photocatalysts, repectively (*6*).

As for the influences of morphologies on photocatalytic performance, although the nano-sized semiconductor powder photocatalysts have some distinct advantages in aqueous photocatalytic reactions due to their good dispersion, large surface area, and abundant active-sites, the powder-like nanoparticles in aqueous photocatalytic reactions often suffer from agglomeration, de-activation, and difficulties in separation for catalyst recycling. Compared with nanopowders, the microspherical photocatalysts can exhibit superior properties in aqueous reactions because of their low density, high surface area, easy settlement, good delivering capability, and surface permeability, leading to large improvement in photocatalytic performance. Therefore, design and fabrication of high quality photocatalysts with desired nanostructure or morphology are of great fundamental and technological interest in enhancing photocatalytic activity and stability.

Moveover, it is highly desirable to develo green and low-energy-consumption methods in photocatalyst fabrication. To obtain photocatalysts with high crystallinity or mesoporous structures, high temperature calcination treatment is often needed in order to remove the template or increase the crystallinity in conventional preparation methods. Experimental drawbacks are mainly associated with the calcination process, because during the calcination process the ordered structure of the mesoporous photocatalyts often collapses and/or the surface area decreases significantly. Therefore, other lower temperature synthetic methods for photocatalyst synthesis, such as sonochemical processing, hydrothermal preparation, etc., are expected to be developed.

To illustrate recent progress in designing highly efficient photocatalysts and their applications in decomposing organic pollutants, this Chapter will focus on the design and preparation of heterojunction and microspherical photocatalysts, as well as sonochemical fabrication of photocatalyst with novel nano-structures.

Design of Heterojunction To Enhance Photocatalytic Performance

Since the time that the photocatalytic technology was proposed as a green-chemistry method to deal with environmental problems, significant effort has been devoted to the photocatalytic degradation of organic pollutants in water by semiconductor photocatalysts. However, up to now, there are still seldom large-scale applications in waste water purification. In practical dye waste water treatment, the photocatalytic technique still confronts with some problems, such as the low degradation efficiency, high operating cost, and difficulty in recycling catalysts. At the same time, high-concentration dye solutions always show a thick color, which further decreases the photocatalytic efficiency due to poor light transmittance (*34, 35*). Therefore, how to promote the photocatalytic efficiency is one of the main tasks to meet the urgent needs.

The photocatalytic performance of photocatalysts is dependent on the separation efficiency of photo-generated electrons and holes. To promote the photocatalytic activity, noble metals such as Pt, Ag and Pd are commonly used for deposition over the photocatalyst because the deposited noble metals can effectively reduce the recombination rate of photo-generated hole/electron pairs (*36–40*). In such composite systems with noble metals, the photo-generated electrons accumulate on the metal while the holes remain on the semiconductor photocatalyst surface, thus suppressing the recombination of charge carriers. On the other hand, doping with non-metal anions (e.g., C, N, S, F) (*41–44*) also benefits by restraininh the recombination of photo-generated electrons and holes or extending the titania's light absorption into the visible region.

In recent years, many investigations have shown that the formation of interface or heterojunction between two semiconductors is an effective strategy in enhancing the separation of photo-generated electron/hole pairs (*45*). For example, the photocatalytic activity of TiO_2 is directly related to the surface phase structure (anatase/rutile). The formation of phase junction between the surfaces of anatase and rutile nanoparticles could greatly enhance the photocatalytic activity for H_2 production (*46*). The following Ag_2O/Ag_2CO_3 heterostructure is a typical example of the design of a heterojunction to enhance the photocatalytic performance in degrading dyes or phenol (*6*).

The Ag_2O/Ag_2CO_3 heterostructures were synthetized *via* a facile phase transformation route (*6*). Ag_2CO_3 was decomposed thermally to Ag through two distinct steps with Ag_2O as the intermediate compound. The Ag_2CO_3 begins to gradually decompose to Ag_2O at an elevated temperature (>180 °C). In the temperature range from 250 to 300 °C, Ag_2O is stable. Over 300 °C, Ag_2O was further decomposed to Ag. Therefore, the Ag_2O/Ag_2CO_3 composite phases can be obtained by precise control of the calacination temperature and time. The formation of Ag_2O/Ag_2CO_3 heterostructures is shown in Figure 1. In this Ag_2O/Ag_2CO_3 heterostructure system, Ag_2CO_3 particles were covered by a surface layer of Ag_2O, producing core-shell like Ag_2O/Ag_2CO_3 heterostructures with well-contacted interfaces. The colors of Ag_2CO_3, Ag_2O/Ag_2CO_3, and Ag_2O are yellow green, ash black and deep brown, respectively.

Figure 1. (a) Schematic illustration of the phase transformation: $Ag_2CO_3 \rightarrow Ag_2O/Ag_2CO_3 \rightarrow Ag_2O$; (b) Photographs of the prepared samples. (Reproduced with permission from ref. (6). Copyright 2014 John Wiley & Sons.)

The electronic band structure of the valence band (VB) edge positions of Ag_2CO_3 and Ag_2O were calculated according to the electronegativity (*46, 47*). The band gaps of Ag_2CO_3 and Ag_2O are 2.5 and 1.3 eV, respectively. The top of the VB and the bottom of the CB of Ag_2CO_3 are thus calculated to be 2.77 and 0.27 eV, respectively, and the VB and CB of Ag_2O are estimated to be 1.5 and 0.2 eV, respectively. The details of the electronic band structures of Ag_2CO_3 and Ag_2O are presented in Figure 2.

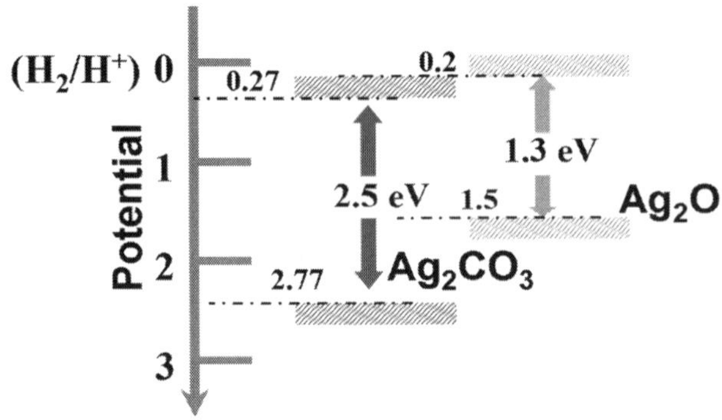

Figure 2. Electronic band structures of Ag_2CO_3 and Ag_2O.

In this unique Ag_2O/Ag_2CO_3 heterostructure system, Ag_2O has a more negative potential of the conduction band (CB: 0.2 eV) and valence band (VB: 1.5 eV) than that of Ag_2CO_3 (CB: 0.27 eV, VB: 2.77 eV). Under visible right irradiation, the photo-generated electrons in the Ag_2O layer can fast transfer to the surface of the Ag_2CO_3; at the same time, the photo-induced holes on the surface of Ag_2CO_3 can migrate to the Ag_2O surface, which increases the effective separation of photo-excited electrons and holes, and the probability of electron–hole recombination is thus largely suppressed. The electrons on the surface of Ag_2CO_3 are trapped by O_2 in solution to form $O_2^{\bullet -}$ (*48–50*), whereas the holes on the surface of Ag_2CO_3 transfer to Ag_2O and react with H_2O to generate ·OH. Both $O_2^{\bullet -}$ and ·OH radicals can effectively decompose methyl orange (MO) dye and phenol. The photocatalytic decomposition rate of MO over the Ag_2O/Ag_2CO_3 catalyst is about 67 times the activity over Ag_2CO_3 and 31 times that over Ag_2O.

The heterostructure effect in promoting photocatalytic performance was also confirmed in H_2 production (*45*). Different amounts of anatase phase TiO_2 (denoted TiO_2(A)) were loaded onto the surface of rutile phase TiO_2 (denoted TiO_2(R)) by an impregnation method, giving rise to anatase TiO_2/rutile TiO_2 heterostructure systems. Compared with the pure rutile TiO_2 support, the photocatalytic activity is remarkably increased by depositing even a small amount of anatase (<10 wt%) on the surface of rutile. This large increase in photocatalytic activity was attributed to the formation of the surface anatase/rutile phase junction.

Over the heterostructure system, one semiconductor is the main body photocatalyst and another semiconductor can be called the cocatalyst (*51*). For example, in MoS_2/CdS with intimate junctions between CdS and MoS_2, MoS_2 acts as the cocatalyst, showing much higher photocatalytic H_2 evolution activity than Pt/CdS in the presence of lactic acid as the sacrificial electron donor even though Pt possesses superior performance than MoS_2 for the activation of H_2 in electrochemical systems. Formation of atomically well-defined junctions between CdS and PdS might facilitate charge transfer between CdS and PdS. Chemically deposited MoS_2/CdS exhibits much higher activity than the mixture of CdS and

MoS_2. Other heterostructure systems such $Bi_5O_7I/BiOI$ (*52*), $Bi_{24}O_{31}Br/BiOBr$ (*53*), $CuO/BiVO_4$ (*54*), WO_3/ZnO (*55*), Bi_2WO_6/ZnO (*56*) further confirmed this conclusion. Figure 3 is a typical example to illustrate the promotion of the separation of photo-excited electrons and holes in a WO_3/ZnO system. Due to the different positions of the valence and conduction bands of WO_3 and ZnO, excited electrons in the conduction band of ZnO can easily transfer to the conduction band of WO_3 because W^{6+} can capture electrons and be reduced to W^{5+}. Meanwhile, the holes in the valence band of WO_3 can transfer to the valence band of ZnO. Therefore, the recombination of the photo-generated carriers is suppressed, leading to an increase in the photo-oxidation efficiency.

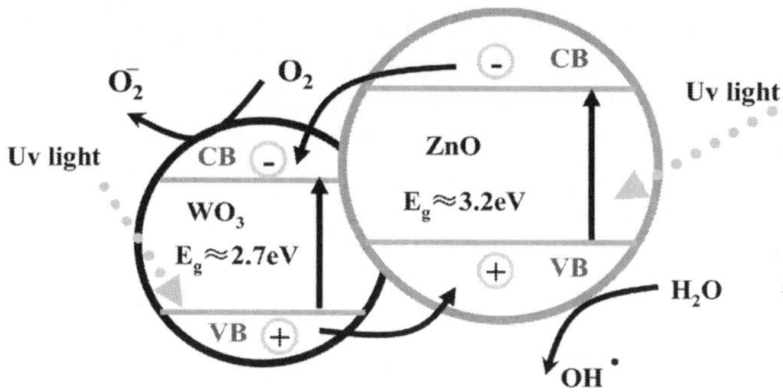

Figure 3. Enhancing the separation of photo-excited electrons and holes in a WO_3/ZnO system.

The above examples demonstrate that the formation of well-defined junctions between two semiconductors with matching electronic band structure can effectively facilitate charge transfer and suppress the recombination of photogenerated electrons and holes, leading to extremely high activity and stability. Therefore, tailoring and designing the interface/junctions between the light-harvesting semiconductor and the cocatalyst is of great importance for improving the activity.

Up to now, different methods, such as phase transformation (*6, 57, 58*), dip-coating (*59*), impregnation-deposition (*60*), high temperature solid-state reaction (*61*), electro-deposition (*62*), sol-gel (*63*), hydrothermal or solvothermal method (*64*), cation exchange (*65*), etc., have been developed to synthesize photocatalyts with heterojunctions.

However, at present the enhancing mechanism of heterostructure in photocatalytic performance is still not very clear. Some key issues remain to be addressed in full, such as the relationships between the content of heterojunctions and photocatalytic activity, the effect of electronic band structure of the two semiconductors on the probability of electron–hole recombination. Moreover, it is difficult to quantify the heterostructure over two semiconductors. New methods to fabricate well-defined junctions should continue to be developed.

Fabrication of Microspherical Photocatalysts

A survey of the literature indicates that most of the photocatalysts are powders, and the reaction matrixes of suspension are accordingly designed. The suspension systems have a fine contact between the catalyst particles and the substrate, and illumination of the particles is ensured. However, the suspension reaction matrix suffers from the problem of recovering the photocatalyst after use, because photocatalysts are usually fabricated as nano-sized ultrafine particles, which are easy to lose, and are difficult to separate them from the reaction systems due to the formation of a milky dispersion of TiO_2 particles. To accelerate the powder settling, some additional treatments are needed. For example, a coagulate such as ferrous sulfate is added, but the photocatalyst is fouled unfortunately and can no longer be reused. Sometimes, the powder photocatalysts are immobilized onto supports such as glass, fiber, or stainless steel. Unfortunately, a big loss in the contact area between the immobilized photocatalyst and light occurs, which lowers the efficiency of photocatalytic degradation of organic substrate.

Clearly, the problem of photocatalyst recycling requires a novel concept of photocatalyst fabrication. Since the suspension system has high quantum efficiency due to the sufficient contact between the photocatalyst particles and light, the advantage of this system should be kept. On the other hand, to ease the operation of recycling, the powder-like photocatalyst should be modified. Therefore, microsphere-based photocalysts appear to be ideal in order to overcome the disadvantages of the powder-like photocatalysts. Some works in this regard show that such microsphere photocatalysts can be successfully fabricated (*17, 18*). If the size of micorspheres is appropriate, they can be suspended by air bubbling, thus the good contact of catalyst with light is still maintained. Meanwhile, once air bubbling stops, the microspheres can quickly settle down to the reactor's bottom under the force of gravity, thus the treated aqueous phase can be decanted and easily separated from the microspheres, and the microspheres can be readily reused.

TiO_2 microspheres can be fabricated by different methods. Yu et al (*17*) have developed a novel sonochemical method to fabricate pure and fluorinated TiO_2 microspheres with perfect mesoporous structures and uniform particles size (1.5 μm). These TiO_2 microspheres were produced by an ultrasound-induced aggregation mechanism under ultrasonic irradiation conditions because these TiO_2 microspheres were composed of TiO_2 nanospheres. The specific surface area of the pure TiO_2 is 180 m^2/g. The pore diameter of the pure TiO_2 is 6.4 nm with a very narrow pore size distribution. The fluorinated TiO_2 possesses an average pore diameter of 7.2 nm and has a relatively low specific surface area (160 m^2/g). The pore volume is virtually identical for pure (0.35 cm^3/g) and fluorinated TiO_2 (0.34 cm^3/g). These TiO_2 microspheres displayed high photocatalytic performance in degradation of methylene blue (MB). Moreover, these microspheres are easily separated from treated aqueous phase and are readily reused.

TiO_2 microsphsere photocatalyst can be prepared by reconstruction of the mixture of TiO_2 sol and TiO_2 nano-sized powders with a spray drier (*12*). Then, the as-prepared microspheres are thermally treated to obtain an anatase-dominant sample. A typical image of the TiO_2 microspehres is shown in Figure 4.

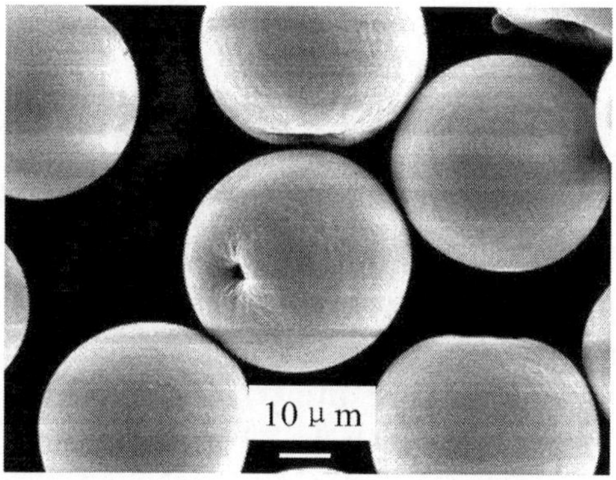

Figure 4. The SEM image of a TiO_2 microsphere sample. (Reproduced with permission from ref. (12). Copyright 2003 American Chemical Society.)

It can be seen that the TiO_2 particles have perfect spherical shape, and the surface appears to be smooth at the magnification of 600x (Figure 5). The size of TiO_2 microspheres was determined to be in the range of 30 - 160 μm with a mean value of around 80 μm. Such TiO_2 microspheres exhibit a quick settling time. The settling velocity of 150 mm min^{-1} for the TiO_2 microspheres is much faster than those values for the other two TiO_2 powders (3.3 and 3.8 mm min^{-1}). The activity was evaluated in terms of the degradations of salicylic acid (SA) and sulfosalicylic acid (SSA) with an illumination of 365-nm UV light (*12*). The experimental results showed that the reaction rate using the TiO_2 microspheres was similar to that using the TiO_2 powder counterparts in the SA suspensions and even higher than that in the SSA suspensions. Moreover, the prepared TiO_2 microsphere samples could be reused in the photocatalytic oxidation reaction for more than 50 times, and there was no loss of the photocatalyst and significant weakening in its activity, nor any change in their particle shape. The excellent performance of the TiO_2 microspheres in the water recycled treatment can be attributed to their good stability and adsorption ability which contributes to the relatively high photocatalytic activity.

Other microsphere photoctalyst were fabricated by different methods and obvious advantages were achieved compared to nano-sized powder photocatalyst. For example, $PbWO_4$ microspheres (about 2–3μm) with hierarchical nanostructures were prepared by hydrothermal method, and showed high photocatalytic activity in degradation of the acid orange II dye (*66*). Figure 5 shows the $PbWO_4$ crystals with different morphologies (micro-polyhedron, microsphere, and irregular morphology) which were fabricated by adjusting the pH value under hydrothermal conditions. After 5 h UV light irradiation, the acid orange II dye degradation rates are 98%, 47% and 72% for $PbWO_4$ microspheres, micro-polyhedrons and irregular particles, respectively, demonstrating that the photocatalytic activity of $PbWO_4$ strongly depends on the microstructure. The hierarchical structure of $PbWO_4$ microsphere photocatalyst gives rise to an improvement in activity due to their suitable adsorption capability, high surface area, and rich surface -OH groups. Besides the excellent activity, these $PbWO_4$

microspheres are quite stable during the liquid-phase photocatalysis since no significant decrease in activity was observed even after being recycled for five times (Table 1) showing promise in applications of environmental remediation.

Hollow Pt-ZnO nanocomposite microspheres with hierarchical structure can be fabricated by another solvothermal strategy under mild conditions using Zn $(CH_3COO)_2 \cdot H_2O$ and $HPtCl_4$ as the precursors, and polyethylene glycol-6000 (PEG-6000) and ethylene glycol as the reducing agent and solvent, respectively (*67*). Suitable concentration of $HPtCl_4$ in the reaction solution system can produce hierarchically hollow Pt-ZnO nanocomposite microspheres, which are composed of assembled fine Pt-ZnO nanocrystals (Figure 6). Table 2 shows that, compared to pure ZnO nano-powder and corresponding Pt-deposited ZnO nano-powder, Pt-ZnO microspheres exhibited extremely high photocatalytic activity and stability in degradation of the acid orange II dye. The remarkable photocatalytic performance of hollow Pt-ZnO microspheres mainly originated from their unique nanostructures and low recombination rate of the e^-/h^+ pairs by the platinum nanoparticles embedded in ZnO nanocrystals. These novel hollow Pt-ZnO nanocomposite microspheres will be promising materials for degrading organic pollutants and other environmental purification applications.

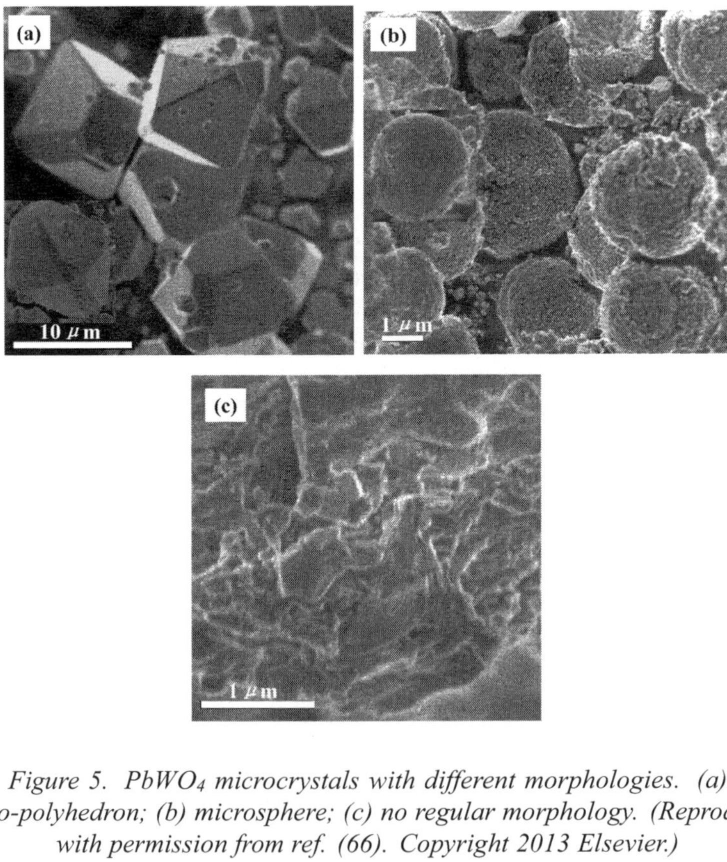

Figure 5. $PbWO_4$ microcrystals with different morphologies. (a) micro-polyhedron; (b) microsphere; (c) no regular morphology. (Reproduced with permission from ref. (66). Copyright 2013 Elsevier.)

Table 1. Recycling tests of $PbWO_4$ microspheres in acid orange II dye degradation.

Recycle times	*Degradation rate (%)*
New	96
Recycle 1	95
Recycle 2	94
Recycle 3	94
Recycle 4	92
Recycle 5	91

Figure 6. SEM images of the Pt(2%)-ZnO microspheres (a) and a zoom-in individual microsphere. (Reproduced with permission from ref. (67). Copyright 2013 RSC Publishing.)

Table 2. Photocatalytic recycling tests of pure ZnO nano-powder, Pt(2%)-ZnO composite microspheres and Pt(2%)/ZnO nano-powder.

Recycle times	*Degradation rate (%)*		
	ZnO nano-powder	*Pt(2%)-ZnO microspheres*	*Pt(2%)/ZnO nano-powder*
New	41	86	55
Recycle 1	30	84	50
Recycle 2	25	82	44
Recycle 3	20	81	40
Recycle 4	16	80	36
Recycle 5	10	75	30

It should be noted that once the photocatalysis is applied on a large scale, the operating cost should be emphasized, and recycling of photocatalyst becomes critically important to cut down the operation cost to an affordable level. The immobilization of powder photocatalysts onto supports results to a big loss in the contact between the catalyst and light, which further leads to a decreased efficiency of photocatalytic degradation of pullutants. The microsphere photocatalyst appears to be an effective alternative for the photocatalyst recycling.

Therefore, photocatalysts can be fabricated as microspheres rather than common powders (e.g. nano-sized powders). The microspheres should have a moderate size to ensure their suspension under air bubbling and quick settlement under gravity. Also they should have a suitable hardness to resist the collision occurring in the reactor, thereby to prevent any rupture or loss of finer particles from the surface. Moreover, the microspheres should have a high surface area and strong adsorption capability towards the organic substrate, thus the overall photocatalytic efficiency of degrading the organic substrate can be guaranteed.

Some methods have been developed to synthetize microsphere photocatalysts, including ultrasound-induced aggregation (*17*), thermal hydrolysis (*68*), sol-gel method (*69*), aerosol-spraying (*70*), hydrothermal (*71*) or solvothermal method (*72*), liquids-assisted hydrothermal method (*73*), template method (*74*), etc.

Sonochemical Fabrication Photocatalysts with Novel Nanostructures

The photocatalytic performance is strongly dependent on the fabrication method of photocatalyst. Recently, great effort is devoted to developing mild and green methods in fabrication of photocatalysts. The high crystallinity of semiconductor photocatalyst is necessary for good photocatalytic activity because the poor crystallinity produces many defect centers for the recombination of photo-generated electrons and holes. To obtain high crystallinity, high temperature calcination is always needed, especially for mesoporous photocatalysts because the mesopores are produced by burning off the template. On the other hand, the high temperature calcination step readily causes the collapse of the ordered structure and loss of surface.

Recently, much investigation indicates that the sonochemical method is a useful technique to synthesize novel photocatalyts under mild conditions (*75, 76*). In sonochemical fabrication, the chemical effects of ultrasound irradiation arise from acoustic cavitation, that is, the formation, growth and collapse of bubbles in a liquid. The extremely high temperatures in local sites (e.g. ~5000 °C), pressures (e.g. >20 M pa), and cooling rates (e.g. $>10^9$ degrees s^{-1}) attained during cavitation collapse lead to many unique properties in the irradiated solution, and these extreme conditions can benefit the formation of photocatalyst with unique morphology and high crystallinity. The sonochemical fabrication reactor is shown in Figure 7. The ultrasonic probe was connected with the ultrasonic generator.

The power and frequency can be controlled automatically. An electric fan was installed on the top of ultrasonic probe to dissipate the heat produced from the ultrasonic reactor, which can ensure the normal work of ultrasonic reactor with high power for a long time. A box was used to isolate the ultrasonic noise. In the synthesis of photocatalysts, the ultrasonic probe was dipped into the reaction solution, and cooling water was applied to control the reaction solution.

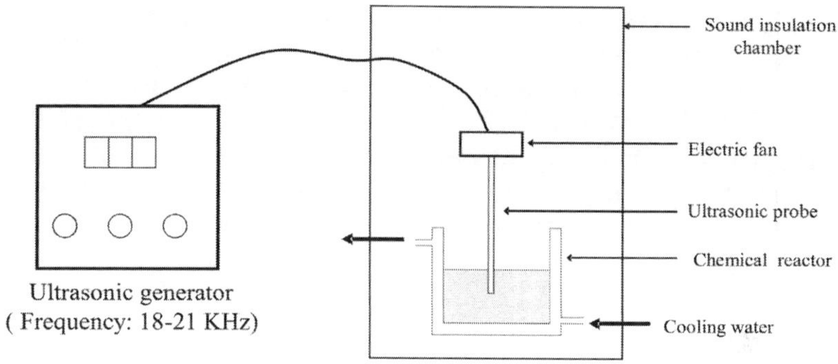

Figure 7. Schematic illustration for the sonochemical reactor.

The acoustic cavitation can generate the role of ultrasound-induced aggregation which benefits the formation of mesoporous structures of TiO_2 or F-doped TiO_2 even without templates (*17*). The process for the formation of mesoporous structures is as follows. Titanium isopropoxide was dissolved in ethanol to form a homogeneous solution. When this solution was added dropwise to the F^- aqueous solution under sonication, the prime amorphous sol particles (titanium oxo clusters) were produced by hydrolyzation of alkoxides. Cavitations and shock waves created by ultrasound can accelerate the prime sol particles to high velocity, leading to inter-particle collisions and effective fusion at the point of collision, finally producing the mesoporous structure by particle aggregation (Table 3). The energy generated during collision accelerates the crystallization of the amorphous particles. At the same time, the localized high temperature and pressure regions produced by the effects of acoustic cavitation could make F^- simultaneously doped into the lattice of TiO_2 during the crystallization course of the amorphous TiO_2 particles. The produced F-doped TiO_2 with mesoporous structures exhibited high activity in degradation of phenol under UV light irradiation (Table 3). The degradation rate constant over F(1.3%)/ TiO_2 (0.172 (mim^{-1}) is 3.8 times that of commercial P25 TiO_2 obtained from the Degussa Co. (Dossenheim, Germany).

The effect of acoustic cavitation can be applied to fabricate other mesoporous semiconductors, such as CdS nanospheres (*77*), $ZnWO_4$ nanorods (*78*), etc. Figure 8 illustrates the formation mechanism of mesoporous CdS nanocrystals. The surfactant P123 acts as a dispersion stabilizer and template for the synthesis of mesoporous CdS. P123 possesses a long poly(propylene oxide) chain and two medium length poly(ethylene oxide) chains, and has a relatively high molecular weight (M_{av}=5800). Such a structure favors the formation of micelles which have hydrophobic poly(propylene oxide) blocks in the core and hydrophilic poly(ethylene oxide) segments on the surface at low surfactant concentrations (low critical micellization concentration, cmc) and at low temperatures (low critical micellization temperature, cmt). In P123-rich solution, Cd^{2+} ions could first interact with the blocks of triblock copolymer templates and form the Cd-(PEO-PPO-PEO) units. When the system is irradiated with ultrasound, the added S^{2-} ions might attack the Cd-(PEO-PPO-PEO) units and form the prime CdS nanoparticles. Cavitations and shock waves created by ultrasound can accelerate the prime particles to high velocities, leading to particle collisions and inducing effective fusion. Therefore, ultrasound induces the attachment of primary particles onto the intermediate polyhedral or spherical aggregates to form sphere-like structures with irregular pores. This explanation can also explain the high pore volume and broad pore-size distribution of the sonicated CdS nanocrystal.

Table 3. Mesoporous property of sonochemically prepared samples and their pseudo-first-order rate constants in phenol degradation.

Sample	*Mesoporous property*			*Degradation rate constants (mim⁻¹)*
	S_{BET} (m^2/g)	*Pore volume (cm^3/g)*	*Pore size (nm)*	
TiO_2	88.20	0.27	13.03	0.027
F(0.7%)/TiO_2	71.03	0.21	12.80	0.102
F(1.3%)/TiO_2	68.30	0.22	12.96	0.172
F(2.4%)/TiO_2	71.85	0.20	11.53	0.060
P25	54.72	-	-	0.045

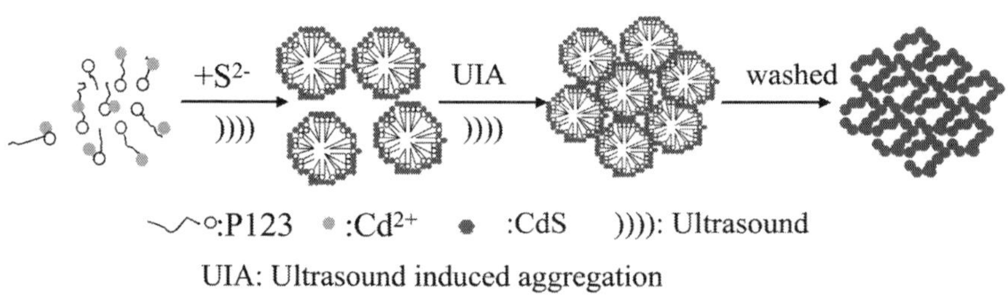

Figure 8. Schematic illustration for the formation of the mesoporous CdS nanocrystals. Adapted with permission from ref (77). Copyright 2012 Springer.

Moreover, the effect of acoustic cavitation can benefit the doping of nonmetal or transition metal elements (*78–81*). Using ethylenediamine as the precursor, the ultrasonic irradiation can bring a 2.2 times increase in N doping efficiency than that under common conditions, and the produced TiO_2 nanocrystals have mesoporous structures (*79*). In Figure 9, it can be seen that pure TiO_2 practically does not absorb visible-light, whereas the doping of nitrogen greatly extends the optical absorption edges to the visible-light region. N/TiO_2 prepared under ultrasonic irradiation exhibits the largest visible-light absorption and the N/TiO_2 (U-N/TiO_2) gives much higher efficiency in degradating dimethyl phthalate (DMP). After 5 h of reaction, the degradation rates of DMP are 41% and 58% over N/TiO_2 and U-N/TiO_2, respectively. The degradation rate for U-N/TiO_2 is ~1.4 times that of N/TiO_2. The high photocatalytic degradation activity of the ultrasonic fabricated N-doped TiO_2 is mainly attributed to that high N-content doping effectively increases visible light absorption.

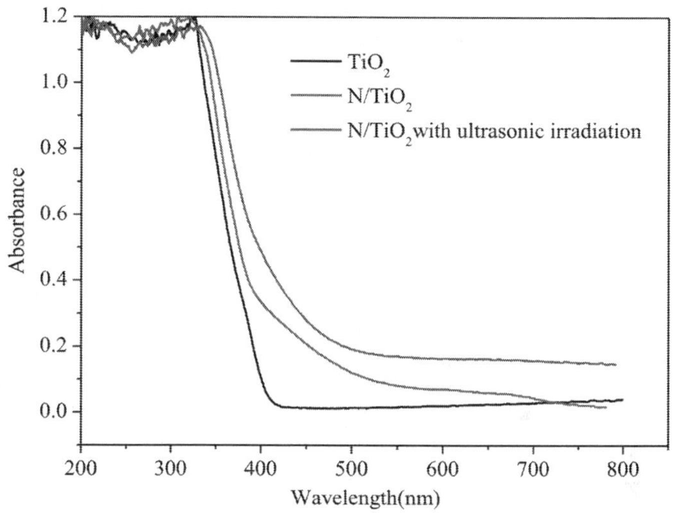

Figure 9. The UV-Vis diffuse reflectance absorption spectra of the pure TiO_2, N/TiO_2 prepared without ultrasonic irradiation and N/TiO_2 prepared under ultrasonic irradiation. Adapted with permission from ref (79). Copyright 2013 Elsevier.

The sonication method can be applied to synthesize composite photocatalysys. At room temperature, under intense ultrasound irradiation, the precursors of copper acetate and cobaltous acetate could transform into CuO and Co_3O_4 respectively and the amorphous $BiVO_4$ can easily crystallize to highly crystalline $BiVO_4$ (*81*). The as-produced Co_3O_4 and CuO were instantaneously doped into $BiVO_4$ and produced p-n heterojunction of Co_3O_4/$BiVO_4$ or CuO/$BiVO_4$. The formation mechanism is as follows. The intense ultrasonic irradiation introduces a series of chemical reactions, including hydrolysis, oxidation, reduction, dissolution, and decomposition. The elevated temperatures and pressures produced by the effects of acoustic cavitation could cause pyrolysis of water into H· and OH· radicals. The OH· radicals can combine with each other to form H_2O_2. The formation of Co_3O_4 occurs from cobaltous acetate, which could be produced by sonochemical hydrolysis with oxidation, whereas sonochemical hydrolysis alone was responsible for the formation of CuO from its acetate.

The produced Co_3O_4 and CuO were uniformly dispersed over $BiVO_4$ by the dispersion role of ultrasound. At the same time, under intense ultrasonic irradiation, the amorphous $BiVO_4$ rapidly crystallized to highly crystalline $BiVO_4$. The produced Co_3O_4/ $BiVO_4$ or CuO/$BiVO_4$ composite photocatalysts exhibited much higher photocatalytic activity than that of pure $BiVO_4$ (Table 4). The enhanced photocatalytic performance could be attributed to the high crystallinity of $BiVO_4$ and the formed p-n heterojunction of Co_3O_4/$BiVO_4$ or CuO/$BiVO_4$. These two factors can effectively suppress the recombination of photogenerated hole-electron pairs.

Table 4. The degradation rate over pure $BiVO_4$, Co/$BiVO_4$ and Cu/$BiVO_4$ samples after visible light irradiation for 5 h.

Samples	*Degradation rate (%)*
$BiVO_4$	48
Co(1%)/$BiVO_4$	52
Co(3%)/$BiVO_4$	86
Cu(1%)/$BiVO_4$	78
Cu(3%)/$BiVO_4$	64

The reported research (*80*) showed that Ce can be successfully doped into TiO_2 under ultrasonic irradiation conditions, and such Ce-doped TiO_2 exhibited good mesoporous structures, which can be contributed to the ultrasound-induced aggregation effect.

Moreover, sonication can be used to assist deposition of metal nanoparticles on photocatalysts, producing highly active photocatalysts. The following is an example to incorporate highly dispersed gold nanoclusters into mesoporous TiO_2 films (*82*). In the fabrication process, the first step involves sonication of a TiO_2 film immersed in a gold chloride solution. This effectively removes the air trapped in the porous film matrix and drives gold chloride into the pore channels, leading to a homogeneous adsorption of ionic gold in the TiO_2 mesoporous matrix. The second step takes advantage of the photocatalytic property of TiO_2 to reduce the adsorbed Au ions to Au^0. As the produced gold nanoclusters are stabilized by the TiO_2 meso-network, no organic capping molecules are required. The as-produced highly dispersed Au/TiO_2 nanoheterojunction arrays can be used as interesting photocatalytic materials. Similarly, this sonication method can be used to fabricate the robust three-dimensional mesoporous Ag/TiO_2 nanohybrid film (*82*).

The sonochemical method is also effective in control over morphologies of semiconductor nanostructures, such as ZnS hollow nanospheres (*83*), BiOCl nanoplates (*84*), ZnO nanorods and nanodisks (*85*), etc. Another advantage of sonochemical method is that ultrasonic irradiation can quickly promote the crystallization of semiconductor. This effect is illustrated in Figure 10. Compared with the sonicated BiOCl, only weak reflection peaks are observed for the BiOCl obtained by stirring 24 h. Similar changes were observed for BiOBr samples prepared under different conditions. The XRD analysis demonstrates that BiOCl and BiOBr with high crystallinity can be fast prepared by powerful ultrasonic irradiation at low temperature. The extreme chemical conditions arising from acoustic cavitation could accelerate the formation of BiOCl and BiOBr crystals because the transient high-temperature (~5000 °C) and high-pressure (>20 MPa) field produced during ultrasound irradiation could provide a favorable environment for the crystallization process of the amorphous BiOCl and BiOBr. Furthermore, the energy generated during collision can induce the crystallization of the amorphous particles. Table 5 shows the comparison of the photocatalytic activity of the samples under UV light irradiation. The sonochemistry-fabricated BiOCl shows the highest photocatalytic activity; almost 100% degradation rate of acid orange II can be obtained after 1 h of light irradiation, which is higher than that of commercial TiO_2 (P25, Degussa). The sonochemistry-fabricated BiOBr also shows the higher degradation rate (64%) than that (56%) of BiOBr prepared by conventional stirring. Moreover, sonochemistry-fabricated BiOCl and BiOBr exhibit high stability in recycling tests.

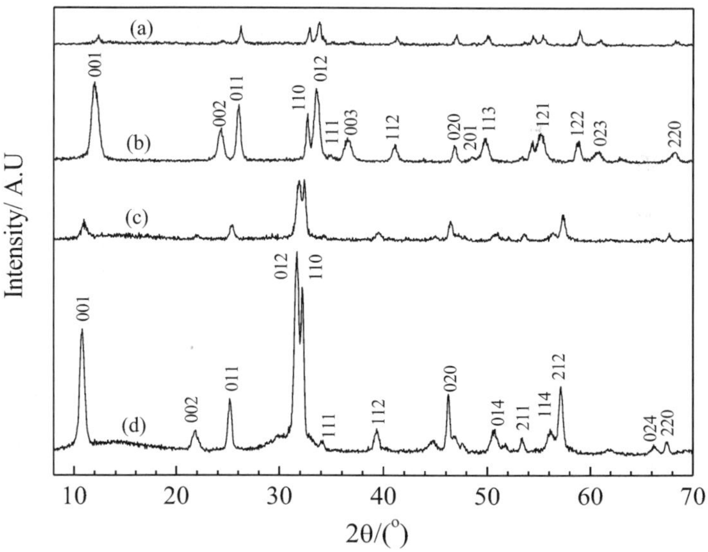

Figure 10. XRD patterns of samples prepared under different conditions. (a) BiOCl stirred for 24 h; (b) BiOCl sonicated for 10 min; (c) BiOBr stirred for 24 h; (d) BiOBr sonicated for 10 min. Adapted with permission from ref (84). Copyright 2011 Chinese Electronic Periodical Services.

Table 5. The degradation rate of acid orange II over various samples after UV or visible light irradiation for 60 min.

Samples	*Degradation rate (%)*	
	UV light	*visible light*
BiOCl (stirred for 24 h)	73	10
BiOCl (sonicated for 10 min)	100	11
BiOBr (stirred for 24 h)	56	35
BiOCBr (sonicated for 10 min)	64	43
P25-TiO_2	91	5

Conclusion

To advance the green photocatalysis into large-scale application of water purification and degradation of organic pollutants, more efforts are still required in terms of further increasing the photocatalytic efficiency and reducing the operating cost.

First, tailoring and designing the interface/junction between the light-harvesting semiconductor and the cocatalyst is important and also very prosmising for largely improving the activity because the formation of well-defined junctions between two semiconductors with matching electronic band structures can effectively facilitate charge transfer and suppress the recombination of photogenerated electrons and holes, resulting in extremely high activity and stability. Further effort should be devoted to clarifying the enhancing mechanism of heterojunction. The relationship between the content of heterojunction and the photocatalytic activity is still unclear. New methods to be developed are expected to quantify each semiconductor phase over the surface of photocatalyst.

Secondly, sometimes, a photocatalyst may have a relatively low efficiency in a single run, but it has the advantage of being recycled for multiple times of usage—which significantly reduces the operating cost. Compared to the powder-like counterparts, the microsphere photocatalysts have some obvious advantages in recycling reactions. Further evaluation of the total cost for the application of microspheres in commercial water treatment, including the cost in the preparation, operation, etc, with the reference of the powder-like counterparts, is also necessary. In this case, the balance between the efficiency and the economics should be evaluated carefully.

Finally, mild and green methods are expected for fabrication of photocatalysts. To obtain high crystallinity of semiconductor photocatalysts, high-temperature calcination treatment is always needed. The high temperature calcination causes huge energy consumption. Moreover, calcination easily causes the collapse of the ordered structure and the loss of surface. Sonochemical techniques can meet the requirement in the synthesis of novel photocatalyts under mild conditions.

References

1. Fujishim, A.; Honda, K. *Nature* **1972**, *238*, 37–38.
2. Hoffmann, M. R.; Martin, S. T.; Choi, W. Y.; Bahnemann, D. W. *Chem. Rev.* **1995**, *95*, 69–96.
3. Herrmann, J. M. *Catal. Today* **1999**, *53*, 115–129.
4. Wang, P.; Huang, B. B.; Zhang, X. Y.; Qin, X. Y.; Jin, H.; Dai, Y.; Wang, Z. Y.; Wei, J. R.; Zhan, J.; Wang, S. Y.; Wang, J. P.; Whangbo, M. H. *Chem. Eur. J.* **2009**, *15*, 1821–1824.
5. Yu, C. L.; Wei, L. F.; Chen, J. C.; Xie, Y.; Fan, Q. Z. *Ind. Eng. Chem. Res.* **2014**, *53*, 5759–5766.
6. Yu, C. L.; Li, G.; Kumar, S.; Yang, K.; Jin, R. C. *Adv. Mater.* **2014**, *26*, 892–898.
7. Byrne, J. A.; Eggins, B. R.; Brown, N. M. D.; McKinney, B.; Rouse, M. *Appl. Catal., B* **1998**, *17*, 25–36.
8. Feitz, A. J.; Waite, T. D.; Boyden, B. H.; Jones, G. J. *J. Adv. Oxid. Technol.* **2002**, *5*, 22–26.
9. Malato, S.; Caceres, J.; Aguera, A.; Mezcua, M.; Hernando, D.; Vial, J.; Fernandez-Alba, A. R. *Environ. Sci. Technol.* **2001**, *35*, 4359–4366.
10. Fernandez-Alba, A. R.; Hernando, D.; Aguera, A.; Caceres, J.; Malato, S. *Water Res.* **2002**, *36*, 4255–4262.
11. Asahi, R.; Morikawa, T.; Ohwaki, T.; Aoki, K.; Taga, Y. *Science* **2001**, *293*, 269–271.
12. Li, X. Z.; Liu, H.; Cheng, L. F.; Tong, H. J. *Environ. Sci. Technol.* **2003**, *37*, 3989–3994.
13. Zhang, J.; Yu, J. G.; Jaroniec, M.; Gong, J. R. *Nano Lett.* **2012**, *12*, 4584–4589.
14. Lin, F.; Wang, D.; Jiang, Z.; Ma, Y.; Li, J.; Li, R.; Li, C. *Energy Environ. Sci.* **2012**, *5*, 6400–6406.
15. Tong, H.; Ouyang, S. X.; Bi, Y. P.; Umezawa, N.; Oshikiri, M.; Ye, J. H. *Adv. Mater.* **2012**, *24*, 229–251.
16. Kubacka, A.; Fernández-García, M.; Colón, G. *Chem. Rev.* **2012**, *112*, 1555–1614.
17. Yu, C. L.; Yu, J. C.; Chan, M. *J. Solid State Chem.* **2009**, *182*, 1061–1069.
18. Yu, J. C.; Yu, J. G.; Zhang, L. Z.; Ho, W. K. *J. Photochem. Photobiol., A: Chem.* **2002**, *148*, 263–271.
19. Boppana, V. B. R.; Lobo, R. F. *J. Catal* **2011**, *281*, 156–168.
20. Yu, C. L.; Yu, J. C.; Zhou, W. Q.; Yang, K. *Catal. Lett.* **2010**, *140*, 172–183.
21. Zhang, M.; Chen, C. C.; Ma, W. H.; Zhao, J. C. *Angew. Chem. Int. Ed.* **2008**, *47*, 9730–9733.
22. Grätzel, M. *J. Photochem. Photobiol., C* **2003**, *4*, 145–153.
23. Gong, X. Q.; Selloni, A.; Dulub, O.; Jacobson, P.; Diebold, U. *J. Am. Chem. Soc.* **2008**, *130*, 370–381.
24. Rodrigues, S.; Ranjit, K. T.; Uma, S.; Martyanov, I. N.; Klabunde, K. J. *Adv. Mater.* **2005**, *17*, 2467–2471.
25. Liu, G.; Zhao, Y. N.; Sun, C. H.; Li, F.; Lu, G. Q.; Cheng, H. M. *Angew. Chem., Int. Ed.* **2008**, *47*, 1–6.

26. Chen, X. F.; Wang, X. C.; Hou, Y. D.; Huang, J. H.; Wu, L.; Fu, X. Z. *J. Catal.* **2008**, *255*, 59–67.
27. Yu, C. L.; Li, G.; Kumar, S.; Kawasaki, H.; Jin, R. C. *J. Phys. Chem. Lett.* **2013**, *4*, 2847–2852.
28. Asahi, R.; Morikawa, T.; Ohwaki, T.; Aoki, K.; Taga, Y. *Science* **2001**, *293*, 269–271.
29. Yu, C. L.; Cao, F. F.; Li, G.; Wei, L. F.; Yu, J. C.; Jin, R. C.; Fan, Q. Z.; Wang, C. Y. *Sep. Purif. Technol.* **2013**, *120*, 110–122.
30. Wang, X. C.; Maeda, K.; Thomas, A.; Takanabe, K.; Xin, G.; Carlsson, J. M.; Domen, K.; Antonietti, M. *Nat. Mater.* **2009**, *8*, 76–80.
31. Zhou, X. M.; Liu, G.; Yu, J. G.; Fan, W. H. *J. Mater. Chem.* **2012**, *22*, 21337–21354.
32. Zong, X.; Yan, H. J.; Wu, G. P.; Ma, G. J.; Wen, F. Y.; Wang, L.; Li, C. *J. Am. Chem. Soc.* **2008**, *130*, 7176–7177.
33. Zhang, J.; Xu, Q.; Feng, Z. C.; Li, M. J.; Li, C. *Angew. Chem., Int. Ed.* **2008**, *47*, 1766–1769.
34. Hameed, B.; Ahmad, A.; Aziz, N. *Chem. Eng. J.* **2007**, *133*, 195–203.
35. Yu, C. L.; Li, G.; Wei, L. F.; Shu, Q.; Fan, Q. Z.; Yu, J. C. *Catal. Today* **2014**, *224*, 154–162.
36. Zhou, X. M.; Liu, G.; Yu, J. G.; Fan, W. H. *J. Mater. Chem.* **2012**, *22*, 21337–21354.
37. Yu, C. L.; Wei, L. F.; Li, X.; Chen, J. C.; Fan, Q. Z.; Yu, J. C. *Mater. Sci. Eng., B* **2013**, *178*, 344–348.
38. Yu, C. L.; Yang, K.; Zhou, W. Q.; Fan, Q. Z.; Wei, L. F.; Yu, J. C. *J. Phys. Chem. Solids* **2013**, *74*, 1714–1720.
39. Yu, C. L.; Fan, C. F.; Meng, X. J.; Yang, K; Cao, F. F.; Li, X. *React. Kinet. Catal. Lett.* **2011**, *103*, 141–151.
40. Yu, C. L.; Yu, J. C.; Fan, C. F.; Wen, H. R.; Hu, S. *J. Mater. Sci. Eng.* **2010**, *166*, 213–219.
41. Yu, C. L.; Fan, Q. Z.; Xie, Y.; Chen, J . C.; Shu, Q.; Yu, J. C. *J. Hazard. Mater.* **2012**, *237/238*, 38–45.
42. Yu, C. L.; Zhou, W. Q.; Yang, K.; Rong, G. *J. Mater. Sci.* **2010**, *45*, 5756–5761.
43. Yu, C. L.; Cai, D. J.; Yang, K.; Yu, J. C.; Zhou, Y.; Fan, C. F. *J. Phys. Chem. Solids* **2010**, *71*, 1337–1343.
44. Yu, C. L.; Yu, J. C. *Catal. Lett.* **2009**, *129*, 462–470.
45. Zhang, J.; Xu, Q.; Feng, Z. C.; Li, M. J.; Li, C. *Angew. Chem., Int. Ed.* **2008**, *47*, 1766–1769.
46. Kim, Y.; Atherton, S. J.; Brigham, E. S.; Mallouk, T. E. *J. Phys. Chem.* **1993**, *97*, 11802–11810.
47. Yu, C. L.; Yang, K.; Shu, Q.; Yu, J. C.; Cao, F. F.; Li, X.; Zhou, X. Z. *Sci. China, Chem.* **2012**, *55*, 1802–1810.
48. Wang, P.; Huang, B.; Zhang, X.; Qin, X.; Jin, H.; Dai, Y.; Wang, Z.; Wei, J.; Zhan, J.; Wang, S.; Wang, J.; Whangbo, M. H. *Chem. Eur. J.* **2009**, *15*, 1821–1824.
49. Wang, P.; Huang, B. B.; Qin, X. Y.; Zhang, X. Y.; Dai, Y.; Wei, J. Y.; Whangbo, M. H. *Angew. Chem., Int. Ed.* **2008**, *47*, 7931–7933.

50. Yamada, H.; Bhattacharyya, A. J.; Maier, J. *Adv. Funct. Mater.* **2006**, *16*, 525–530.
51. Zong, X.; Yan, H. J.; Wu, G. P.; Ma, G. J.; Wen, F. Y.; Wang, L.; Li, C. *J. Am. Chem. Soc.* **2008**, *130*, 7176–7177.
52. Yu, C. L.; Fan, C. F.; Yu, J. C.; Zhou, W. Q.; Yang, K. *Mater. Res. Bull.* **2011**, *46*, 140–146.
53. Yu, C. L.; Zhou, W. Q.; Yu, J. C.; Cao, F. F.; Li, X. *Chin. J. Chem.* **2012**, *30*, 721–726.
54. Yu, C. L.; Yang, K.; Yu, J. C.; Cao, F. F.; Li, X.; Zhou, X. Z. *J. Alloys Compd.* **2011**, *509*, 4547–4552.
55. Yu, C. L.; Yang, K.; Shu, Q.; Yu, J. C.; Cao, F. F.; Li, X. *Chin. J. Catal.* **2011**, *32*, 555–565.
56. Yu, C. L.; Yang, K.; Yu, J. C.; Cao, F. F.; Li, X.; Zhou, X. Z. *J. Inorg. Mater.* **2011**, *26*, 1157–1163.
57. Yu, C. L.; Fan, C. F.; Yu, J. C.; Zhou, W. Q.; Yang, K. *Mater. Res. Bull.* **2011**, *46*, 140–146.
58. Yu, C. L.; Zhou, W. Q.; Yu, J. C.; Cao, F. F.; Li, X. *Chin. J .Chem.* **2012**, *30*, 721–726.
59. Xu, Q. C.; Wellia, D. V.; Ng, Y. H.; Amal, R.; Tan, T. T. Y. *J. Phys. Chem. C* **2011**, *115*, 7419–7428.
60. Reddy, K. H.; Martha, S.; Parida, K. M. *Inorg. Chem.* **2013**, *52*, 6390–6401.
61. Lin, X. P.; Xing, J. C.; Wang, W. D.; Shan, Z. C.; Xu, F. F.; Huang, F. Q. *J. Phys. Chem. C* **2007**, *111*, 18288–18293.
62. Yang, L. X.; Luo, S. L.; Li, Y.; Xiao, Y.; Kang, Q.; Cai, Q. Y. *Environ. Sci. Technol.* **2010**, *44*, 7641–7646.
63. Huang, H. J.; Li, D. Z.; Lin, Q.; Zhang, W. J.; Yu, S. Y.; Chen, Y. B.; Sun, M.; Fu, X. Z. *Environ. Sci. Technol.* **2009**, *43*, 4164–4168.
64. Chang, C.; Zhu, L. Y.; Wang, S. F.; Chu, X. L.; Yue, L. F. *ACS Appl. Mater. Interfaces* **2014**, *6*, 5083–5093.
65. Zhang, J.; Yu, J. G.; Zhang, Y. M.; Li, Q.; Gong, J. R. *Nano Lett.* **2011**, *11*, 4774–4779.
66. Yu, C. L.; Cao, F. F.; Li, X.; Li, G.; Xie, Y.; Yu, J. C.; Shu, Q.; Fan, Q. Z.; Chen, J. C. *Chem. Eng. J.* **2013**, *219*, 86–95.
67. Yu, C. L.; Yang, K.; Xie, Y.; Fan, Q. Z.; Yu, J. C.; Shu, Q.; Wang, C. Y. *Nanoscale* **2013**, *5*, 2142–2151.
68. Cho, C. H.; Kim, D. K.; Kim, D. H. *J. Am. Ceram. Soc.* **2003**, *86*, 1138–1145.
69. Rengifo-Herrera, J. A.; Blanco, M. N.; Pizzio, L. R. *Appl. Catal., B* **2011**, *110*, 126–132.
70. Huo, Y. N.; Miao, M.; Zhang, Y.; Zhu, J.; Li, H. X. *Chem. Commun.* **2011**, *47*, 2089–2091.
71. Zheng, Z. K.; Huang, B. B.; Lu, J. B.; Qin, X. Y.; Zhang, X. Y.; Dai, Y. *Chem. Eur. J.* **2011**, *17/52*, 15032–15038.
72. Yu, C. L.; Yang, K.; Xie, Y.; Fan, Q. Z.; Yu, J. C.; Shu, Q.; Wang, C. Y. *Nanoscale* **2013**, *5*, 2142–2151.
73. Lou, Z. Z.; Huang, B. B.; Wang, P.; Wang, Z. Y.; Qin, X. Y.; Zhang, X. Y.; Cheng, H. F.; Zheng, Z. K.; Dai, Y. *Dalton Trans.* **2011**, *40* (16), 4104–4110.

74. Yu, J. G.; Liu, W.; Yu, H. G. *Cryst. Growth Des.* **2008**, *8* (3), 930–934.
75. Yu, J. C.; Zhang, L. Z.; Yu, J. G. *New J. Chem.* **2002**, *26*, 416–420.
76. Yu, J. C.; Wang, X. C.; Wu, L.; Ho, W. K.; Zhang, L. Z.; Zhou, G. T. *Adv. Funct. Mater.* **2004**, *14*, 1178–1183.
77. Yu, C. L.; Zhou, W. Q.; Yu, J. C.; Yang, J. G.; Fan, Q. Z. *Chem. Res. Chin. Univ.* **2012**, *28*, 124–128.
78. Yu, C. L.; Yu, J. C. *Mater. Sci. Eng.* **2009**, *164*, 16–22.
79. Zhou, W. Q.; Yu, C. L.; Fan, Q. Z.; Wei, L. F.; Chen, J. C.; Yu, J. C. *Chin. J. Catal.* **2013**, *34*, 1250–1255.
80. Yu, C. L.; Shu, Q.; Zhang, C. X.; Xie, Z. P.; Yu, J. C. *J. Porous Mater.* **2012**, *19*, 903–911.
81. Yu, C. L.; Yang, K.; Yu, J. C.; Cao, F. F.; Li, X.; Zhou, X. Z. *J. Alloys Compd.* **2011**, *509*, 4547–4552.
82. Yu, J. C.; Wang, X. C.; Wu, L.; Ho, W. K.; Zhang, L. Z.; Zhou, G. T. *Adv. Funct. Mater.* **2004**, *14*, 1178–1183.
83. Zhou, H.; Fan, T. X.; Zhang, D.; Guo, Q. X.; Ogawa, H. *Chem. Mater.* **2007**, *19*, 2144–2146.
84. Yu, C. L.; Zhou, W. Q.; Yu, J. C. J. *Chin. Inorg. Chem.* **2011**, *27*, 2033–2038.
85. Jung, S. H.; Oh, E.; Lee, K. H.; Yang, Y.; Park, C. G.; Park, W. J.; Jeong, S. H. *Cryst. Growth Des.* **2008**, *8*, 265–269.

Chapter 9

Environmentally-Friendly Designs and Syntheses of Metal-Organic Frameworks (MOFs)

Tomislav Friščić,* Patrick A. Julien, and Cristina Mottillo

Department of Chemistry and FRQNT Centre for Green Chemistry and Catalysis, McGill University, 801 Sherbrooke St. W., Montreal, Canada
***E-mail: tomislav.friscic@mcgill.ca**

Over the past 15 years, the area of metal-organic frameworks (MOFs) has been the subject of intense efforts in research and development, both on the laboratory and on the industrial scale. Such persistent interest, fueled largely by the desire to develop new functional materials for storage of fuel gases (e.g. hydrocarbons, hydrogen) has led to MOFs being recognized as enabling materials for a number of technologies. Such rising importance of MOFs, accompanied by their recent commercialization, has begun to highlight previously unknown challenges associated with safe, environmentally-friendly synthesis of metal-organic materials on a large scale. These challenges have recently been summarized in a set of evaluative criteria for industrial MOF synthesis, which kindle the need to develop clean and sustainable methods of their manufacture. However, in contrast to organic synthesis, concepts of green and sustainable chemistry have been slow to adopt in inorganic synthesis. This lack of suitable "green" and industrially acceptable synthetic concepts can now be recognized as an outstanding challenge in the MOF area. Focusing on MOF synthesis, we now highlight three recent experimental developments which advance sustainable and environmentally-friendly synthesis with respect to the reaction environment, reactant choice and the synthetic design: (1) using (near) supercritical water as the reaction medium, (2) using biocompatible building blocks and (3) direct conversion of metal oxides into MOFs. These developments signal an important and necessary shift of green inorganic chemistry from designing "green" materials to conducting syntheses mindful of sustainability and environmental impact. In the context of recently established criteria for MOF manufacture these developments also serve to illustrate how embracing aspects of green chemistry and sustainability can be compatible with the requirements of industrial production.

Introduction

Rising concerns about the energy efficiency, toxicity and environmental impact of chemical processes have highlighted the importance of discovering “green” methods for synthesizing compounds and materials (*1–3*). Recent efforts in process chemistry and synthetic organic chemistry have thus focused towards more sustainable and environmentally-friendly synthesis, and the need to clearly establish the environmental impact of a chemical process has inspired the development of green metrics, exemplified by the concepts of the E-factor (*4*) and atom efficiency (*5*). While these, and other elements of green metrics, have been introduced to and are being actively applied in organic synthesis (*6*), the “green” perspective of research and development in inorganic and metal-organic synthesis has traditionally focused on developing materials for applications recognized as green and environmentally-friendly, such as energy storage (*7*), carbon sequestration (*8*), hydrogen storage (*9*), heterogeneous (*10*) (e.g. zeolites) and homogeneous (*11*) catalysts. In other words, inorganic and metal-organic chemistry (coordination chemistry), have dedicated much less attention to the environmentally-friendly nature of the synthetic process than, for example, synthetic organic chemistry. However, it is clear that broad applications of all these functional materials will also require synthetic or recycling approaches that are efficient, environmentally-friendly and economically viable on a large scale. Unfortunately, a large gap is often observed between the "green" properties of an advanced material and the "greenness" of its synthesis, meaning. that extensive research is necessary for the development of "greener" synthetic concepts, methodologies and techniques in inorganic and metal-organic chemistry.

In this contribution, we aim to illustrate several such innovative synthetic concepts, as applied to the synthesis of metal-organic frameworks (MOFs), one of the most popular and rapidly developing areas of modern metal-organic materials chemistry. Whereas MOFs have opened a route to metal-organic materials and their syntheses on large industrial scales, the exploration of underlying literature reveals that research in the MOF area has only recently begun to address the challenges of their environmentally sustainable and commercially viable synthesis (*12–14*). Indeed, the area of MOF synthesis has so far focused largely, and fervently, on making frameworks with improved properties, in particular stability and porosity. This led to impressive developments, including record porosities of > 7000 m^2 g^{-1} (*15*), theoretical modelling (*16*) and even prediction (*17*) of gas sorption properties of MOFs.

Acknowledging that tuning intrinsic properties of MOFs requires continued exploration of novel frameworks, we believe that the time is now ripe to shift the focus towards improvements in synthetic methods. Indeed, a revision of current, mostly laboratory-based, methods of MOF synthesis has been highlighted by industrial researchers as one of the prerequisites for commercialization of MOFs. Recent summaries by the team of BASF, who have pioneered the industrial manufacture of MOFs (*11*), have outlined the following critical evaluative criteria for MOF commercialization and large-scale production: availability/purity of chemicals, raw material costs, toxicity of reagents and solvents, and safety. Whereas these criteria (*18*) for successful industrial-scale synthesis of MOFs were recently outlined for individual cases, *e.g.* for the water-based, ton-scale synthesis of the aluminum fumarate framework Basolite A520 (*19*), it is clear that the diversity of MOFs, and their general novelty, are preventing the development of a general means to address them. Consequently, we suggest that the improvement of existing synthetic methodologies and the development of new ones should remain at the forefront of MOF research. The availability of synthetic criteria (*11*, *18*) desirable for large-scale MOF synthesis, as outlined by industrial researchers, has now empowered academic researchers to make synthetic innovations compatible both with the requirements of green synthesis and the needs of industrial manufacture. In this overview we will outlining these evaluative criteria for industrial synthesis of MOFs and highlight four (*12–14*) recent developments that demonstrate how MOF research is turning towards "greener" synthetic concepts. We note that these concepts remain fully compatible with one or more criteria for industrial manufacture.

Prospective MOF Applications

MOFs are a wide class of materials with highly diverse applications, that have been addressed in a number of comprehensive reviews (*20*). Thus, we will now provide only a brief summary of prospective applications (*21*), so as to link the emergent synthetic approaches to expected uses. High surface areas and microporous (<2nm) design make MOFs ideal for rational development of materials for gas storage and separation. Extensive effort has been put in the design of MOFs that would meet the requirements for hydrogen and natural gas energy storage density put forward by the United States Department of Energy (DOE), while minimising dangerous and coslty use of high pressures (*21*). Significant increases in hydrogen sorption have been reported and promising work on MOFs for hydrogen storage continues (*22*). While DOE goals for methane storage have already been met and exceeded by MOFs (*23*, *24*). However, more data on methane absorption and packing density is needed to evaluate the practical application of MOFs in methane storage.

Separation of gases, such as O_2, N_2, H_2, and simple hydrocarbons using either the selective surface-absorbate or size exclusion properties of MOFs has yielded recyclable frameworks with separations comparable with or superior to currently used zeolites. The pore size tunability of MOF-based mesh-adjustable molecular sieves in different temperatures has shown promising results in otherwise difficult gas separations (*25–27*).

Designs and uses of MOFs in catalysis have been extensively tested and reviewed (*28*). The high crystallinity of MOFs produces uniform catalytic sites, ideally contained in an insoluble matrix. As such, MOFs are promising as heterogenous catalysts. Strategies include incorporating catalytically active open transition metal centers (*29*), catalytically active linker ligands (*30*), and the immobilization of catalytic molecular or nanoparticle-based (*31*) species. Homochiral porous MOFs have been synthesized and used as stereoselective and size-selective catalysts (*32*).

The inability to control drug release rate in orally administered medication has promoted the development of MOFs as a vehicle for drug delivery (*33*). Materials with pore sizes in the mesoporous range (2 nm - 50 nm) are needed to store and release an appreciable number of drug molecules. A long and steady delivery of ibuprofen with 20 wt% loading has been reported although research into MOFs as drug delivery agents remains in its infancy (*34*).

MOFs with a diverse range of magnetic properties have been synthesized and are being explored for use in memory storage (*35*) and other electronic applications (*36*). Luminescent metal ions (*37*) and linkers (*38*) have been used to generate MOFs with solvent-controlled tunable emission for multifunctional luminescent materials. Some of these exhibit strong quenching and selectivity which could be further developed into highly selective and sensitive sensors (*39*).

Criteria for the Industrial Manufacture of MOFs

Evaluative criteria for MOF synthesis were recently described by the team that pioneered industrial MOF production (*11*, *18*, *19*), and address the availability, cost, toxicity, and safety for all components (reagents, solvents, products and byproducts) of the synthesis. These clearly outline the challenges facing the green and scalable synthesis of MOFs, especially for applications in gas storage and separation. In order to facilitate the examination of the efficiency and applicability of a MOF synthesis, the evaluation is broken down into three main components: (1) the metal source(s); (2) the linker(s) and (3) the reaction conditions. The latter include solvent, reaction time, temperature, and pressure. Each of these evaluation components is analysed with respect to availability, cost, toxicity, and safety. For any industrial synthesis, processing steps must also be included in the evaluation of MOF manufacture. An overview of both synthetic and processing steps is given in Scheme 1, and is based on work by the team that pioneered industrial MOF production (*11*, *18*, *19*).

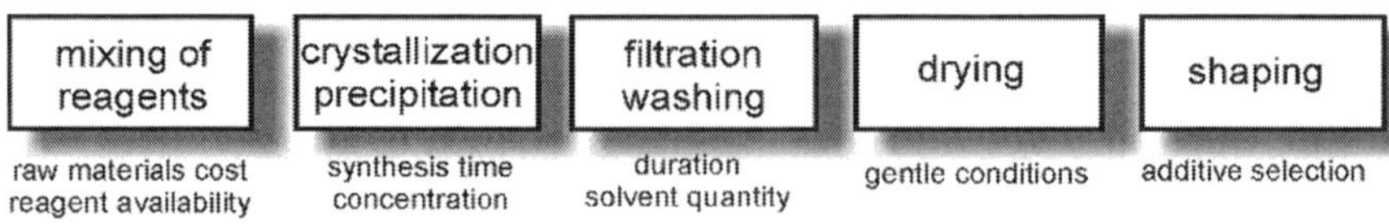

Scheme 1. A simplified overview of processing steps important in industrial MOF manufacture. Criteria in the development of each step are highlighted below.

In contrast to most laboratory solution-based syntheses reported in the research literature, the industrial manufacturing criteria strongly advise against the use of soluble nitrate salts as metal precursors. This is principally due to the risks associated with the counterions: whereas metal nitrates are immensely popular as metal sources in the research environment, due to their solubility and relatively poor coordination ability of the NO_3^- ion, in the context of industrial manufacture the same properties bring about serious problems of toxicity and oxidizing properties. Similarly, the use of metal halides is strongly discouraged due to their strong tendency for metal coordination and corrosion, which puts additional concerns on the design of large-scale metal reactors. It can, therefore, be assumed that any proposed green synthesis of MOFs which depends on soluble metal chlorides or nitrates will need to be revised before being considered for industrial applications. In contrast, these adverse issues are not encountered with metal oxides and sulfates as precursors. Such precursors, however, often exhibit much lower solubilities and the development of a green and industrially-attractive MOF synthesis will need to address the efficient transformation of slightly soluble reagents. As an example, the recently illustrated large-scale synthesis of the porous copper(II) trimesate framework (*40*), known as HKUST-1 and commercialized by BASF under the name Basolite C300®, has opted for electrochemical generation of Cu^{2+} ions from copper metal in order to avoid the use of toxic or corrosive counterions.

Aside from the selection of metal precursor is the choice of ligand for MOF synthesis. This selection criterion will favor the large-scale production of materials based on simple, commercially available ligands or ligands that can be readily generated in a small number of steps involving least toxic reagents, intermediates or byproducts. Simple commercially available ligands involve carboxylic acids such as succinic, terephthalic or fumaric acid. Obviously, ligands need to be stable under the conditions of synthesis, otherwise the formation of byproducts for completely unexpected materials can interfere with the process. An example of the latter is the formation of a complex mixed-ligand bimetallic frameworks involving oxalate ions, reported by Feller *et al.* (*41*). Oxalates formed in the solvothermal reaction system by unexpected redox reaction between Cu^{2+} and one of the organic ligands became incorporated as a stoichiometric component in the product (Figure 1).

While a large number of conventional MOF syntheses depend on the use of high boiling organic solvents with potential for templating the porous framework structure, such as N,N′-dimethylformamide (DMF) or N,N′-diethylformamide (DEF), industrial synthesis avoids these materials due to toxicity, safety concerns and cost. For example, the synthesis of a porous (~1500 $m^2\ g^{-1}$) aluminium fumarate MOF (Basolite A520) (*42*) was optimized with the key improvement consisting of using water as a solvent substitute for DMF, a known teratogen that is also sensitive to thermal treatment and hydrolysis following the reaction (Eq. 1) (*43*):

$$HCON(CH_3)_2 + H_2O \rightarrow HCO_2^- + H_2N(CH_3)_2^+ \text{ (Equation 1)}$$

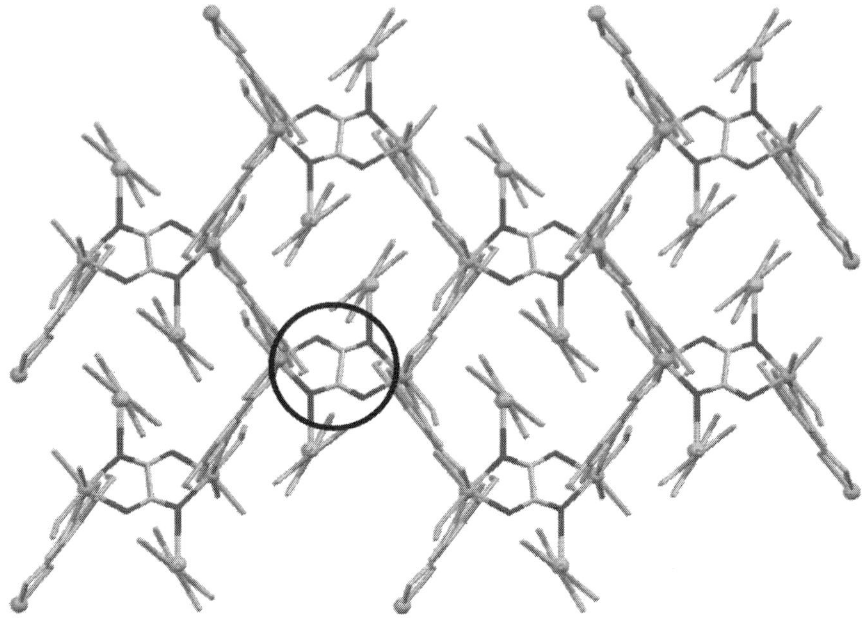

Figure 1. View of the complex mixed-ligand bimetallic (Cu and Zn) framework formed by unexpected oxidation of ligands to oxalate. The oxalate byproduct (circled) became incorporated in the final MOF as a structural building block (41).

Solvent substitution also improved on the cost and availability of the solvent. In addition, the synthesis was conducted at ambient pressure to reduce cost and safety concerns. Anion-related concerns of safety and corrosion were addressed by using aluminum sulfate instead of a nitrate or a chloride. Fumaric acid was chosen as a readily available and inexpensive linker with only moderate toxicity and few safety concerns. The synthesis of Basolite A520 by this route substantially increased the space-time-yield from 7 kg/m^3/day for conventional synthesis to >3600 kg/m^3/day, and the yield percentage from 92 to 98 mol%. The use of inexpensive and environmentally compatible materials with emphasis on cost, availability, toxicity, and safety enabled the production of several tons of Basolite A520.

Green Innovations in MOF Synthesis

While the above illustrated industrial synthesis of the aluminum fumarate MOF represents a breakthrough in the scalability of MOF synthesis, it is not very likely that one scale-up methodology will be generally applicable to all MOFs of industrial and commercial interest, leaving considerable room for the development of innovative and green synthetic methodologies.

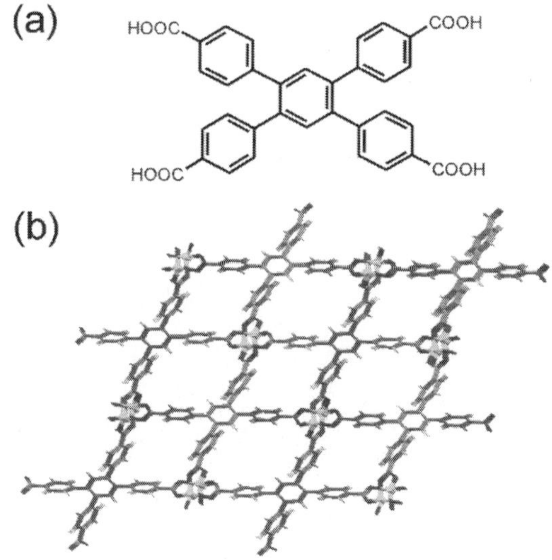

*Figure 2. (a) The ligand H_4**tcpb** used by Ibarra et al. in MOF synthesis using near-critical water and (b) the resulting porous MOF (12).*

Super-Critical Water as a Solvent

Besides a careful selection of metals and linkers, other factors such as topology, porosity, pore size and shape, thermal and chemical stability are equally critical for practical MOF applications. This introduces further restraints on synthetic design. Switching to water as the solvent might not be always possible, as the topology of many MOFs is solvent-dependent. Although hydrothermal synthesis has been employed early in the development of porous MOFs (*44*), many potential ligands and MOFs are either not stable or not soluble in water, limiting the scope of potential room temperature aqueous or hydrothermal MOF syntheses. Solubility or stability concerns may make the synthesis of MOFs with multiple linkers in water especially challenging.

The potential of near-supercritical fluids as reaction media in MOF synthesis has been explored by Ibarra *et al.*, who illustrated a different approach to using water as a solvent for MOF synthesis (*12*), by using near-critical high-temperature water as the reaction medium. As the physicochemical properties of water change dramatically when water approaches its critical point (374 °C, 220 bar), it is expected that the potential of water in MOF synthesis will also change. In particular, the dielectric constant of water decreases to values comparable to those of non-polar solvents (*e.g.* acetone, acetonitrile), allowing the solvation of organic ligands (*45*, *46*). At the same time, the ionic product of water increases by three orders of magnitude to reach a maximum at 280°C, providing a unique environment with simultaneously increased concentrations of H^+ and OH^- ions. These properties enabled the authors to use near-critical water at 300°C and 80 bar as a solvent for the reaction of the ligand 1,2,4,5-tetrakis(4-carboxyphenyl)benzene (H_4**tcpb**) with zinc nitrate. This synthesis yielded a novel MOF Zn_2(**tcpb**)·$3H_2O$ (Figure 2). Despite the use of water as the reaction medium, the product does not incorporate water molecules bound directly to metal sites. Instead, the water was found to reside as a guest in the pores of the generated MOF. The unique benefits arising from the properties of near-critical water are illustrated by the observation that, when the reaction is run at 150 °C under otherwise identical conditions, an unidentified product with poor crystallinity is produced.

While the use of near-critical water can broaden the range of MOF syntheses utilising water as a solvent, it should also be noted that such a methodology may place synthetic restraints with respect to the thermal stabilities of MOFs and ligands, as well as cost and safety concerns arising from the need for high pressures and temperatures. The potential of supercritical solvents in large-scale MOF synthesis has recently been highlighted by the same group, who applied supercritical ethanol as a solvent in developing a continuous flow process for the manufacture of the aluminum MOF known as MIL-53 (*47*).

Renewable Linkers for Edible MOFs

The use of organic solvents is not the only area of environmental, cost, and safety concern. Much of the cost of many MOFs can be attributed to the use of petrochemical feedstocks for linker synthesis. In an attempt to discover renewable, inexpensive, and non-toxic feedstocks for MOF synthesis, Smaldone *et al.* (*48*) used γ-cyclodextrin, a cyclic oligosaccharide (Figure 3a) that is mass produced enzymatically from starch. Reaction of γ-cyclodextrin with aqueous potassium hydroxide solution, followed by vapour-phase diffusion of methanol, yielded crystalline material of composition $[(C_{48}H_{80}O_{40})(KOH)_2(H_2O)_8\text{-}(CH_3OH)_8]_n$. Using single crystal X-ray diffraction, the material was found to consist of cube-like hydrogen-bonded motifs of γ-cyclodextrin. Individual cyclodextrin molecules from adjacent cube-like motifs are bridged by coordinated potassium ions, leading to a three-dimensional open structure filled with guest methanol molecules (Figure 3b-c). Upon removing the alcohol guests, the cyclodextrin-based material was thermally stable up to 175 °C and displayed a porous area of >1200 $m^2\ g^{-1}$. Analogous porous or open structures were also obtained by using hydroxides of other alkaline metals (Na, Rb, Cs), as well as salts of other anions (*49*). The most notable in this context was the ability to synthesize the porous framework using food grade γ-cyclodextrin, potassium chloride (a known salt substitute) and commercial bottled water. Subsequent crystallization of the product MOF was performed with 190-proof grain alcohol to yield completely edible, food-grade MOFs.

This approach highlights the possibility of using naturally derived renewable sources for MOF linkers. In this synthesis, the metal ion, the organic linker, and the solvent are all considered non-toxic for a majority of purposes, readily available, and inexpensive. This provides a promising source of inexpensive, environmentally benign MOFs. Although cyclodextrin-based materials are not likely to rapidly find applications in industrial catalysis or hydrogen storage, primarily due to potential reactivity of the highly hydroxylated cyclodextrin fragment and its solubility in water, the demonstrated synthesis of "food-grade MOFs" could be a potentially important development for MOF use in biological or medicinal applications such as the aforementioned drug delivery, as it mitigates any toxicity concerns.

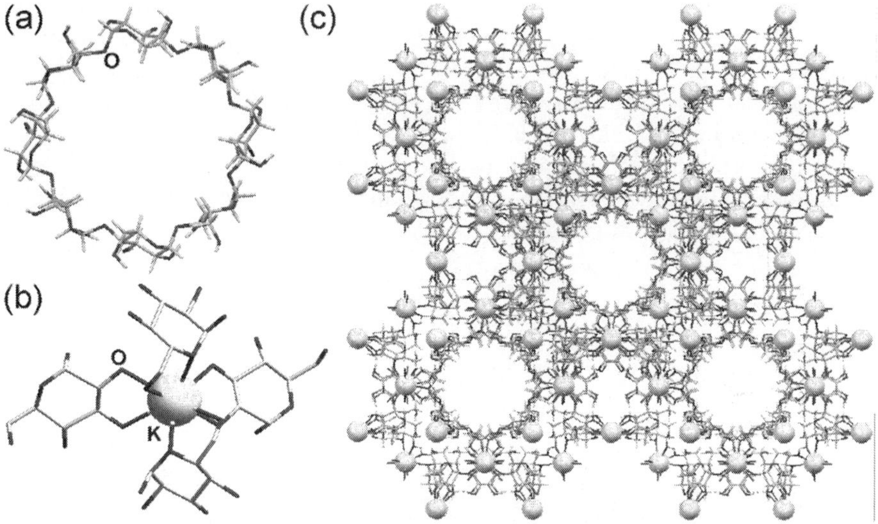

Figure 3. (a) A single molecule of γ-cyclodextrin as found in the crystal structure; (b) the coordination environment of the potassium ion in the crystal structure of a potassium-based "edible MOF" and (c) view of the crystal structure of an "edible MOF" based on γ-cyclodextrin and KOH, with included solvent molecules omitted for clarity (48, 49).

Solvent-Free Synthesis of MOFs: Thermochemical, Mechanochemical and Aging Reactions

Solvent-free reactions are an obvious response to difficulties imposed upon MOF synthesis from cleaner, safer, but poorly soluble oxides or sulfates. As solubility-related issues are a consequence of using solutions as reaction media, switching to a solvent-free environment is expected to allow circumventing such issues. Currently, there are three principal routes for the solvent-free synthesis of porous MOFs. These generally differ in the energy input required for the synthesis, as well as in how energy is introduced to the reaction. These are thermochemical (*50*), mechanochemical (*51*) and diffusion-controlled "accelerated aging" reactions (*52*). Whereas these methodologies have been developed by different groups focused on a variety of reactions, involving diverse selections of starting materials, this overview will focus on the synthesis of MOFs from simple and basic metal precursors, largely metal oxides.

Thermochemical Reactions

The potential of thermochemical solvent-free chemistry for clean synthesis of MOFs has been illustrated by Lin *et al.* (*50*) in 2011 who reported a multi-gram synthesis of the microporous framework ZIF-8 (also known as MAF-4), commercially available as Basolite Z1200®, via a thermochemical melt reaction of ZnO and 2-methylimidazole. The reaction completely avoids bulk solvents, can be conducted in quantitative fashion and can provide shaped objects. As the reaction proceeds at an appreciable rate only above 160 °C, it is likely that large-scale application will need to address energy efficiency. Thermochemical reactivity does not seem to be limited to zinc-based frameworks, but is also applicable to cobalt(II) frameworks (*53*), as demonstrated by Lanchas and co-workers who conducted thermochemical synthesis of a variety of cobalt(II) based imidazolate frameworks from cobalt(II) oxide or hydroxide.

Mechanochemical Reactions

Mechanochemical synthesis (*54–56*) entails the mechanical grinding or milling of reactants, nominally at or near room temperature. Whereas the simplest approach to mechanochemical synthesis is by manual grinding using a mortar and pestle, such a methodology is clearly not readily applicable for syntheses at a scale exceeding a few grams. More importantly, the open design of mortar-and-pestle reactions (sometimes also called "grindstone chemistry") makes control of a number of important reaction parameters (intensity of grinding, temperature, atmosphere, and humidity) impossible to control, resulting in sometimes irreproducible results. All these hindrances are readily eliminated by the application of an electronically controlled automated ball mill (*57*), wherein reacting materials are enclosed in a rapidly oscillating or rotating closed vessel (the milling jar) and ground by the abrasive motion of the milling media. Moreover, automated milling equipment is available in a number of designs (shaker mills, planetary mills, attritors, etc.) and sizes suitable for applications in laboratory exploration and development (grams to hundreds of grams), as well as industrial manufacturing (kilograms to hundreds of kilograms). Although mechanochemical reactivity is sometimes thought to proceed through the appearance of hot-spots caused by grinding or milling, these are likely to be extremely short lived (on the order of nanoseconds) (*58*) and the overall transformation proceeds close to room temperature.

Developments over the past half a decade have revealed that a number of coordination polymers and microporous MOFs can be synthesized at nominally room temperature, by grinding together a metal oxide and a suitable ligand (*59*), sometimes in the presence of suitable additives. Indeed, the application of small amounts of additives in milling reactions has provided an important breakthrough for a more general application of mechanochemistry in the synthesis of molecules and materials (*60*). In the context of making porous MOFs, the additive is typically an organic liquid which acts as a stoichiometric space-filling template and ends up being incorporated in the final reaction product as a guest. An example (*51*) of such liquid-assisted grinding (LAG) (*61*) methodology in the synthesis of MOFs is the synthesis of pillared MOFs (*51*, *62*), based on a design which metal carboxylate sheets are organized into a porous three-dimensional structure by small molecules acting as pillars, such as 4,4′-dipyridyl (**bipy**) or 1,2-*trans*-bis(4-pyridyl)ethylene (**bpe**) (Figure 4) (*51*). The synthesis of pillared MOFs normally requires several days of solvothermal reactivity in water or an organic solvent, at temperatures >120 °C. In contrast, mechanochemical milling of zinc oxide, a suitable carboxylic acid and the pillaring ligand in the exact 2:2:1 stoichiometric ratio was found to readily provide the expected pillared MOFs as long as a stoichiometric amount of a liquid phase was present in the mixture. Interestingly, the application of LAG enabled the synthesis of pillared MOFs not only in DMF, which is one of the most popular solvents for MOF synthesis as well as a known teratogen, but also using more environmentally-acceptable liquid phases such as methanol or ethanol (*51*).

It is thought that the liquid phase acts both as a means to enhance the reaction rate, as well as a space-filling agent that becomes incorporated into the final pillared MOF product as a guest. This view is largely based on systematic investigations conducted in the context of organic materials, such a pharmaceutical cocrystals and hydrogen-bonded lattice hosts (*63*, *64*). Indeed, if the above described reactions are attempted by neat grinding, *i.e.* milling without added liquid, no pillared MOFs are produced, suggesting that the added liquid might actively be templating the formation of the porous structure. It is important to note that the amount of liquid present in a LAG reaction is stoichiometrically equivalent to or smaller than the reactants, *i.e.* it is thousands of times smaller than the amount of liquid present in a conventional solvothermal synthesis (*65*).

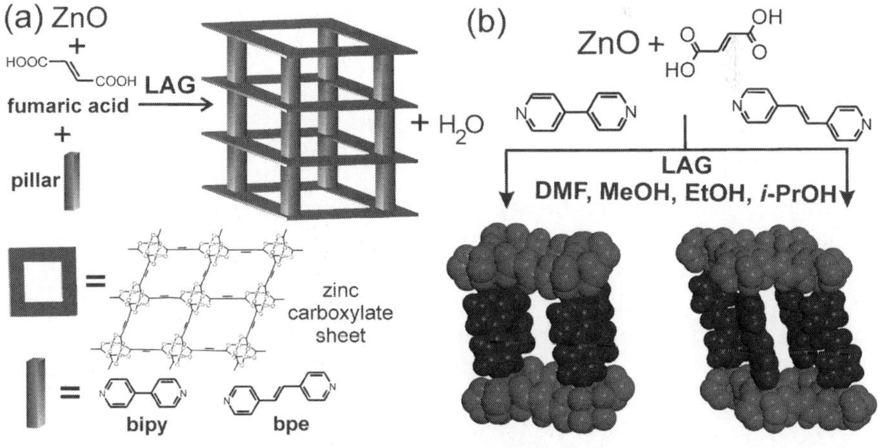

*Figure 4. (a) General scheme for the mechanochemical assembly of zinc fumarate pillared MOFs from a metal oxide using LAG and (b) the LAG synthesis of two pillared MOFs using **bipy** and **bpe** as pillaring ligands (51).*

The scope of mechanochemical LAG reactivity is further expanded through the addition of catalytic amounts of inorganic salts that appear necessary for the synthesis of highly open metal-organic structures and can also direct their formation (*66*). This ion- and liquid-assisted grinding (ILAG) methodology was first utilized to provide access to high-porosity pillared MOFs directly from zinc oxide (Figure 5). Whereas milling together of zinc oxide, terephthalic acid (H_2**ta**) and **dabco** failed to provide any new metal-organic materials, either by neat milling or LAG, conducting the reaction in the presence of catalytic amounts of nitrate salts (ca. 0.1 mol% with respect to zinc) results in the rapid assembly of the popular square-grind pillared MOF Zn_2(**ta**)$_2$(**dabco**). The application of mechanochemistry for the synthesis of pillared MOFs also revealed structure templating effects that have not been observed in conventional solution-based approaches. If the a sulfate salt was used in the above described ILAG synthesis, the product was another, hexagonal isomer of the Zn_2(**ta**)$_2$(**dabco**) framework, based on carboxylate sheets adopting a Kagome topology, rather than a square-grid one (*66*). The effect of anion in controlling the porous MOF structure is surprising as most laboratory studies focused on soluble nitrates or acetates as precursors.

Another application of ILAG in MOF synthesis is for the mechanosynthesis of frameworks with zeolitic topologies, obtained by combining a tetrahedrally-coordinated metal center (*e.g.* Zn^{2+}, Co^{2+}) with a bridging azolate ligand (*67*, *68*). In particular ILAG of zinc oxide and different 2-substituted imidazoles gave rise to a number of zeolitic imidazolate frameworks (ZIFs, Figure 5b), including zinc 2-methylimidazolate, known as ZIF-8 and commercially manufactured as Basolite Z1200® (Figure 5c) (*69*). Whereas neat grinding of ZnO and 2-methylimidazole (**HMeIm**) leads to the slow and incomplete formation of ZIF-8 within 72 hours (*70*), almost complete conversion is achieved by LAG within 30 minutes (*69*). Switching from LAG to ILAG provided further reaction acceleration and allowed complete conversion (Figure 6a). In contrast to the ILAG synthesis of pillared MOFs, where the role of salt in activating the reaction mixture is still unclear, in the ILAG synthesis of ZIFs the salt additive is thought to participate in an acid-base reaction with either zinc oxide or the imidazole ligand (*52*, *69*).

Understanding Reaction Mechanisms and Kinetics

Wider industrial application of mechanochemical reactivity mandates the development of reliable approaches for scaling up the gram-size laboratory processes to industrial sizes of hundreds of kilograms. Unfortunately, there are as yet no clear guidelines for converting reaction conditions accessible in laboratory-size shaker or planetary mills into those associated with industrial attrition or media mills. Consequently, the exploration of the impact of scaling-up to energy and heat transfer, as well as reaction yield of mechanochemical reactions, is an area of active exploration (*57*, *71*). An important requirement for such exploration is the understanding of mechanisms and kinetics of reactions taking place in a ball mill. Such investigations have only recently been initiated, through the introduction of first techniques for *in situ* and real-time monitoring of mechanochemical reactions in a ball mill (*72*, *73*). Direct observation of transformations in a ball mill is hindered by the rapid movement of the reaction vessel, as well as the violent motion of the grinding media. It was, however, recently demonstrated that a sufficiently high energy (short wavelength) X-rays can be used to perform a powder X-ray diffraction (PXRD) measurement on the mechanochemical reaction mixture as it is being milled (*72, 73*). These *in situ* and real-time measurements can be performed in conventional milling jars of steel or aluminum, although transparent poly(methylmethacrylate) (PMMA) plastic was found just as applicable and, indeed, the preferred choice for mechanochemical processes involving only poorly scattering elements of low atomic number. Using this *in situ* and real-time scattering technique readily provided qualitative information on the reaction rate, reaction course and the appearance of intermediate phases (Figure 6a,b) (*72, 74*). Importantly, this work revealed that mechanochemical formation of MOFs can take place within minutes, *i.e.* often faster than observed for solution-phase reactions.

Quantitative information on mechanochemical reaction kinetics are accessible if the measurements are conducted in the presence of a small amount of crystalline silicon as the internal diffraction standard (*75*). Such quantitative measurements revealed that the kinetics of mechanochemical reactions are surprisingly similar to the rate laws established for 1st- and 2nd-order reaction in solution chemistry. While it is not yet clear why mechanochemical reactions between rapidly agitated powders kinetically resemble solution reactions, similar conclusions have also been drawn by the James group (*76*) who monitored mechanochemical formation of ZIFs using Raman spectroscopy. It has been proposed that rapid agitation of reactant powders leads to a "pseudo-fluid" state in which the reaction rate depends on the number of reactive collisions between particles (*76*).

Accelerated Aging Reactions

"Accelerated aging" is a recently introduced (*52*) approach to the synthesis of metal-organic materials, inspired by the processes of biological mineralization, generally known as mineral weathering or mineral neogenesis (*77, 78*), in which simple binary minerals (oxides or sulfides) are converted into industrially significantly more tractable secondary metal-organic (also known as "organic") (*79*) minerals. Well known examples of mineral neogenesis involve the transformation of metal ores into metal oxalates by extended exposure to oxalic acid which is excreted by lichens or to ammonium oxalate found in deposits of guano (*80*). It was recently demonstrated by Qi and co-workers (*81*) that these chemically simple processes can be readily reproduced in the laboratory, by exposing physical mixtures of a metal oxide and oxalic acid, in a 1:1 stoichiometric ratio, to suitable conditions of temperature and humidity. In particular, increasing relative air humidity (R.H.) to 98% and temperatures up to 45 °C allowed acceleration of the aging process and the complete conversion of metal oxides into corresponding metal oxalate coordination polymers within several days. This facile conversion of normally inert and high-melting metal oxides into metal-organic materials was found to be generally applicable to oxides of main group (*e.g.* MgO, PbO), as well as transition (e.g. MnO, CoO, NiO, CuO, CdO) metals (*81*). Surprisingly, this simple diffusion-controlled reactivity was readily applicable to samples sizes of at least 20 grams, leading to complete conversions of metal oxides even without mechanical stirring. Consequently, the operationally simple accelerated aging procedures provide a means to convert poorly reactive substances, normally activated at high temperatures or extended milling, into significantly more active metal-organic materials without the need for high temperatures, solvents or extensive input of mechanical energy. Importantly, the accelerated aging processes could readily be directed towards the formation of open two- and three-dimensional frameworks by the addition of cationic templates in the form of alkyl-ammonium oxalates (Figure 7b) (*81*). The selection of alkyl-ammonium templates for the formation of a particular MOF topology was based on the results previously established in solvothermal synthesis by the groups of Rao and Cheetham (*82*, *83*).

In a step beyond mimicking and accelerating mineral weathering processes in nature, the accelerated aging technique was readily adapted for the synthesis of microporous ZIFs, by aging either ZnO or CoO with a suitable imidazole ligand and a catalytic acid salt (*84*). The exploration of different reaction conditions revealed that the selection of the catalytic salt could be used to modify the outcomes of the aging reactions. For example, simple ammonium salts, such as NH_4NO_3, readily induced the formation of the ZIF-8 structure from ZnO and HMeIm. However, the product also rapidly collapsed into the non-porous polymorph of the framework, with a diamondoid topology. In contrast, the use of caffeinium sulfate or benzimidazolium sulfate as catalytic salt additives led to the formation of the ZIF-8, but without the subsequent decomposition. In that way, variation of different catalytic salt additives allowed the design of a low-temperature, low-energy approach for the direct conversion of ZnO into microporous ZIF-8 (*84*).

The reactivity underlying accelerated aging processes also opens up an interesting opportunity for reducing the input of energy and reagents in mineral processing. Namely, the differences in rates of accelerated aging reactions of oxalic acid with different metal oxides allowed the demonstration of a solvent-free segregation of metal oxides in a mixture. Specifically, accelerated of a mixture of zinc and copper(II) oxides to a limited amount of oxalic acid led to the preferred conversion of ZnO into the corresponding oxalate, leaving behind unreacted CuO. After several days, the resulting mixture of zinc oxalate (density 2.2 g cm^{-3}) and copper(II) oxide (density 6.3 g cm^{-3}) could be readily separated by gravity, using flotation (*81*). The demonstrated separation of CuO from ZnO is a technologically important process, as well as the separation of ZnO from PbO, and of CuO from PbO, which were all readily achieved in the same manner. However, although the demonstrated accelerated aging approaches for metal oxide separation are highly promising as a laboratory proof-of-principle for developing new, solvent-free and mild techniques for mineral processing, their applicability to real mineral concentrates still needs to be explored.

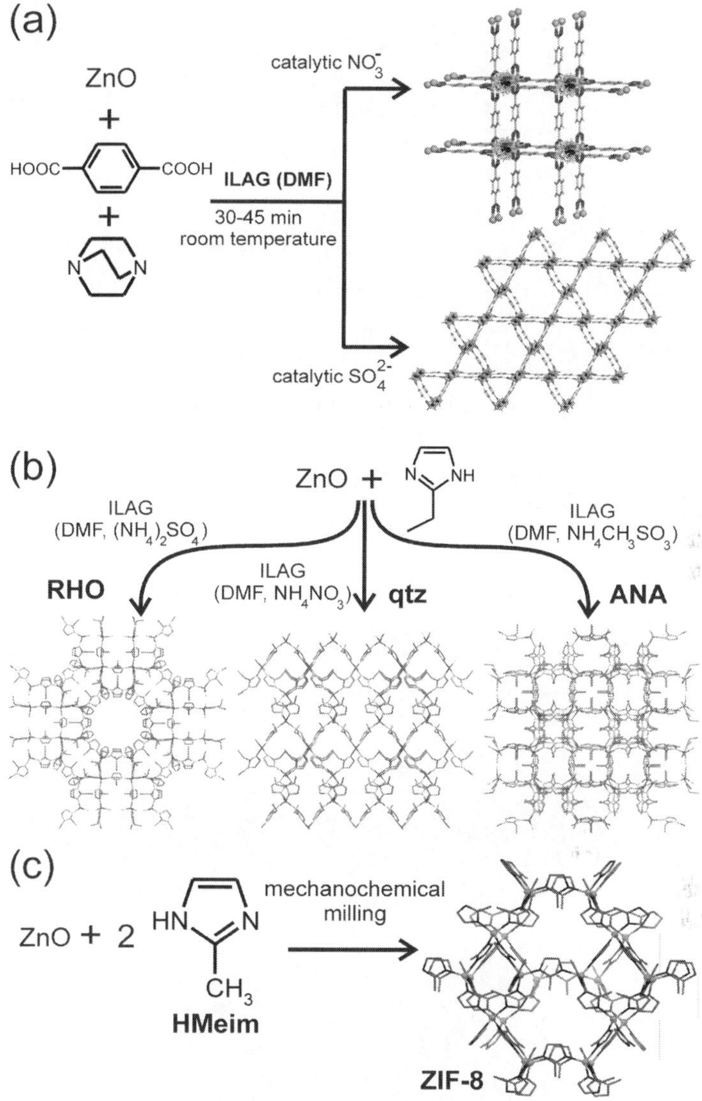

Figure 5. Applications of ILAG in the synthesis of microporous MOFs: (a) the synthesis of large-pore pillared MOFs directed by the choice of salt additive (66); (b) the ILAG synthesis of zeolitic imidazolate frameworks (ZIFs) with different topologies and (c) the mechanochemical synthesis of ZIF-8 (Basolite Z1200®) (69).

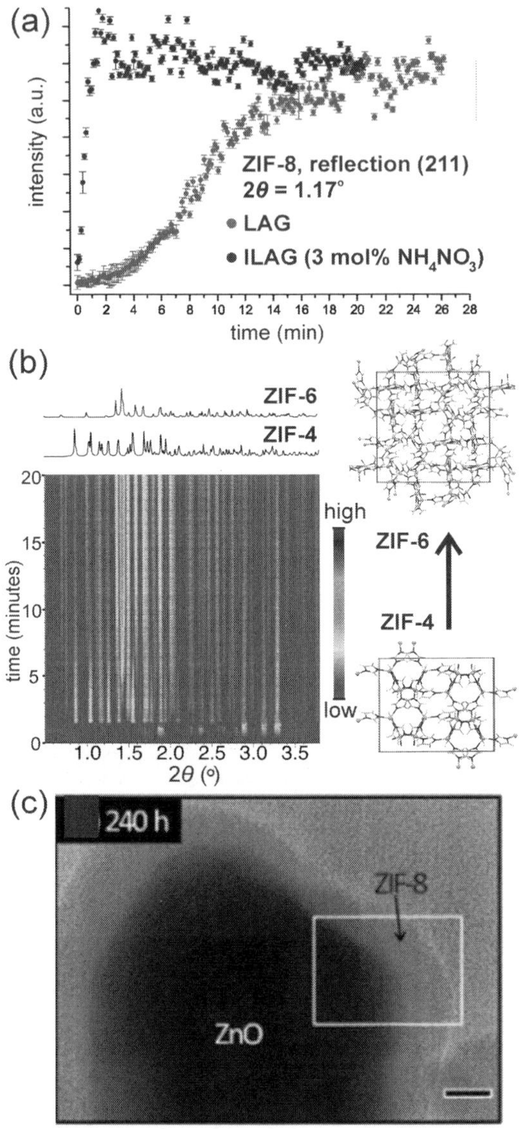

Figure 6. (a) Comparison of rates mechanochemical formation of ZIF-8 by LAG and ILAG, demonstrating higher reaction yield and faster product formation using ILAG (72); (b) an example of a time-resolved X-ray powder diffractogram demonstrating the mechanochemical formation of a ZIF product (ZIF-6) via a short-lived intermediate (ZIF-4) (72) and (c) Scanning Electron Microscopy image of a particle in the mechanochemical neat grinding reaction of ZnO and HMeIm after 10 days milling, revealing a core-shell structure (70). Figure (c) has been adapted from reference (70). Copyright (2013, Royal Society of Chemistry).

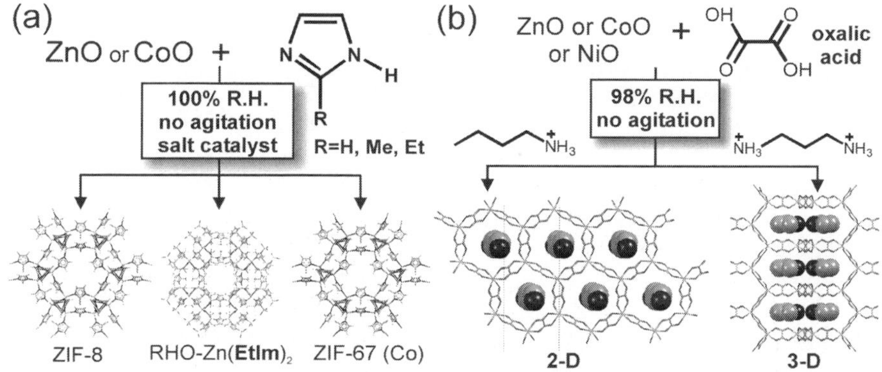

Figure 7. The synthesis of metal-organic frameworks by accelerated aging: (a) zeolitic imidazolate frameworks from zinc or cobalt(II) oxides and (b) templated assembly of two- and three-dimensional metal(II) oxalate frameworks from zinc, nickel(II) or cobalt(II) oxides (81, 83).

Beyond MOFs: Metal-Organic OLEDs and Metallodrugs

Whereas MOFs are certainly among the most popular and rapidly developing families of modern metal-organic materials, it is important to note that there are also other ones, for which the application of oxide-based synthesis can be just as beneficial in the context of green and sustainable synthesis. One example of such synthesis was reported by Ma and co-workers (*85*) who described the synthesis of the important, archetypal OLED material (*86*) aluminium(8-hydroxoquinolinate) (**Alq$_3$**) directly from basic aluminium acetate and the organic ligand 8-hydroxyquinoline (Figure 8a). Besides being a pioneering (and successful) effort in the solvent-free manufacture of a commercial OLED material, this work also recognized the presence of an intermediate phase, characterized via structure solution from X-ray powder diffraction data as a previously not known acetic acid solvate of **Alq$_3$**, **Alq$_3$**·AcOH (*85*). Importantly, the solvate **Alq$_3$**·AcOH is also luminescent, demonstrating the ability to discover and characterize novel potentially functional materials by a combination of mechanosynthesis and structure determination using X-ray powder diffraction. The authors also highlighted the ability to easily manufacture the OLED material **Alq$_3$** on a scale of at least 50 grams and, if high purity of the product is necessary, perform direct sublimation from the reaction mixture in that way achieving an entirely solvent-free access to a technologically important metal-organic material.

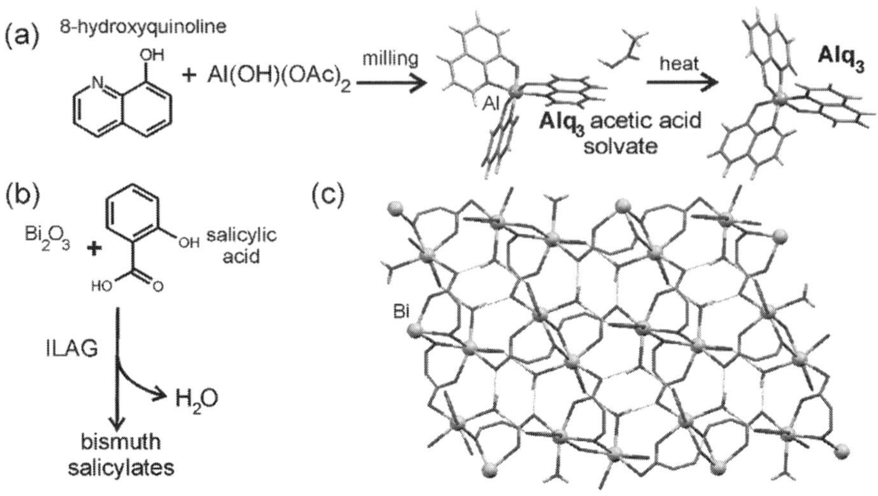

Figure 8. (a) Mechano- and thermochemical synthesis of the OLED material Alq3 from basic aluminum acetate, as reported by the James group (85); (b) the mechanosynthesis of bismuth salicylates (87), including bismuth subsalicylate, the active ingredient of Pepto-Bismol®, from a metal oxide and (c) the crystal structure of mechanochemically prepared bismuth disalicylate is the first structure of a bismuth salicylate without auxiliary organic ligands (87).

Mechanochemical oxide-based reactions are also applicable for the synthesis of metal-based drugs and pharmaceuticals directly from metal oxides. For example, the use of ILAG enabled the rapid and environmentally-friendly synthesis of the metallodrug bismuth subsalicylate, the active component of the popular gastrointestinal drug Pepto-Bismol®, as well as of two further bismuth salicylates directly from Bi_2O_3 (Figure 8b) (*87*). The outcome of the mechanochemical reactions, which are performed by ball milling in the presence of water and salt additive, was determined by the stoichiometric ratio of bismuth and salicylic acid in the reaction mixture. The quantitative reaction yield and purity of the products also facilitated the structural characterization of mechanochemical products using PXRD data, leading to the structural characterization of bismuth disalicylate monohydrate, an outstanding target from the 2002 Second Structure Determination by Powder Diffractometry Round Robin challenge (SDPDRR-2) (*88*). The crystal structure revealed a non-porous 2-D metal-organic framework and corrected the previously proposed chemical composition which involved hydroxide ligands (Figure 8c) (*89*). The structure of bismuth disalicylate hydrate is currently the only known crystal structure for a bismuth salicylate complex without auxiliary organic ligands and, as a potential model for the bismuth subsalicylate active drug ingredient, it is in striking contrast to previously proposed models based on discrete oxo-cluster structures (*90, 91*).

Conclusions and Outlook

The introduction of MOFs as enabling materials for a number of modern technologies and the development of first approaches to their commercial, large-scale manufacture have highlighted green and sustainable synthesis as a challenge that still has to be met by inorganic and metal-organic chemists. The recent reports from BASF, describing the criteria of availability, cost, toxicity, and safety, have provided a good first roadmap towards developing a commercially viable and environmentally acceptable MOF synthesis on a large scale. In this brief and critical overview we have attempted to illustrate several aspects of developing more sustainable approaches to MOFs, such as improvements in solvent systems achieved by employing (near) supercritical conditions, changes in design to incorporate biocompatible building blocks and the development of a range of solvent-free methodologies permitting MOF assembly from the simplest possible feedstocks. Certainly, there have been significant advances in other areas of MOF design and synthesis that are outside the scope of this brief overview. Possibly most important and promising among these is the rapid development of theoretical methods that allow the modelling of guest behavior in MOFs (*92*), as well as the prediction of structures of porous solids, including MOFs (*93*) but also related organic solids (*94*). These theoretical tools are expected to facilitate the design of MOFs for selected applications and replace experimental resources currently being dedicated to the discovery of new MOFs with computational screening of hypothetical MOF structures catered towards specific applications (*95*). Clearly, while the application of the principles of Green Chemistry to synthesis of metal-organic compounds is a relatively recent development, it appears that its importance has been realized by both industry and academia.

References

1. Izatt, R. M.; Izatt, S. R.; Bruening, R. L.; Izatt, N. E.; Moyer, B. A. *Chem. Soc. Rev.* **2014**, *43*, 2451–2475.
2. Federsel, H.-J. *Green Chem.* **2013**, *15*, 3105–3115.
3. Dunn, P. J. *Chem. Soc. Rev.* **2012**, *41*, 1452–1461.
4. Sheldon, R. A. *Chem. Soc. Rev.* **2012**, *41*, 1437–1451.
5. Trost, B. M. *Science* **1991**, *254*, 1471–1477.
6. Constable, D. J. C.; Dunn, P. J.; Hayler, J. D.; Humphrey, G. R.; Leazer, J. L., Jr.; Linderman, R. J.; Lorenz, K.; Manley, J.; Pearlman, B. A.; Wells, A.; Zaks, A.; Zhang, T. Y. *Green Chem.* **2007**, *9*, 411–420.
7. Lu, J.; Li, L.; Park, J.-B.; Sun, Y.-K.; Wu, F.; Amine, K. *Chem. Rev.* **2014**, *114*, 5611–5640.
8. Sathre, R.; Masanet, E. *RSC Adv.* **2013**, *3*, 4964–4975.
9. Seayad, A. M.; Antonelli, D. M. *Adv. Mater.* **2004**, *16*, 765–777.
10. Weitkamp, J. *Solid State Ionics* **2000**, *131*, 175–188.
11. Czaja, A.; Leung, E.; Trukhan, N.; Müller, U. Industrial MOF synthesis. In *Metal-Organic Frameworks: Applications from Catalysis to Gas Storage*, 1st ed.; Farrusseng, D., Ed.; Wiley-VCH Verlag & Co: Weinheim, Germany, 2011.

12. Ibarra, I. A.; Bayliss, P. A.; Pérez, E.; Yang, S.; Blake, A. J.; Nowell, H.; Allan, D. R.; Poliakoff, M.; Schröder, M. *Green Chem.* **2012**, *14*, 117–122.
13. Friščić, T.; Halasz, I.; Štrukil, V.; Eckert-Maksić, M.; Dinnebier, R. E. *Croat. Chem. Acta* **2012**, *85*, 367–378.
14. Saravade, P.; Tan, H.; Polshettiwar, V. *ACS Sustainable Chem. Eng.* **2013**, *1*, 66–74.
15. Farha, O. K.; Eryacizi, I.; Jeong, N. C.; Hauser, B. G.; Wilmer, C. E.; Sarjeant, A. A.; Snurr, R. Q.; Nguyen, S. T.; Yazaydin, A. Ö.; Hupp, J. T. *J. Am. Chem. Soc.* **2012**, *134*, 15016–15021.
16. Vaidyanathan, R.; Iremonger, S. S.; Shimizu, G. K. H.; Boyd, P. G.; Alavi, S.; Woo, T. K. *Science* **2010**, *330*, 650–653.
17. Wilmer, C. E.; Leaf, M.; Lee, C. Y.; Farha, O. K.; Hauser, B. G.; Hupp, J. T.; Snurr, R. Q. *Nat. Chem.* **2012**, *4*, 83–89.
18. Yilmaz, B.; Trukhan, N.; Müller, U. *Chin. J. Catal.* **2012**, *33*, 3–10.
19. Gaab, M.; Trukhan, N.; Maurer, S.; Gummaraju, R.; Müller, U. *Microporous Mesoporous Mater.* **2012**, *157*, 131–136.
20. Kuppler, R. J.; Timmons, D. J.; Fang, Q.-R.; Li, J.-R.; Makal, T. A.; Young, M. D.; Yuan, D.; Zhao, D.; Zhuang, W.; Zhou, H.-C. *Coord. Chem. Rev.* **2009**, *253*, 3042–3066.
21. Czaja, A. U.; Trukhan, N.; Müller, U. *Chem. Soc. Rev.* **2009**, *38*, 1284–1293.
22. Rowsell, J. L. C.; Yaghi, O. M. *Angew. Chem., Int. Ed.* **2005**, *44*, 4670–4679.
23. Peng, Y.; Krungleviciute, V.; Eryacizi, I.; Hupp, J. T.; Farha, O. K.; Yildirim, T. *J. Am. Chem. Soc.* **2013**, *135*, 11887–11894.
24. Mason, J. A.; Veenstra, M.; Long, J. R. *Chem. Sci.* **2014**, *5*, 32–51.
25. Li, J.-R.; Kuppler, R. J.; Zhou, H.-C. *Chem. Soc. Rev.* **2009**, *38*, 1477–1504.
26. Herm, Z. R.; Wiers, B. M.; Mason, J. A.; van Baten, J. M.; Hudson, M. R.; Zajdel, P.; Brown, C. M.; Masciocchi, N.; Krishna, R.; Long, J. R. *Science* **2013**, *340*, 960–964.
27. Herm, Z. R.; Bloch, E. D.; Long, J. R. *Chem. Mater.* **2014**, *26*, 323–328.
28. Lee, J. Y.; Farha, O. K.; Roberts, J.; Scheidt, K. A.; Nguyen, S. T.; Hupp, J. T. *Chem. Soc. Rev.* **2009**, *38*, 1450–1459.
29. Liu, J.; Chen, L.; Cui, H.; Zhang, J.; Zhang, L.; Su, C.-Y. *Chem. Soc. Rev.* **2014**, *43*, 6011–6061.
30. Lu, W.; Wei, Z.; Gu, Z.-Y.; Liu, T.-F.; Park, J.; Park, J.; Tian, J.; Zhang, M.; Zhang, Q.; Gentle, T., III; Bosch, M.; Zhou, H.-C. *Chem. Soc. Rev.* **2014**, *43*, 5561–5593.
31. Liu, Y.; Tang, Z. *Adv. Mater.* **2013**, *25*, 5819–5825.
32. Liu, Y.; Xuan, W.; Cui, Y. *Adv. Mater.* **2010**, *22*, 4112–4135.
33. Huxford, R. C.; Della Rocca, J.; Lin, W. *Curr. Opin. Chem. Biol.* **2010**, *14*, 262–268.
34. Sun, C.-Y.; Qin, C.; Wang, X.-L.; Su, Z.-M. *Exp. Opin. Drug Delivery* **2013**, *10*, 89–101.
35. Zeng, M.-H.; Yin, Z.; Tan, Y.-X.; Zhang, W.-X.; He, Y.-P.; Kurmoo, M. *J. Am. Chem. Soc.* **2014**, *136*, 4680–4688.
36. Kurmoo, M. *Chem. Soc. Rev.* **2009**, *38*, 1353–1379.
37. Heine, J.; Müller-Buschbaum, K. *Chem. Soc. Rev.* **2013**, *42*, 9232–9242.

38. Allendorf, M. D.; Bauer, C. A.; Bhakta, R. K.; Houk, R. J. T. *Chem. Soc. Rev.* **2009**, *38*, 1330–1352.
39. Banerjee, D.; Hu, Z.; Li, J. *Dalton Trans.* **2014**, *43*, 10668–10685.
40. Mueller, U.; Schubert, M.; Teich, F.; Puetter, H.; Schierle-Arndt, K.; Pastré, J. *J. Mater. Chem.* **2006**, *16*, 626–636.
41. Feller, R. K.; Cheetham, A. K. *CrystEngComm* **2009**, *11*, 980–985.
42. Gaab, M.; Trukhan, N.; Maurer, S.; Gummaraju, R.; Müller, U. *Microporous Mesoporous Mater.* **2012**, *157*, 131–136.
43. Hawxwell, S. M.; Brammer, L. *CrystEngComm* **2006**, *8*, 473–476.
44. Stock, N.; Biswas, S. *Chem. Rev.* **2012**, *112*, 933–969.
45. Dalinger, D.; Kappe, C. O. *Chem. Rev.* **2007**, *107*, 2563.
46. Patrick, H. R.; Griffith, K.; Liotta, C. L.; Eckert, C. A.; Gläser, R. *Ind. Eng. Chem. Res.* **2001**, *40*, 6063–6067.
47. Bayliss, P. A.; Ibarra, I. A.; Pérez, E.; Yang, S.; Tang, C. C.; Poliakoff, M.; Schröder, M. *Green Chem.* **2014**, *16*, 3796–3802.
48. Smaldone, R. A.; Forgan, R. S.; Furukawa, H.; Gassensmith, J. J.; Slawin, A. M. Z.; Yaghi, O. M.; Stoddart, J. F. *Angew. Chem., Int. Ed.* **2010**, *49*, 8630–8634.
49. Forgan, R. S.; Smaldone, R. A.; Gassensmith, J. J.; Furukawa, H.; Cordes, D. B.; Li, Q.; Wilmer, C. E.; Botros, Y. Y.; Snurr, R. Q.; Slawin, A. M. Z.; Stoddart, J. F. *J. Am. Chem. Soc.* **2012**, *134*, 406–417.
50. Lin, J.-B.; Lin, R.-B.; Cheng, X.-N.; Zhang, J.-P.; Chen, X.-M. *Chem. Commun.* **2011**, *47*, 9185–9187.
51. Friščić, T.; Fábián, L. *CrystEngComm* **2009**, *11*, 743–745.
52. Cliffe, M. J.; Mottillo, C.; Stein, R. S.; Bučar, D.-K.; Friščić, T. *Chem. Sci.* **2012**, *3*, 2495–2500.
53. Lanchas, M.; Vallejo-Sánchez, D.; Beobide, G.; Castillo, O.; Aguayo, A. T.; Luque, A.; Román, P. *Chem. Commun.* **2012**, *48*, 9930–9932.
54. James, S. L.; Adams, C. J.; Bolm, C; Braga, D.; Collier, P.; Friščić, T.; Grepioni, F.; Harris, K. D. M.; Hyett, G.; Jones, W.; Krebs, A.; Mack, J.; Maini, L.; Orpen, A. G.; Parkin, I. P.; Shearouse, W. C.; Steed, J. W.; Waddell, D. C. *Chem. Soc. Rev.* **2012**, *41*, 413–447.
55. Braga, D.; Giaffreda, S. L.; Grepioni, F.; Pettersen, A.; Maini, L.; Curzi, M.; Polito, M. *Dalton Trans.* **2006**, 1249–1263.
56. Boldyreva, E. *Chem. Soc. Rev.* **2013**, *42*, 7719–7738.
57. Burmeister, C. F.; Kwade, A. *Chem. Soc. Rev.* **2013**, *42*, 7660–7667.
58. Urakaev, F. Kh.; Boldyrev, V. V. *Powder Technol.* **2000**, *107*, 197–206.
59. Friščić, T. *Chem. Soc. Rev.* **2012**, *41*, 3493–3510.
60. Friščić, T. *J. Mater. Chem.* **2010**, *20*, 7599–7605.
61. Friščić, T.; Jones, W. *Cryst. Growth Des.* **2009**, *9*, 1621–1637.
62. Dybtsev, D. N.; Chun, H.; Kim, K. *Angew. Chem., Int. Ed.* **2004**, *43*, 5033–5036.
63. Friščić, T.; Trask, A. V.; Jones, W.; Motherwell, W. D. S. *Angew. Chem., Int. Ed.* **2006**, *45*, 7546–7550.
64. Friščić, T.; Trask, A. V.; Jones, W.; Motherwell, W. D. S. *Cryst. Growth Des.* **2008**, *8*, 1605–1609.

65. Friščić, T.; Childs, S. L.; Rizvi, S. A. A.; Jones, W. *CrystEngComm* **2009**, *11*, 418–426.
66. Friščić, T.; Reid, D. G.; Halasz, I.; Stein, R. S.; Dinnebier, R. E.; Duer, M. J. *Angew. Chem., Int. Ed.* **2010**, *49*, 712–715.
67. Phan, A.; Doonan, C. J.; Uribe-Romo, F. J.; Knobler, C. J.; O'Keeffe, M.; Yaghi, O. M. *Acc. Chem. Res.* **2010**, *43*, 58–67.
68. Zhang, J.-P.; Zhang, Y.-B.; Lin, J.-B.; Chen, X.-M. *Chem. Rev.* **2012**, *112*, 1001–1033.
69. Beldon, P. J.; Fábián, L.; Stein, R. S.; Thirumurugan, A.; Cheetham, A. K.; Friščić, T. *Angew. Chem., Int. Ed.* **2010**, *49*, 9640–9643.
70. Tanaka, S.; Kida, K.; Nagaoka, T.; Ota, T.; Miyake, Y. *Chem. Commun.* **2013**, *49*, 7884–7886.
71. Tumanov, I. A.; Achkasov, A. F.; Boldyreva, E. V.; Boldyrev, V. V. *CrystEngComm* **2011**, *13*, 2213–2216.
72. Friščić, T.; Halasz, I.; Beldon, P. J.; Belenguer, A. M.; Adams, F.; Kimber, S. A. J.; Honkimäki, V.; Dinnebier, R. E. *Nat. Chem.* **2013**, *5*, 66–73.
73. Halasz, I.; Kimber, S. A. J.; Beldon, P. J.; Belenguer, A. M.; Adams, F.; Honkimäki, V.; Nightingale, R. C.; Dinnebier, R. E.; Friščić, T. *Nature Protoc.* **2013**, *8*, 1718–1729.
74. Halasz, I.; Puškarić, A.; Kimber, S. A. J.; Beldon, P. J.; Belenguer, A. M.; Adams, F.; Honkimäki, V.; Dinnebier, R. E.; Patel, B.; Jones, W.; Štrukil, V.; Friščić, T. *Angew. Chem., Int. Ed.* **2013**, *52*, 11538–11541.
75. Halasz, I.; Friščić, T.; Kimber, S. A. J.; Užarević, K.; Puškarić, A.; Mottillo, C.; Julien, P.; Štrukil, V.; Honkimäki, V.; Dinnebier, R. E. *Faraday Discuss.* **2014**, *170*, 203–221.
76. Ma, X.; Yuan, W.; Bell, S. E. J.; James, S. L. *Chem. Commun.* **2014**, *50*, 1585–1587.
77. Adamo, P.; Violante, P. *Appl. Clay Sci.* **2000**, *16*, 229–256.
78. Chen, J.; Blume, H.-P.; Beyer, L. *Catena* **2000**, *39*, 121–146.
79. Echigo, T.; Kimata, M. *Can. Mineral.* **2010**, *48*, 1329–1357.
80. Clarke, R. M.; Williams, I. R. *Mineral. Mag.* **1986**, *50*, 295–298.
81. Qi, F.; Stein, R. S.; Friščić, T. *Green Chem.* **2014**, *16*, 121–132.
82. Cheetham, A. K.; Rao, C. N. R. *MRS Bull.* **2005**, *30*, 93–98.
83. Vaidhyanathan, R.; Natarajan, S.; Rao, C. N. R. *Dalton Trans.* **2001**, 699–706.
84. Mottillo, C.; Lu, Y.; Pham, M.-H.; Cliffe, M. J.; Do, T.-O.; Friščić, T. *Green Chem.* **2013**, *15*, 2121–2131.
85. Ma, X.; Lim, G. K.; Harris, K. D. M.; Apperley, D. C.; Horton, P. N.; Hursthouse, M. B.; James, S. L. *Cryst. Growth Des.* **2012**, *12*, 5869–5872.
86. Tang, C. W.; van Slyke, A. *Appl. Phys. Lett.* **1987**, *51*, 913–915.
87. André, V.; Hardeman, A.; Halasz, I.; Stein, R. S.; Jackson, G. J.; Reid, D. G.; Duer, M. J.; Curfs, C.; Duarte, M. T.; Friščić, T. *Angew. Chem., Int. Ed.* **2011**, *50*, 7858–7861.
88. Le Bail, A.; Cranswick, L. M. D. SDPD Round Robin 2002 results. *Commission on Powder Diffraction IUCr Newsletter* **2003**, *29*, 31–34.
89. Timakova, E. V.; Udalova, T. A.; Yukhin, Yu. M. *Russ. J. Inorg. Chem.* **2009**, *54*, 873–880.

90. Andrews, P. C.; Deacon, G. B.; Forsyth, C. M.; Junk, P. C.; Kumar, I.; Maguire, M. *Angew. Chem., Int. Ed.* **2006**, *45*, 5638–5642.
91. Boyd, T. D.; Kumar, I.; Wagner, E. E.; Whitmire, K. H. *Chem. Commun.* **2014**, *50*, 3556–3559.
92. Kadantsev, E. S.; Boyd, P. G.; Daff, T. D.; Woo, T. K. *Phys. Chem. Lett.* **2013**, *4*, 3056–3061.
93. Wilmer, C. A.; Leaf, M.; Lee, C. Y.; Farha, O. K.; Hauser, B. G.; Hupp, J. T.; Snurr, R. Q. *Nat. Chem.* **2012**, *4*, 83–89.
94. Jones, J. T. A.; Hassel, T.; Wu, X.; Bacsa, J.; Jelfs, K. E.; Schmidtmann, M.; Chong, S. Y.; Adams, D. J.; Trewin, A.; Schiffmann, F.; Cora, F.; Slater, B.; Steiner, A.; Day, G. M.; Cooper, A. I. *Nature* **2011**, *474*, 367–371.
95. Snurr, R. Q.; Colón, Y. J. *Chem. Soc. Rev.* **2014**, *43*, 5735–5749.

Biomass and Waste Valorization to Chemicals, Materials and Fuels

Chapter 10

Food Waste Valorisation for High Value Chemicals and Energy Production

Wan Chi Lam and Carol Sze Ki Lin*

School of Energy and Environment, City University of Hong Kong, Tat Chee Avenue, Kowloon, Hong Kong
***E-mail: carollin@cityu.edu.hk**

Significant amount of food and industrial wastes is being produced every year at alarming rate. The conventional ways to manage the excessive amount of waste generated include landfilling and incineration. However, these approaches often lead to uncontrolled emission of greenhouse gas or toxic compounds such as carbon dioxide, nitric oxides, methane and dioxin in large quantity during decomoposition and causing adverse environmental consequences such as air pollution, smog, acid rain and greenhouse effect. Therefore, converting waste to energy in a sustainable way using green technologies would be a desirable approach for waste management. In this chapter, conversion of food waste to energy and high value chemicals through different green technologies are presented.

Introduction

Approximately, 1.3 billion tons of food produced for human consumption is lost or wasted worldwide every year (*1*). In developing countries, approximately 15-30% food is wasted and it occurs mostly during food production stage in the farm due to poor agricultural techniques or other reasons such as droughts, flooding, and pests. Food wastage in these cases is often inevitable. In contrast, food waste occurs mostly during food processing, transportation, the retail and commercial sectors (supermarkets and restaurants) and households in developed countries, which accounts for approximately 30% of the food produced (*2*). In many cases, however, food wastage is avoidable. Food waste problem in developed cities such as Hong Kong is imminent. Alarming amounts of food waste of around 3000-3600 tons have been produced in Hong Kong every day for over the past 10 years between 2003 and 2012. The amount of food waste produced in Hong Kong per capita per day is 0.4 kg (*3*) which is significantly higher than other nearby Asian cities such as Tapei city and Seoul. In both cases, the food waste produced per capita per day is 0.2 kg (*3*).

Food waste including bread waste (*4*, *5*), pastry (*6*), cake (*6*) and restaurant food waste (*7*), generated from household, retail and commercial sectors are often rich in nutrients such as starch, proteins and lipid, see Table 1, making it a potential feedstock for fermentative production of chemicals, materials and biofuels.

Table 1. Composition of various food wastes from retail and commercial sectors (per 100g).

	Bread waste	*Pastry waste*	*Cake waste*	*Restaurant food waste*
Moisture (g)	22.3	34.5	45.0	N/A
Carbohydrate	46.8	33.5	62.0	33.3
Starch	59.8	N/A	N/A	N/A
Lipids	0.9	35.2	19.0	15.0
Sucrose	3.0	4.5	22.7	N/A
Fructose	N/A	2.3	11.9	N/A
Protein (TN × 5.7)	8.9	7.1	17.0	10.4
Total phosphorous	Trace	1.7	1.5	N/A
Ash	N/A	2.5	1.6	N/A

Through enzymatic hydrolysis, significant amount of fermentable nutrients such as glucose, amino acids and fatty acids can be produced and used for chemicals or biofuels production. Food waste hydrolysis could be effectively achieved via chemical and enzymatic approaches for nutrients recovery, various enzymes such as glucoamylase, protease and lipase. The overall food waste biorefinery concept is illustrated in Figure 1. Valorisation of food waste for energy and chemicals production using green technologies could be a sustainable and environmental friendly approach to the problem.

Fermentative Productions of High Value Chemicals from Food Waste

Solid state fermentation (SSF) and submerged (SmF) fermentation are two commonly used techniques for fermentative chemicals production. A wide range of high value chemicals, bio-polymers, enzymes, bio-pigments as well as anti-biotics could be produced in large scale through SSF or SmF. In SSF, fermentation is conducted on moist solid substrates in the absence of free water. Agro-industrial food residues such as crops residues, bakery waste and restaurant food waste which contains complex carbon source in form of starch and cellulose are frequently used as feedstock for fermentative chemicals production. In contrast, SmF is conducted in free flowing liquid substrate, therefore, soluble sugars such as glucose, molasses, maltose or sucrose are commonly used as the carbon sources.

Direct Use of Food Waste for Enzymes and Bio-Colorant Production

Fungi such as *Aspergillus*, *Rhizopus* and *Monascus* are capable of secreting significant amounts of extracellular amylases and proteases that can digest starch and proteins in food waste to fermentable sugars and amino acids to support its growth. In general, food waste which contains 30–60% starch and 5–10% proteins could serve as a promising substrate for fermentation (*8*). Since solid state fermentation (SSF) only require little substrate handling procedures to start fermentation, consequently, direct use of bakery and restaurant food wastes for fermentative chemicals production through SSF could be a promising approach to produce valuable chemicals. Food waste is first blended in a mixer to obtain homogeneous mixture, the moisture content and pH of the homogenised food waste are adjusted and then re-shaped to small pellets. Afterwards, spore suspension of the fungal host is inoculated to the surface of the food waste and keeps in an incubator for fermentation in order to produce desirable products. At the end of fermentation, target products are isolated and purified. Direct use of food waste as feedstock for fermentative chemicals production is summarised in Figure 2.

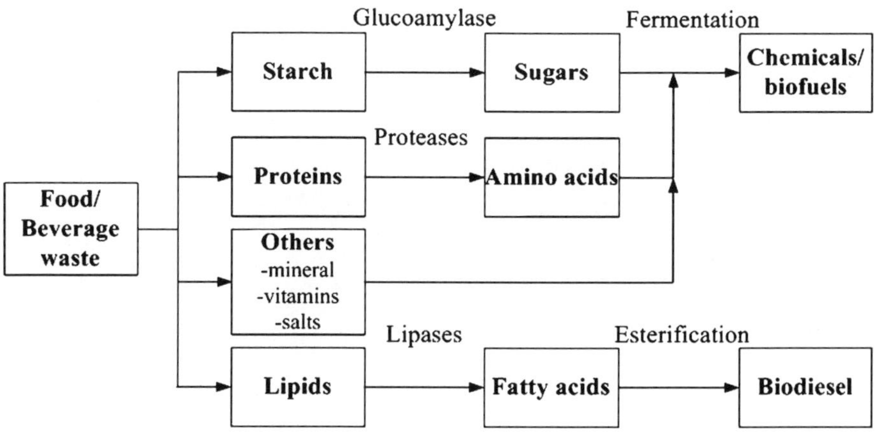

Figure 1. Food waste based biorefiney for chemicals and biofuels production.

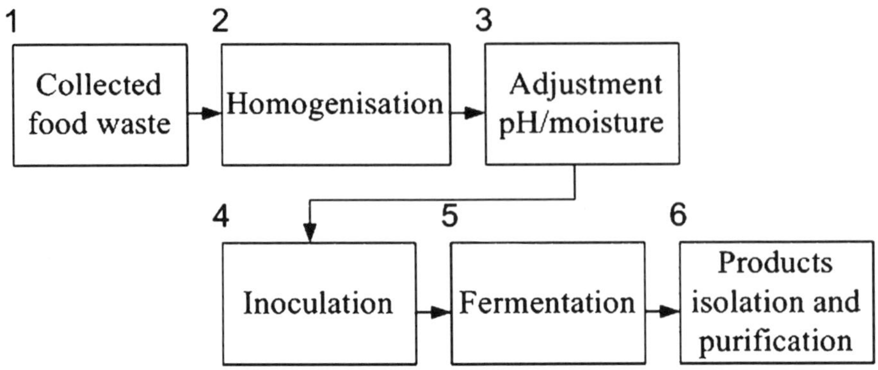

Figure 2. Overall process for direct use of food waste for fermentative chemicals production.

Production of Glucoamylase and Protease from Food Wastes

Glucoamylase is one of most industrially important enzymes. The enzyme catalysed the cleavage of α–(1,4) glycosidic bonds and α–(1,6) glycosidic bonds of polysaccharides from the non-reducing ends, giving glucose as product (*9, 10*). It is used for crystalline glucose or glucose syrup manufacturing from starch while the glucose can be used subsequently for wine, biofuels and chemicals production. Proteases refer to a diverse group of enzymes which hydrolysed peptide bonds of proteins and giving shorter peptides or amino acids as products (*11*). It is often used for animal feed, fish sauce or soy sauce production. Both glucoamylase and proteases could be produced from different origins including animals, plants and microorganisms. Microbial enzymes remain the predominant enzyme source, however, due to susceptible to genetic manipulation for target enzymes over production, rapid growth and limited space requirement for cultivation. Commercial glucoamylase and protease are commonly produced by filamentous fungi from *Aspergillus* and *Rhizopus* genera (*10, 12–14*), due to its ability to produce extracellular thermostable glucoamylases in large quantity and the host is generally regarded as safe (*15*).

Direct use of food waste to produce glucoamylase and proteases through solid state fermentation (SSF) has been studied widely (*10, 15–28*), the enzyme yields from various wastes are summarised in Table 2.

Table 2. Production of glucoamylase and protease from agro-industrial waste[a]

Substrate	*Concentration (U mL^{-1})*	*Yield (Ug^{-1})*	*Fungus*	*References*
Glucoamylase				
Rice powder	N/A	71.3 ± 2.34	*A. niger*	(*16*)
Wheat bran	N/A	110 ± 1.32	*A. niger*	(*16*)
Mixed food waste	137	N/A	*A. niger*	(*17*)
Cowpea waste	970	N/A	*A. oryzae*	(*18*)
Wheat bran	4.4	48.0	*A. awamori*	(*19*)
Wheat piece	3.32	81.3	*A. awamori*	(*20*)
Waste bread	3.94	78.4	*A. awamori*	(*10*)
Waste bread	N/A	114	*A. awamori*	(*21*)
Pastry waste	76 ± 6.1	253.7 ± 20.4	*A. awamori*	(*22*)
Proteases				
Wheat bran	N/A	[b]31.2	*A. oryzae* NRRL 1808	(*15*)
Mixed wheat and rice bran	N/A	[c]1256	*A. oryzae* Ozykat-1	(*23*)
Rice bran	N/A	[c]1400	*A. oryzae* Ozykat-1	(*23*)
Canola cake	N/A	[d]371	*A. oryzae* CCBP 001	(*24*)
Rice broken	N/A	[c]44.7 - 67.7	*A. niger* MTCC 281	(*25*)
Hulled grain of wheat	[c]183.1	N/A	*A. niger* I1	(*26*)
Wine stillage	[c]200	N/A	*A.* ***niger*** **THA**001	(*27*)
Jatropha seed cake	N/A	3260.5	*A. versicolour* CJS-98	(*28*)

[a] One unit of protease activity was defined as the amount of enzyme that liberated. [b] 1μM tyrosine min^{-1}. [c] 1μg tyrosine min^{-1}. [d] 0.01 absorbance at 660 nm min^{-1}.

Bio-Pigments

Monascus pigments are produced by the fungal species *Monascus*. The pigments produced have been used as a source of natural colouring agents and preservatives in food and beverage industries in many Asian countries including China, Japan, Indonesia and Thailand long time ago due to the pigments produced are non-toxic and possess anti-microbial activity (*29*, *30*). Over 50 *Monascus* pigments (*31*) have been identified so far, while the pigments rubropunctamine and ascorubramine (red), monascin and ankaflavin (yellow) and rubropunctatin and monascorubrin (orange) are the most studied pigments since 1932.

In addition to its applications in food and beverage industry, *Monascus* pigments could potentially be used in pharmaceutical industry owning to its biological activities. Mostafa et al. have reviewed the application of *Monascus* pigments according to its therapeutic properties which include anticancer, anti-cardiovascular diseases, antimicrobial and anti-cholesterols (*32*). More recently, novel applications of *Monascus* pigments as textile dye (*33*, *34*) and low cost dye-sensitised solar cells (*35*, *36*) have also been examined.

Optimisation of submerged fermentation for red *Monascus* pigments has been investigated by Lee et al. (*37*), the reported optimal carbon and nitrogen sources were 30 g L^{-1} of glucose and 1.5 g L^{-1} of monosodium glutamate (MSG) at pH 6.5. Moreover, the effect of metal ions on red pigments induction was also studied, iron (Fe^{3+}) exhibited the strongest stimulatory effect on red pigment production among other tested ions (Zn^{2+} and Mn^{2+}). Under the optimal conditions, the specific productivity is reported as 32.5 OD g DCW^{-1} h^{-1}. In addition to defined medium, the feasibility of using agricultural residues such as corncob (*38*) and jackfruit seed wastes (*39*) for *Monascus* pigments production through solid state fermentation has also been explored. Approximately, 25 OD g^{-1} dried waste was produced in both cases.

Utilisation of Hydrolysed Food Waste for High Value Chemicals Production

Since not all microbes are able to produce extracellular hydrolytic enzymes such as amylases and proteases, especially for bacteria. Moreover, it is also difficult to up-scaling solid state fermentation, in these cases, submerged fermentation (SmF) can be applied. To conduct SmF using food wastes as feedstock, it requires a hydrolysis step in order to convert the solid food waste into a nutrient-complete liquid medium. In this case, the blended food waste are subjected to enzymatic digestion in a bioreactor under controlled temperature. Commercial glucoamylase and protease could be utilised to recover glucose and amino acids from food waste. The major limitation of this approach is the high cost of the commercial enzymes. To overcome the cost problem, crude fungal enzymes (enzymes mixture secreted by or extracted from fungal culture without purification) could be a low-cost option. As discussed in the previous section, *Aspergillus*, are commonly used for enzyme production. *A. awamori* and *A. oryzae* are capable of secreting significant amount of glucoamylase and proteases, respectively. Utilisations of both fungi to recover glucose and free amino acids (FAN) have been reported (*4*, *5*, *7*).

Two methods i) addition of crude enzyme extract from fungal culture (*21, 22, 40*) and ii) direct addition of fermented solid (fungal mashes) (*4, 7, 20, 41*), could be applied to recover glucose and FAN from food waste. The first method require enzymes extraction and quantification (*22*) before digestion reaction. Although it involves more procedures, this method allows a consistent hydrolysis method. Furthermore, a concentrated enzyme can be prepared in order to shorten the digestion time. In contrast, the second method involves less procedures which only require addition of fungal mashes inoculum to homogenised food waste in a bioreactor at controlled temperature of 55°C for overnight (*4, 7, 42*). Therefore, hydrolytic activities of this method depend greatly on the degree of fungal growth.

Succinic Acid Production from Food Waste Hydrolysate

Succinic acid (SA) is a platform molecule with four carbons which is selected as one among the top twelve chemical building blocks by the U.S. Department of Energy (USDOE) in 2004 and 2011. In 2011, the global market of succinic acid is estimated to 40,000 metric ton while the market value was estimated at $ 63.2 million (*43*). The succinic acid market is predicted to grow rapidly and reach approximately $ 699,449 metric ton by 2020, with an estimated market value of $ 538.8 million. The major industrial application of succinic acid includes resins, coatings and pigments which account for 19.3% of the global succinic acid market (*43*). Succinic acid is produced as an intermediate of the tricarboxylic acid (TCA) cycle which is part of the central metabolism of organisms. Therefore, it can be produced through fermentation. Currently, bio-based succinic acid is produced in industrial scale in various companies. BioAmber, Succinity and Reverdia have produced 350, 500 and 300 metric ton succinic acid, respectively in 2011 (*43*).

Anaerobiospirillum succiniciproducens, *Actinobacillus succinogenes*, *Mannheimia succiniciproducens*, *Basfia succiniciproducens* and recombinant *Escherichia coli* are regarded as promising succinic acid producers, due to their high succinic acid productivity and ability to utilise a wide range of C5 and C6 sugars for fermentative succinic acid production (*44*). Moreover, utilisation of different agricultural residues and waste from diary industry as feedstock for fermentative succinic acid production (*19, 45–48*) through bacteria have also been demonstrated (Table 3).

Poly(3-hydroxybutyrate) (PHB) from Food Waste

Poly(3-hydroxybutyrate) (PHB) is a biopolymer that belongs to the group of polyhydroxyalkanoates (PHAs). It was first isolated and characterised in 1925 by Lemoigne at the Pasteur Institute in Paris. A wide range of bacteria can synthesis it as intracellular carbon and energy storage in large quantity under nutrient limitation conditions such as O, P, N, S, or trace elements like Mg, Ca and Fe but in excess carbon (*49*). *Alcaligenes latus*, *Azotobacter vinelandii*, *Cupriavidus necator* and recombinant *Escherichia coli* have been investigated as PHB producers with high productivities (*50–54*). Due to the chemical properties of PHB are similar to polypropylene and can be modulated easily by co-polymerisation, biodegradable, biocompatible with mammalian cells, PHB is applicable as petroleum based plastic substitutes and valuable biomaterial in tissue engineering such as surgical implants and scaffolds, biodegradable screws and staples and wound dressings (*55*). Recently, *Halomonas boliviensis* has been identified as a potential PHB producer for industrial applications due to its high productivity and high salt tolerance, implying its potential for PHB production using wastes as feedstock. Table 4 shows PHB production with different wastes substrates using *H. boliviensis*. Since fermentative PHB production in large-scale using glucose as carbon source in medium is very expensive and that accounts for approximately 30% of the total production costs (*56*). Utilisation of various wastes material as alternative carbon source were investigated and the feasibility of using various wastes for fermentative PHB production were demonstrated (Table 4) (*6, 57, 58*).

Lactic Acid

Lactic acid ($C_3H_6O_3$) was first discovered from sour milk by the Swedish scientist C. W. Scheele in 1780. Lactic acid exists in two optically active stereo-isomers, the L(+) and the D(−) lactic acid (see Figure 3), while D(−) lactic acid is harmful to human. L(+) lactic acid is a non-volatile, odorless organic acid which has been approved by the U.S. Food and Drug Adminstration (FDA) as food additive for general purpose.

In addition to its traditional application in food and beverage industry as flavouring, preserving, pH buffering and emulsifying agents (*59*), this three-carbon molecule serves as an excellent building block for production of biodegradable polymer and other value-added chemicals such as polylactic acid (PLA), ester, acrylic acid, pyruvic acid, propylene glycol, and 2, 3 pentandione. Therefore, lactic acid is also selected as one of the top ten platform chemicals in 2010 (*60*). Between 2010 and 2013, the annual production of lactic acid is estimated as 260,000 tons while the future market of lactic acid is predicted to be 329,000 and 600,000 tons by 2015 and 2020, respectively (*60*). Similar to succinic acid, it is also one of the intermediate products from the TCA cycle in microorganisms. Nowadays, over 90% of commercial lactic acid is produced through microbial fermentation (*61*) due to the distinctive advantage of possibility of pure L(+) lactic acid production (*62–64*). In contrast to chemical synthesis, which always give a racemic D,L-lactic acid (*65*).

Lactic acid bacteria (LAB) refers to a group bacteria which produce lactic acid as their major anaerobic product of glycolysis. Therefore, lactic acid production is typically carried out under anaerobic conditions. The genera *Lactobacillus*, *Lactococcus, Streptococcus*, *Leuconostoc*, *Pediococcus*, *Aerococcus*, *Carnobacterium*, *Enterococcus*, *Oenococcus*, *Tetragenococcus*, *Vagococcus* and *Weisella* are the major LAB. In addition to bacteria, a variety of microorganisms such as fungus, yeast and microalgae (*63*) are capable of producing lactic acid. Whereas the LAB, *Lactobacillus* remains the major lactic acid producer used in the industry due to its long history of industrial-scale production and the product safety (*63*). The feasibility of utilising different food wastes for lactic acid production have been demonstrated (*66–70*) and summarised in Table 5.

Table 3. Fermentative succinic acid production from glucose and other wastes substrates

Host	*Substrate*	*Yield* $g\ g^{-1}$	*Productivity* $g\ L^{-1}\ h^{-1}$	*SA conc.* $g\ L^{-1}$	*References*
A. succinogenes 130Z	Bread hydrolysate	1.16	1.12	47.3	(*45*)
E. coli SD121 (mutant)	Glucose	1.13	1.55	116.2	(*46*)
A. succinogenes 130Z	Wheat hydrolysate	0.81	1.19	64.2	(*19*)
A. succinogenes 130Z	Cane molasses	N/A	1.15	55.2	(*47*)
A. succinogenes 130Z	Cheese whey	0.57	0.44	21.2	(*48*)

Table 4. PHB production from defined medium and low cost substrates with *H. boliviensis*

Medium/ Carbon source	***Process***	***PHB Conc. (g L^{-1})***	***PHB content (%)***	***References***
Defined	Batch	1.0	45	(*57*)
Wheat bran hydrolysate	Batch	1.08	33.8 ± 2.17	(*57*)
Starch hydrolysate	Batch	N/A	56	(*58*)
Pastry hydrolysate	Fed-batch	N/A	3.6	(*6*)

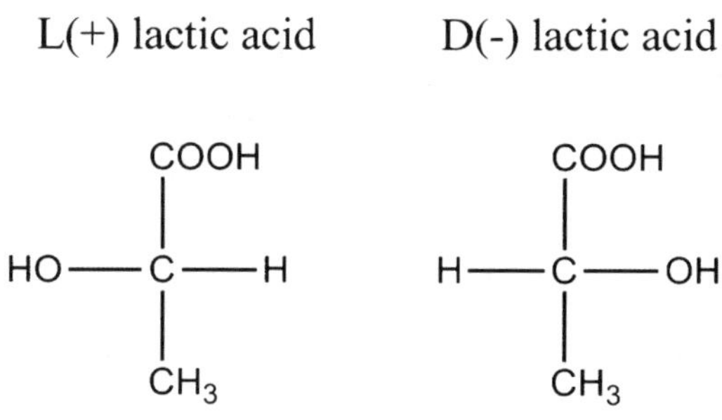

Figure 3. Chemical structures of L(+) and D(-) lactic acid.

Table 5. Production of lactic acid from different waste substrates

Substrate	*Yield ($g\ g^{-1}$)*	*Concentration ($g\ L^{-1}$)*	*Bacterial strain*	*References*
Cafeteria food waste	1.11	48.7	*Lactobacillus manihotivorans* LMG18011	(*66*)
Sugar cane baggage	0.36	10.9	*Lactococcus. lactis* IO-1	(*67*)
Kitchen wastes	0.62	64.0	*Lactic acid bacteria and Clostridium sp.*	(*68*)
Mango peel	N/A	17.4	Indigenous microorganisms	(*69*)
Vine-trimming wastes	0.77	21.8	*Lactobacillus pentosus* ATCC 8041	(*70*)

Valorisation of Food Waste for Biofuel Production

Biofuel is a general term describing fuels such as biodiesel, biogas and bioethanol produced from biological materials. Through various conversion processes including chemical reactions, anaerobic digestion and fermentation, biofuel can be produced from biomass. As described by Lee et al. (*71*), biofuels is usually classified into three generations as follow:

- First-generation biofuels are directly related to a biomass that is generally edible.
- Second-generation biofuels are defined as fuels produced from a wide array of different feedstock, ranging from lignocellulosic feedstocks to municipal solid wastes.
- Third-generation biofuels are related to algal biomass but could to a certain extent be linked to utilisation of CO_2 as feedstock.

Since the first generation biofuel production is primarily food crops based, therefore, this restricted the large-scale production to meet the demand due to competition for land and water used for cultivation of food. In view of this, the second and third generations of biofuels are better alternatives.

Production of Biodiesel from Food Waste Lipid

Biodiesel is defined as a fuel comprised of mono-alkyl esters of long chain fatty acids derived from vegetable oils or animal fats. It is produced through the trans-esterification reaction of vegetable oil or animal fat in the presence of catalyst (Figure 4). Both alkaline (NaOH, KOH) and acidic (H_2SO_4) catalysts are the most commonly used catalysts for biodiesel production. Whereas, lipase could also be used as a catalyst for biodiesel production as a greener approach.

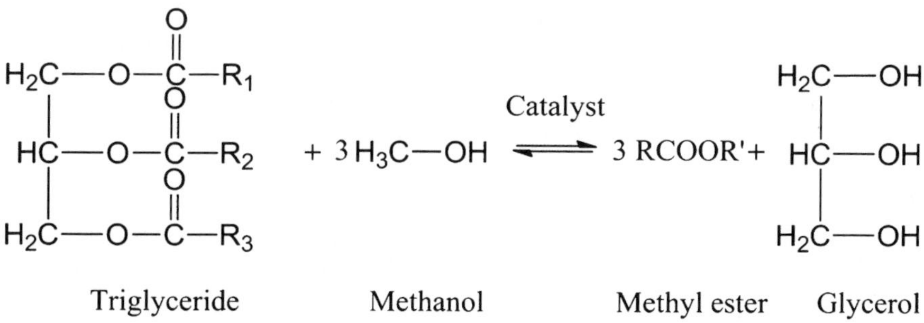

Figure 4. Trans-esterification of triglyceride for biodiesel production

For reaction with 3 moles of methanol and 1 mole of triglyceride, it gives 3 moles of methyl esters and 1 mole of glycerol. The methyl ester formed is biodiesel while glycerol is a by-product. Although the use of biodiesel as fuel is environmentally beneficial in comparison to petroleum diesel as it is biodegradable, nontoxic and resulted a low emission profile. Its production cost is also much higher than the petroleum diesel due to the high cost of feedstock. Since the cost of biodiesel feedstock alone could account for 75% of the overall biodiesel production process (*72*), therefore, utilisation of waste lipid as feedstock could be an economically feasible way of biodiesel production. According to the latest directory in March 2014, there are 136 biodiesel production plants in the U.S. While one-fifth of them utilise waste lipids for biodiesel production (*73*) and it corresponds to an annual production of 286 million gallons biodiesel.

In addition to waste cooking oil, lipid extracted from food waste such as pastry, cake and restaurant waste could potentially be used as feedstock for biodiesel production due to the presence of significant amount of lipid (15-35 %) see Table 1 in the previous section. Moreover, the residue obtained after lipid extraction also contains significant amount of carbohydrate and proteins, which could be used for biogas production through anaerobic digestion or bioethanol production. The benefits of this food waste based approach for biodiesel production is not only limited to the reduction of production cost, but it is also an alternative approach for waste management.

Anaerobic Digestion of Food Waste for Biogas Production

Anaerobic digestion is a commonly used method for methane rich biogas production. The biogas produced is methane rich and also contains carbon dioxide. In anaerobic digestion, a consortium of bacteria is used to degrade pasteurised organic wastes under anaerobic conditions at controlled temperature in order to produce biogas. The key steps involved in anaerobic digestion include: hydrolysis, acetogenesis and methanogenesis as illustrated in Figure 5.

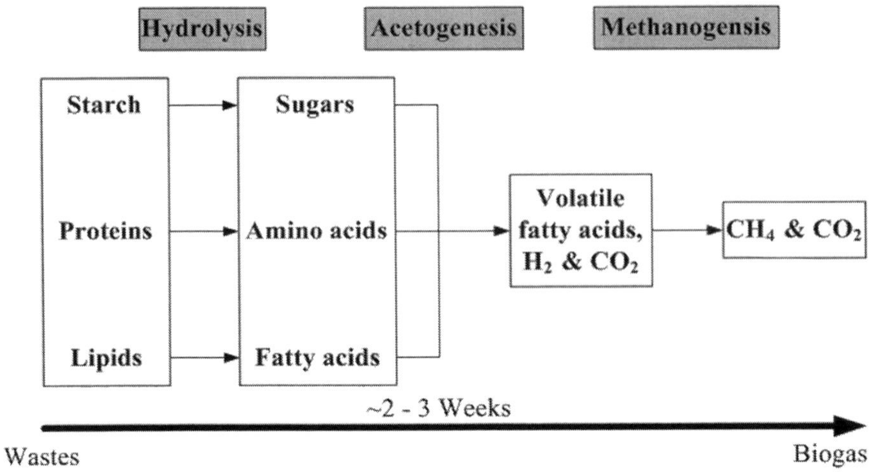

Figure 5. Key steps in organic food waste degradation in anaerobic digestion.

Anaerobic digestion is commonly conducted at either mesophilic (30-40°C) or thermophilic (50-55°C) temperature. Although shorter digestion time is required when anaerobic digestion is conducted at thermophilic temperature, thermophilic anaerobic digestion is also more energy intensive. Consequently, it would be preferred to conduct anaerobic digestion at mesophilic temperature. Grindsted Kommune anaerobic plant in Denmark is one of the benchmark anaerobic digestion facilities which co-digests household waste, sewage sludge, food waste and organic industrial wastewater to produce biogas for heat and electricity generation (*74*). In this facility, anaerobic digestion is carried out at mesophilic temperature for 14 days for biogas production to generate heat and electricity, while the sledge obtained at the end of anaerobic digestion is dewatered and spread on farms as fertiliser.

The Hong Kong Organic Waste Treatment Facilities (OWTF) is a similar anaerobic digestion facility adopted to produce fertiliser and electricity to supply households in Hong Kong. This facility can process approximately 200 tons organic waste (mainly food waste) every day to produce 14 million kWh of surplus electricity per year to the power grid to supply electricity to the household.

Technical Challenges

Utilisation of food waste for high value chemicals and fuel production could be a promising approach leading to sustainable production of chemicals and energy. Currently, food waste sorting policy is lacking in many countries. The 'food waste' collected is likely containing non-biodegradable parts such as, bones, shells or even food packaging. This greatly limits the use of food waste for chemicals and energy conversion. Moreover, food waste may contain high salt concentration and heavy metal ions (*75*) which could inhibit the hydrolytic enzymes used in the hydrolysis step or toxic to the microorganisms for fermentation. Consequently, it makes the waste to energy conversion inefficient and costly. In view of this, appropriate pre-sorting step to remove non-degradable components in food waste is vital to the approach. Furthermore, future development of low-cost technologies for food waste desalting, inhibitors removal or enzyme engineering for producing inhibitors-resistant enzymes could greatly facilitate the application of this waste to energy approach.

Concluding Remarks and Future Prospects

Utilisation of food waste for valuable chemicals and biofuel production through various green methods has been presented in this chapter. Through biological or chemical approaches, it is feasible to convert the carbohydrate, protein and lipid fractions from food waste to a board spectrum of value-added chemicals, bio-polymers and biofuels. In 2013, a European network COST Action TD1203 entitled Food Waste Valorisation for Sustainable Chemicals, Materials and Fuels (*EUBis*) (http://costeubis.org/) was set up. The aim of the action is to provide an integrated alternative and renewable source of carbon for the production of industrially relevant bio-derived chemicals, fuels and materials via the exploration of novel and advanced routes for food supply chain waste valorisation. In the first year, over 100 laboratories from over 30 countries have already joined the network. Such concerted actions could create the critical mass for the development of sustainable technologies based on food waste utilisation.

Acknowledgments

The authors acknowledges the Innovation and Technology Fundings from the Innovation and Technology Commission (ITS/323/11, ITS/353/12, ITP/002/14TP) in Hong Kong. We are also grateful to the support from the Hong Kong Research Institute of Textiles and Apparel (HKRITA). This publication is partially funded by the Applied Research Grant of City University of Hong Kong (project number of 9667101).

References

1. Gustavsson, J.; Cederberg, C.; Sonesson, U.; van Otterdijk, R.; Meybeck, A. *Global food losses and food waste. Extent, causes and prevention*, 2011. http://www.fao.org/docrep/014/mb060e/mb060e00.pdf (accessed May 14, 2014).
2. Morgan, G. Waste Not, Want Not: An Overview of the Food Waste Issue, 2011. http://www.bsr.org/en/our-insights/report-view/waste-not-want-not-an-overview-of-food-waste (accessed May 14, 2014).
3. Hong Kong SAR Environmental Bureau. A Food waste & yard waste plan for Hong Kong 2014-2022, 2012 http://www.enb.gov.hk/en/files/FoodWastePolicyEng.pdf (accessed on May 14, 2014).
4. Leung, C. C. J.; Cheung, A. S. Y.; Zhang, A. Y. Z.; Lam, K. F.; Lin, C. S. K. *Biochem. Eng. J.* **2012**, *65*, 10–15.

5. Zhang, A. Y. Z.; Sun, Z.; Leung, C. C. J.; Han, W.; Lau, K. Y.; Li, M.; Lin, C. S. K. *Green Chem.* **2013**, *15*, 690–695.
6. Arancon, R. A. D.; Lin, C. S. K.; Chan, K. M.; Kwan, T. H.; Luque, R. *Energy Sci. Eng.* **2013**, *1*, 53–71.
7. Pleissner, D.; Lam, W. C.; Sun, Z.; Lin, C. S. K. *Bioresour. Technol.* **2013**, *137*, 139–146.
8. Pleissner, D.; Lin, C. S. K. *Sustain. Chem. Process* **2013**, *1*, 1–6.
9. Sauer, J.; Sigurskjold, B. W.; Christensen, U.; Frandsen, T. P.; Mirgorodskaya, E.; Harrison, M.; Roepstorff, P.; Svensson, B. *Biochim. Biophys. Acta, Protein Struct. Mol. Enzymol.* **2000**, *1543*, 275–293.
10. Norouzian, D.; Akbarzadeh, A.; Scharer, J. M.; Moo Young, M. *Biotechnol. Adv.* **2006**, *24*, 80–85.
11. Sumantha, A.; Larroche, C.; Pandey, A. *Food Technol. Biotech.* **2006**, *44*, 211–220.
12. James, J. A.; Lee, B. H. *J. Food Biochem.* **1997**, *21*, 1–52.
13. Buchholz, K.; Seibel, J. *Carbohydr. Res.* **2008**, *343*, 1966–1979.
14. Rao, M. B.; Tanksale, A. M.; Ghatge, M. S.; Deshpande, V. V. *Microbiol. Mol. Biol. Rev.* **1998**, *62*, 597–635.
15. Sandhya, C.; Sumantha, A.; Szakacs, G.; Pandey, A. *Process Biochem.* **2005**, *40*, 2689–2694.
16. Anto, H.; Trivedi, U. B.; Patel, K. C. *Bioresour. Technol.* **2006**, *97*, 1161–1166.
17. Wang, Q.; Wang, X.; Wang, X.; Ma, H. *Process Biochem.* **2008**, *43*, 280–286.
18. Kareem, S. O.; Akpan, I.; Oduntan, S. B. *Afr. J. Microbiol. Res.* **2009**, *3*, 974–977.
19. Du, C.; Lin, S. K. C.; Koutinas, A.; Wang, R.; Dorado, P.; Webb, C. *Bioresour. Technol.* **2008**, *99*, 8310–8315.
20. Wang, R.; Godoy, L. C.; Shaarani, S. M.; Melikoglu, M.; Koutinas, A.; Webb, C. *Enzyme Microb. Technol.* **2009**, *44*, 223–228.
21. Melikoglu, M.; Lin, C. S. K.; Webb, C. *Biochem. Eng. J.* **2013**, *80*, 76–82.
22. Lam, W. C.; Pleissner, D.; Lin, S. K. C. *Biomolecules* **2013**, *3*, 651–661.
23. Chutmanop, J.; Chuichulcherm, S.; Chisti, Y.; Srinophakun, P. *J. Chem. Technol. Biotechnol.* **2008**, *83*, 1012–1018.
24. Freitas, A. C.; Castro, R. J. S.; Fontenele, M. A.; Egito, A. S.; Farinas, C. S.; Pinto, G. A. S. *ISRN Microbiol.* **2013**, *2013*, 8.
25. Paranthaman, R.; Alagusundaram, K.; Indhumathi, J. *World J. Agric. Sci.* **2009**, *5*, 308–312.
26. Siala, R.; Frikha, F.; Mhamdi, S.; Nasri, M.; Sellami Kamoun, A. *Sci. World J.* **2012**, *2012*, 11.
27. Yang, F. C.; Lin, I. H. *Enzyme Microb. Technol.* **1998**, *23*, 397–402.
28. Veerabhadrappa, M. B.; Shivakumar, S. B.; Devappa, S. *J. Biosci. Bioeng.* **2014**, *117*, 208–214.
29. Wong, H. C.; Koehler, P. E. *J. Food Sci.* **1981**, *46*, 589–592.
30. Kumari, H. P. M.; Naidu, K. A.; Vishwanatha, S.; Narasimhamurthy, K.; Vijayalakshmi, G. *Food Chem. Toxicol.* **2009**, *47*, 1739–1746.
31. Feng, Y.; Shao, Y.; Chen, F. *Appl. Microbiol. Biotechnol.* **2012**, *96*, 1421–1440.

32. Mostafa, M. E.; Abbady, M. S. *Global J. Biotechnol. Biochem.* **2014**, *9*, 1–13.
33. Santis, D. D.; Moresi, M.; Gallo, A. M.; Petruccioli, M. *J. Chem. Technol. Biotechnol.* **2005**, *80*, 1072–1079.
34. European Commision, Guidelines on State aid for environmental protection and energy 2014–2020 Commision Vol. C (2322), 2014. http://ec.europa.eu/smart-regulation/impact/ia_carried_out/docs/ia_2014/c_2014_2322_en.pdf (accessed May 14, 2014).
35. Ito, S.; Saitou, T.; Imahori, H.; Uehara, H.; Hasegawa, N. *Energy Environ. Sci.* **2010**, *3*, 905–909.
36. Lee, J. W.; Kim, T. Y.; Ko, H. S.; Han, S.; Lee, S. H.; Park, K. H. *Spectrochim. Acta, Part A* **2014**, *126*, 76–80.
37. Lee, B. K.; Park, N. H.; Piao, H. Y.; Chung, W. J. *Biotechnol. Bioprocess Eng.* **2001**, *6*, 341–346.
38. Velmurugan, P.; Hur, H.; Balachandar, V.; Kamala-Kannan, S.; Lee, K. J.; Lee, S. M.; Chae, J. C.; Shea, P. J.; Oh, B. T. *J. Biosci. Bioeng.* **2011**, *112*, 590–594.
39. Babitha, S.; Soccol, C. R.; Pandey, A. *Bioresour. Technol.* **2007**, *98*, 1554–1560.
40. Melikoglu, M.; Lin, C. S. K.; Webb, C. *Food Bioprod. Process.* **2013**, *91*, 638–646.
41. Dorado, M. P.; Lin, S. K. C.; Koutinas, A.; Du, C.; Wang, R.; Webb, C. *J. Biotechnol.* **2009**, *143*, 51–59.
42. Pleissner, D.; Kwan, T. H.; Lin, C. *Bioresour. Technol.* **2014**, *158*, 48–54.
43. Weastra. Determination of market potential for selected platform chemicals-Itaconic acid, Succinic acid, 2,5-Furandicarboxylic acid, 2013. http://www.bioconsept.eu/wp-content/uploads/BioConSepT_Market-potential-for-selected-platform-chemicals_report1.pdf (accessed May 14, 2014).
44. Koutinas, A. A.; Vlysidis, A.; Pleissner, D.; Kopsahelis, N.; Lopez Garcia, I.; Kookos, I. K.; Papanikolaou, S.; Kwan, T. H.; Lin, C. S. K. *Chem. Soc. Rev.* **2014**, *43*, 2587–2627.
45. Leung, C.; Cheung, A.; Zhang, A.; Lam, K.; Lin, C. *Biochem. Eng. J.* **2012**, *65*, 10–15.
46. Wang, D.; Li, Q.; Song, Z.; Zhou, W.; Su, Z.; Xing, J. *J. Chem. Technol. Biotechnol.* **2011**, *86*, 512–518.
47. Liu, Y. P.; Zheng, P.; Sun, Z. H.; Ni, Y.; Dong, J. J.; Zhu, L. L. *Bioresour. Technol.* **2008**, *99*, 1736–1742.
48. Wan, C.; Li, Y.; Shahbazi, A.; Xiu, S. *Appl. Biochem. Biotechnol.* **2008**, *145*, 111–119.
49. Anderson, A. J.; Dawes, E. A. *Microbiol. Rev.* **1990**, *54*, 450–472.
50. Yamane, T.; Fukunaga, M.; Lee, Y. W. *Biotechnol. Bioeng.* **1996**, *50*, 197–202.
51. Page, W. J.; Knosp, O. *Appl. Environ. Microbiol.* **1989**, *55*, 1334–1339.
52. Tian, J.; He, A.; Lawrence, A. G.; Liu, P.; Watson, N.; Sinskey, A. J.; Stubbe, J. *J. Bacteriol.* **2005**, *187*, 3825–3832.

53. Koutinas, A. A.; Xu, Y.; Wang, R.; Webb, C. *Enzyme Microb. Technol.* **2007**, *40*, 1035–1044.
54. Lee, S. Y.; Chang, H. N. *J. Environ. Polym. Degrad.* **1994**, *2*, 169–176.
55. Holmes, P. A. *Phys. Technol.* **1985**, *16*, 32–36.
56. Choi, J. I.; Lee, S. Y. *Bioprocess. Eng.* **1997**, *17*, 335–342.
57. Van-Thuoc, D.; Quillaguamán, J.; Mamo, G.; Mattiasson, B. *J. Appl. Microbiol.* **2008**, *104*, 420–428.
58. Quillaguamán, J.; Hashim, S.; Bento, F.; Mattiasson, B.; Hatti-Kaul, R. *J. Appl. Microbiol.* **2005**, *99*, 151–157.
59. Reddy, G.; Altaf, M.; Naveena, B. J.; Venkateshwar, M.; Kumar, E. V. *Biotechnol. Adv.* **2008**, *26*, 22–34.
60. Bozell, J. J.; Petersen, G. R. *Green Chem.* **2010**, *12*, 539–554.
61. Dusselier, M.; Wouwe, P. V.; Dewaele, A.; Makshina, E.; Sels, B. F. *Energy Environ. Sci.* **2013**, *6*, 1415–1442.
62. Ilmén, M.; Koivuranta, K.; Ruohonen, L.; Suominen, P.; Penttilä, M. *Appl. Microbiol. Biotechnol.* **2007**, *73*, 117–123.
63. Abdel-Rahman, M. A.; Tashiro, Y.; Sonomoto, K. *Biotechnol. Adv.* **2013**, *31*, 877–902.
64. Abdel-Rahman, M. A.; Tashiro, Y.; Sonomoto, K. *J. Biotechnol.* **2011**, *156*, 286–301.
65. Wee, Y. J.; Kim, J. N.; Ryu, H. W. *Food Technol. Biotechnol.* **2006**, *44*, 163–172.
66. Ohkouchi, Y.; Inoue, Y. *Bioresour. Technol.* **2006**, *97*, 1554–1562.
67. Laopaiboon, P.; Thani, A.; Leelavatcharamas, V.; Laopaiboon, L. *Bioresour. Technol.* **2010**, *101*, 1036–1043.
68. Zhang, B.; He, P. J.; Ye, N. F.; Shao, L. M. *Bioresour. Technol.* **2008**, *99*, 855–862.
69. Jawad, A. H.; Alkarkhi, A. F. M.; Jason, O. C.; Easa, A. M.; Nik Norulaini, N. A. *J. King Saud Univ., Sci.* **2013**, *25*, 39–45.
70. Bustos, G.; Moldes, A. B.; Cruz, J. M.; Domínguez, J. M. *J. Sci. Food Agric.* **2004**, *84*, 2105–2112.
71. Lee, R. A.; Lavoie, J. M. *Anim. Front.* **2013**, *3*, 6–11.
72. Atabania, A. E.; Silitonga, A. S.; Badruddin, I. A.; Mahlia, T. M. I.; Masjuki, H. H.; Mekhilef, S. *Renewable Sustainable Energy Rev.* **2012**, *16*, 2070–2093.
73. USA Plants. *Biodiesel Magazine*, March 2014. http://www.biodiesel magazine.com/plants/listplants/USA/ (accessed May 14, 2014).
74. European Commission. Grindstead Municipality Wastewater Treatment Plant, 1997. http://ec.europa.eu/energy/renewables/bioenergy/doc/anaero bic/08bm_077_1994.pdf (accessed May 14, 2014)
75. Zhang, R.; EI-Mashad, H. M.; Hartman, K.; Wang, F.; Liu, G.; Choate, C.; Gamble, P. *Bioresour. Technol.* **2007**, *4*, 929–935.

Chapter 11

Lignin Based Carbon Materials for Energy Storage Applications

Sabornie Chatterjee,*[,1] Tomonori Saito,[1] Orlando Rios,[2] and Alexander Johs*[,3]

[1]Chemical Sciences Division, Oak Ridge National Laboratory, Oak Ridge, Tennessee 37831, United States
[2]Materials Science & Technology Division, Oak Ridge National Laboratory, Oak Ridge, Tennessee 37831, United States
[3]Environmental Sciences Division, Oak Ridge National Laboratory, Oak Ridge, Tennessee 37831, United States
***E-mail: chatterjees@ornl.gov (Sabornie Chatterjee); johsa@ornl.gov (Alexander Johs)**

The implementation of Li-ion battery technology into electric and hybrid electric vehicles and portable electronic devices such as smart phones, laptops and tablets, creates a demand for efficient, economic and sustainable materials for energy storage. However, the high cost and long processing time associated with manufacturing battery-grade anode and cathode materials are two big constraints for lowering the total cost of batteries and environmentally friendly electric vehicles. Lignin, a byproduct of the pulp and paper industry and biorefinery, is one of the most abundant and inexpensive natural biopolymers. It can be efficiently converted to low cost carbon fibers with optimal properties for use as anode materials. Recent developments in the preparation of lignin precursors and conversion to carbon fiber-based anode materials have created a new class of anode materials with excellent electrochemical characteristics suitable for immediate use in existing Li- or Na-ion battery technologies.

Introduction

In today's energy intensive world, with rapidly growing use of smart phones and electric cars, there is an increasing demand of high capacity, re-usable, budget friendly batteries. Among commercially available energy storage systems, rechargeable Lithium (Li) ion batteries possess high energy density with favorable discharge characteristics. However, the high cost and long manufacturing time of anode materials are two big constraints to make such batteries. Lignin, one of the most abundant and inexpensive natural biopolymers, can be efficiently converted to low cost carbon fibers for use as a battery anode material (*1–3*). This review will highlight some recent advances in the application of lignin carbon fibers for the preparation of Li-ion anode materials.

Lignin Carbon Fiber

Lignin has a three dimensional complex branched structure based on three basic monomeric units: *p*-hydroxyphenyl (H), Guaiacyl (G) and Syringyl (S) (Figure 1) (*4–6*). The relative distribution of these units and the degree of cross linking in any lignin sample depend primarily on the plant source. The S/G ratios in hardwood and softwood lignins are 2:1 and 1:2-1:3 respectively (*6*, *7*). Typically, the process to produce carbon fiber from purified lignin can be divided into three steps (*1–3*, *8–10*) (Scheme 1); i) spinning of lignin or lignin derivatives; ii) thermal stabilization of lignin fibers and iii) carbonization of lignin fibers to lignin carbon fibers. Lignin is at first converted to a precursor fiber using a variety of spinning techniques (*1*, *8*, *9*, *11*). The lignin fiber is then thermostabilized, which prevents sticking and melting of lignin fiber during carbonization in the next step (*12*, *13*). During thermal stabilization, weak intermolecular bonds are broken and molecules are crosslinked by oxidation and condensation, which increase the glass transition temperature (T_g). At the initial thermostabilization stages (temperatures below 250°C) oxidation reactions dominate and formations of alcohols, phenol, alkyl ethers, aldehydes, ketones, carboxylic acids, etc. occur, which increase the oxygen content of lignin fibers (*14*, *15*). At the later stages of thermostabilization (temperatures above 250°C), the oxygen content of lignin fibers decreases. Then, thermostabilized lignin fibers are converted to carbon fiber by carbonization. During high temperature thermostabilization and the initial stages of carbonization, various functional groups on lignin fibers undergo thermal dehydration, decarboxylation, and condensation reactions resulting in the formation of volatile H_2O, CO_2 and CO, which introduce microscale porosity in the material (*12*, *14*, *16*). Above 900°C, all functional groups of lignin are eliminated and a carbon backbone remains (*15*). During carbonization, lignin is converted into a carbon material with nearly perfect turbostratic disorder. The carbonization temperature determines the degree of turbostratic disorder.

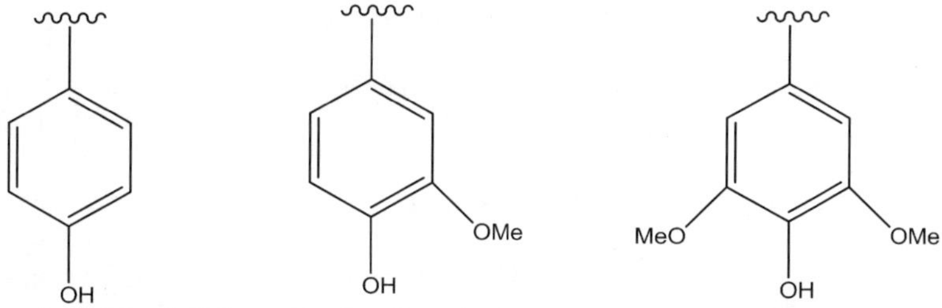

p-Hydroxyphenyl Unit (H) Guaiacyl Unit (G) Syringyl Unit (S)

Figure 1. Lignin monomeric units

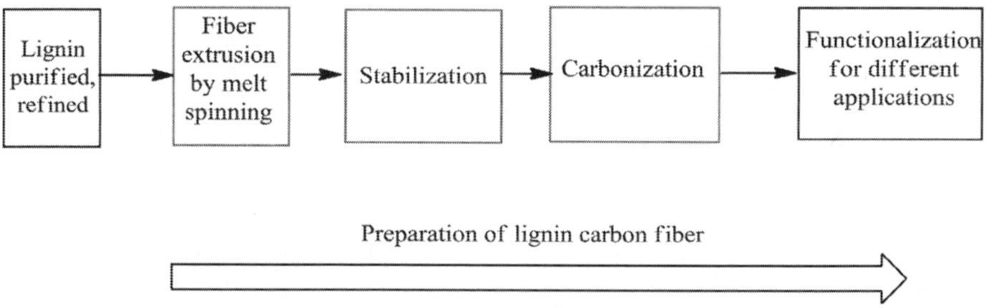

Scheme 1. Preparation of carbon fiber from lignin.

The properties of lignin-based carbon materials including carbon fibers are controlled by the chemical and physical properties of precursor lignin (*17*). Natural inhomogeneity (different distributions of monomeric units) of lignin is an important factor, which modulates the properties of carbon fibers. Additionally, types and purity of lignin play major roles in controlling the properties of carbon materials. Commercial lignins, such as those isolated from alkaline, acidic or organic solvent-based processes, generally undergo extensive fragmentation and degradation (*5, 6, 18*). As a result, a wide variety of lignins, in terms of chemical properties and structures are obtained. Also, most of these industrial lignins contain impurities. In order to control the properties of lignin-derived materials, precursor lignins are sometimes cleaned or pretreated using various methods. A large number of lignin pretreatment methods have been reported in literature where modifications to increase thermal stabilities, conductivity, adhesive and plasticizing properties of lignin derived carbon materials are explored (*19–21*). With proper purification and modification of precursor lignin and optimization of the processing parameters during stabilization and carbonization, stable carbon materials with controllable microstructure can be obtained, which is the key to the favorable properties of lignin carbon fiber as anode material for Li-ion battery.

Challenges with Graphite-Based Anodes in Li-Ion Batteries

The most commonly used anode material for Li-ion batteries is graphite, which can insert lithium reversibly (*22–25*). However, the use of graphite as an anode material has some disadvantages. Production of conventional graphitic anodes requires significant processing involving binders and current collectors, which makes the process expensive and time consuming. Additionally, electrolyte solutions may promote exfoliation and formation of an unstable solid electrolyte interface (SEI) resulting in loss of capacity, eventually resulting in battery failure (*26*).

These limitations drive the development of low cost high capacity anode materials that have increased cycling stability in a wide spectrum of electrolyte solutions. Over the years, many research efforts were aimed at developing alternative carbons that can insert Li ions at higher capacity compared to graphite and also behave more robust during repeated Li ion insertion and de-insertion in a wide spectrum of electrolytic solutions. Several processes for the preparation of new disordered carbon materials have been explored, including gas phase reaction of liquid propane gas (*27*), mesophase pitch based carbon fibers (*28*), chemical vapor deposited carbon cloth (*23*), etc. However, most of these methods require expensive chemicals and material processing and have limited economic viability. Interestingly, there is a broad consensus that a high level of microscale porosity in anode materials helps to make the electrode more robust and less affected by the solvent interactions (*23*, *27–29*). Thus, lignin carbon fibers with inherent microscale porosity and turbostratic disorder, offer some distinct advantages over graphite as an anode material. In addition, carbonization introduces macroscopic fiber-fiber fusions resulting in a monolithic, highly interconnected mat structure eliminating the need for a binder and current collector (*8*). Therefore, the use of lignin carbon fiber as an anode material not only reduces the number of steps and makes the process more economic but also offers stability.

Recently, a few studies reported the use of lignin for electrochemical applications (*30–32*). Initially, lignins were merely used as an additive to prepare composite materials with a lignin ratio of upto 50% (w/w). More recently, the performance of lignin-based carbon fibers as anode materials of Li-ion battery was evaluated. This review focuses recent developments in the use of lignin-based carbon fiber materials for Li-ion batteries.

Lignin Carbon Fiber Based Anode Materials

Wang et al. reported the use of fused free standing lignin carbon fiber mats as high performance anode materials for lithium ion batteries (*31*). In this work, nitrogen-doped, freestanding, fused and highly porous lignin fiber mats were obtained from an organosolv lignin-polyethylene oxide (PEO) (90:10) blend by electrospinning followed by carbonization and thermal annealing in the presence of urea. The addition of PEO facilitated the spinning of the material. The authors report that the addition of heteroatoms into the material further enhances the electrochemical performances of carbon fibers. Scanning electron microscope (SEM) revealed that the electrospun lignin fibers were free of beads with diameters ~ 1 μm (Figure 2, a-c). On thermal stabilization in the air and carbonization under argon, carbon fibers with a diameter ~ 500 nm were obtained (Figure 2, e-g). The use of PEO at high concentrations gave highly fused carbon fibers which offered shorter and continuous pathways decreasing the resistance of the electrode. The electrical conductivities of fused (Figure 2, d) and non fused carbon fibers were 10.53 S/cm and 7.34 S/cm respectively. Nitrogen-doped fused carbon fibers (Figure 2, h) showed even higher conductivity 12.24 S/cm. Interestingly, BET surface area of N-doped fused carbon fibers was lower than that of pure fused carbon fibers, which might be due to the merging of existing micropores rather than generation of new pores on the fiber surface.

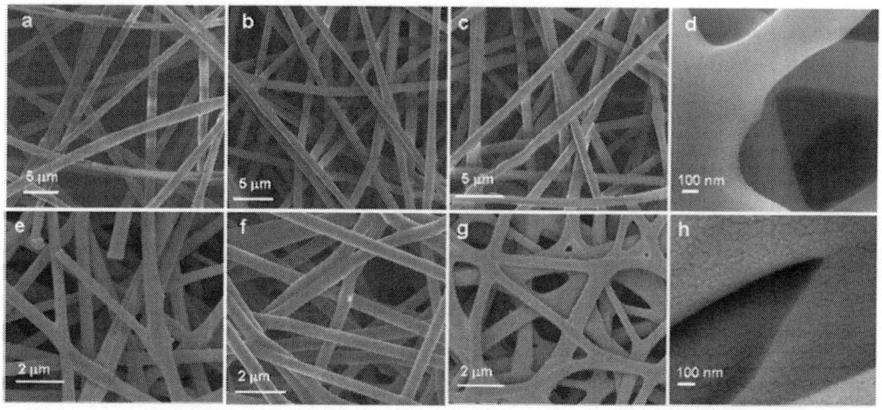

Figure 2. FESEM (Field Emission Scanning Electron Microscopy) images of electrospun nanofibers from lignin−PEO blends with lignin to PEO ratios of (a) 97:3, (b) 95:5, and (c) 90:10. FESEM images of electrospun carbon nanofibers derived from lignin−PEO blends with lignin to PEO ratio of (e) 97:3, (f) 95:5, and (g) 90:10. High resolution FESEM images show the surface of the carbon fibers (d) without and (h) with nitrogen doping. Reproduced from Ref. 31. Copyright (2013, American Chemical Society) (31).

The fused carbon fibers without nitrogen doping exhibited high specific capacities of up to ~ 445 mAh/g at a current density of 30 mA/g and good cyclic stability at different current rates (Figure 3a). The reported capacity is higher than graphite (~320 mAh/g) and comparable to electrospun nanofibers from polyacrylonitrile (~450mAh/g). In the case of the N-doped fibers, the charge capacity was increased to 576 mAh/g and maintained a good capacity of about 200 mAh/g at a high current rate of 2000 mA/g. Most importantly, N-doped carbon fibers showed higher columbic efficiency (extraction capacity/ insertion capacity) than non doped fibers (Figure 3b). The authors proposed that modification of carbon fibers with heteroatoms could prevent the formation of solid electrolyte interfaces (SEI).

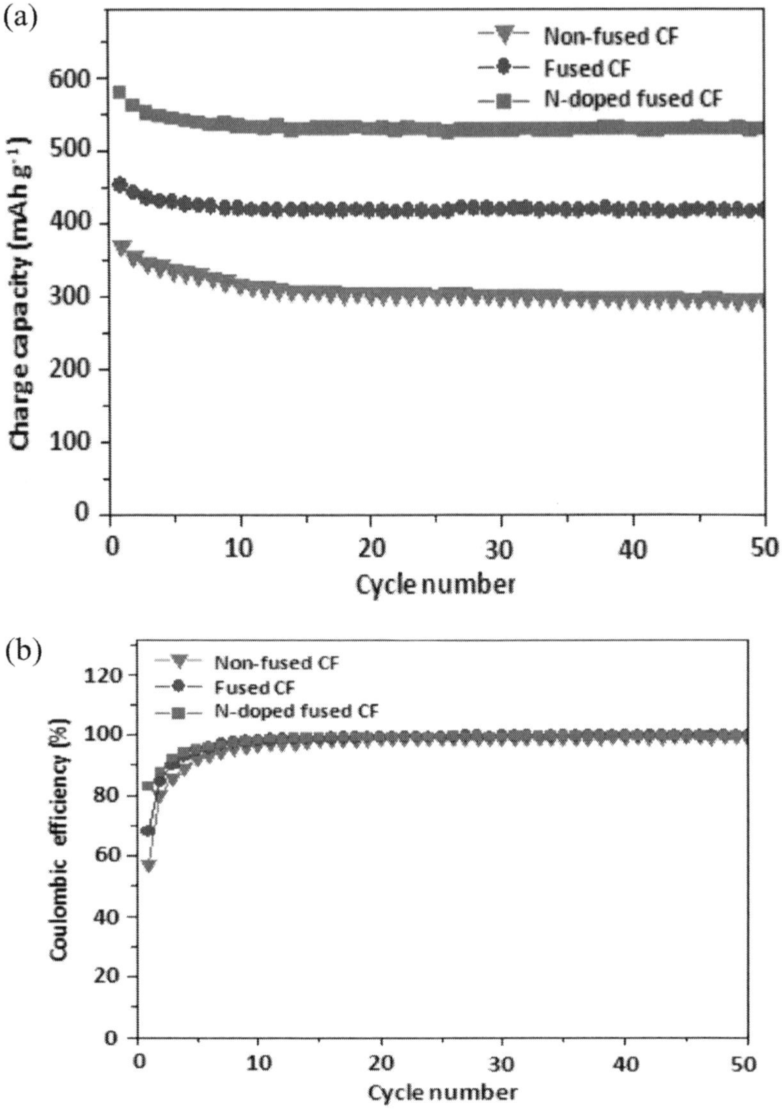

Figure 3. (a) Cyclic performance and (b) Coulombic efficiency of the lignin-derived carbon fiber electrodes cycled at a current density of 30 mA g^{-1}. Reproduced from Ref. 31. Copyright (2013, American Chemical Society) (31).

In another work, Chatterjee et al. developed a method to modify organosolv lignin with optimized properties for melt spinning and production of microstructured lignin carbon fibers suitable to be used as an anode material for lithium ion batteries (*32, 33*). The role of lignin chemistry and the impact of chemical modifications of functional groups on lignin structure and lignin carbon fibers were also investigated. In the work, a method was developed combining chemical modifications with melt processing techniques to obtain functional lignin carbon fibers. In this approach, chemical modification of specific functional groups in lignin was conducted to process lignin into a microstructured carbon material after oxidation and carbonization. Lignin was chemically modified by esterification using dicarboxylic acid anhydrides (acetic, phthalic, succinic and maleic anhydrides), which reacted with hydroxyl groups on lignin (Figure 4).

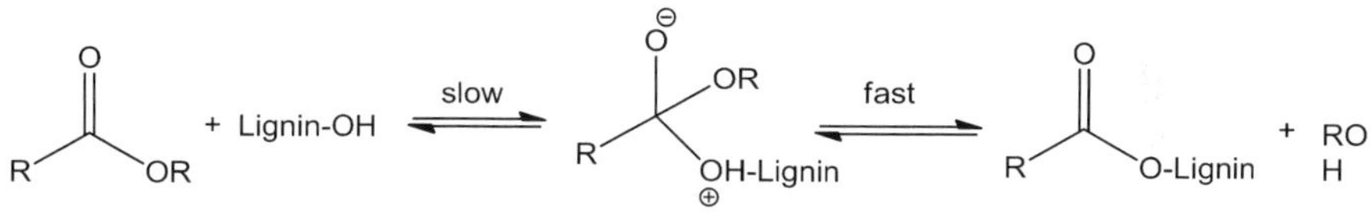

Figure 4. Possible nucleophilic acyl substitution mechanism.

As previously mentioned, during thermal processing, functional groups on lignin undergo thermal dehydration, decarboxylation, and condensation reactions forming volatile H_2O, CO_2 and CO (*12, 14, 16*). With modified lignin, additional ester groups modulate these reactions which in turn facilitate the development of high micro scale porosity and turbostratic disorder in carbon fibers.

The reaction of acid anhydrides and lignin hydroxyl group is an acyl substitution reaction under neutral conditions. Reactions began with a slow equilibrium addition of the nucleophile to the carbonyl carbon (Figure 4). This was subsequently followed by a rapid departure of the acetoxy part of the corresponding acid anhydride to form the nucleophilic acyl substitution product. The structural and thermal properties of lignin resulting from modification with each anhydrides control the microstructure development during oxidation and carbonization. Esterification with different acid anhydrides successfully modified the chemical structure of lignin, which was confirmed by ^{13}C NMR analysis (Figure 5). Characteristic spectral features between 10-40 and 100-200 ppm represented the methyl and carbonyl carbons of resulting esters, respectively.

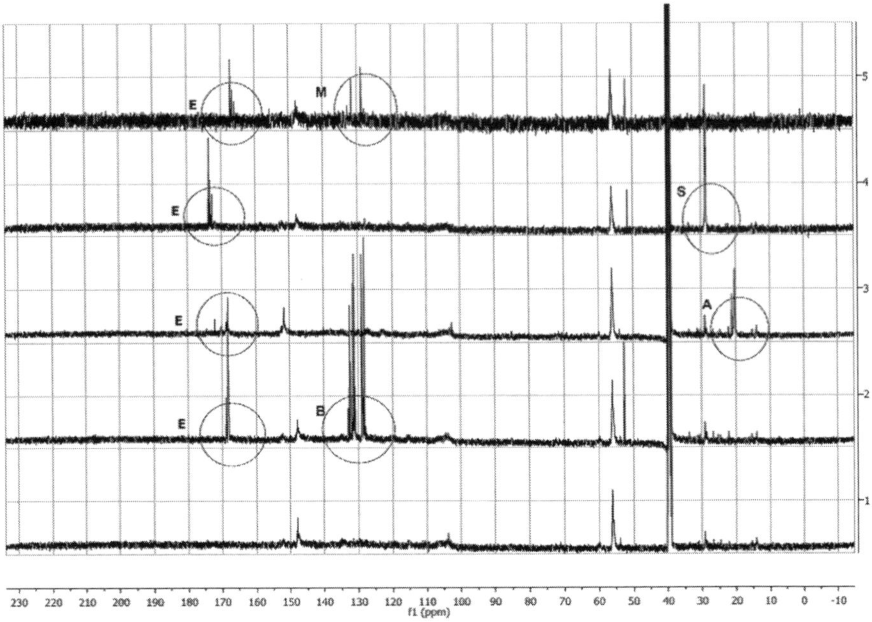

Figure 5. ^{13}C NMR Spectra of unmodified and modified lignin. The spectra are arranged as (1) unmodified lignin, (2) phthalic anhydride modified lignin, (3) acetic anhydride modified lignin, (4) succinic anhydride modified lignin and (5) maleic anhydride modified lignin. All modified lignins showed characteristic ester carbonyl peak, which is indicated by 'E'. In cases of acetic and succinic anhydride-modified lignins, changes in 10-30 ppm are indicated by 'A' and 'S', respectively. In cases of phthalic and maleic anhydride modified lignins, new peaks are shown 'B' and 'M'. Reproduced from Ref. 33. Copyright (2014, The Royal Society of Chemistry) (33).

Esterification of lignin also altered the thermal properties of lignin. Thermogravimetric analysis (TGA) measures the amount and rate of change in the mass of a sample as a function of temperature or time in a controlled atmosphere (Figures 6 and 7). TGA plots of all lignin samples were divided into two ranges; 1) 0 to 250°C corresponds to oxidation in air and 2) 250 to 800°C corresponds to carbonization of lignin samples under an inert atmosphere. During the initial stage of oxidation a small weight loss was observed, which reflected a loss of residual solvent or reagent from the sample. However, temperatures corresponding to 50% weight loss of modified lignin samples during carbonization were higher than that of the unmodified lignin. The higher thermal stability of modified lignin might be a result of partial cross linking of lignin polymers. In addition, the derivatives of weight loss profiles of most modified lignins showed more uniformity during carbonization than that of the unmodified lignin.

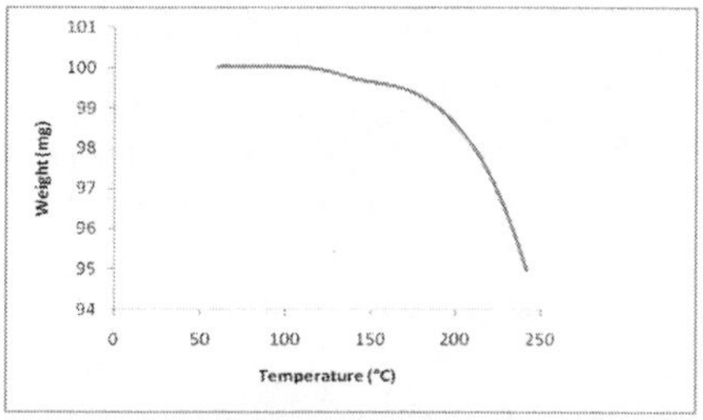

(Unmodified)

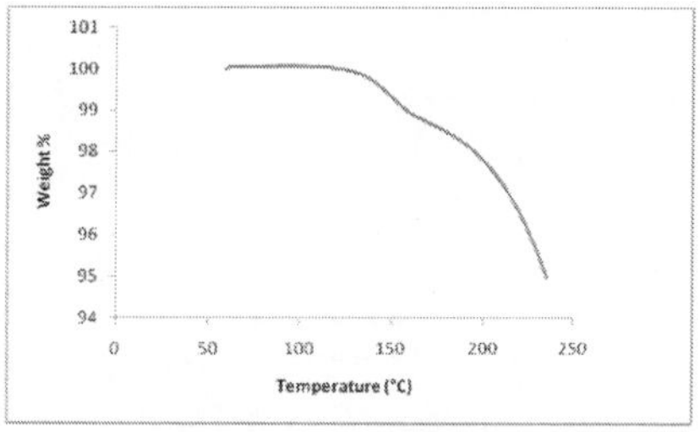

(Phthalic modified)

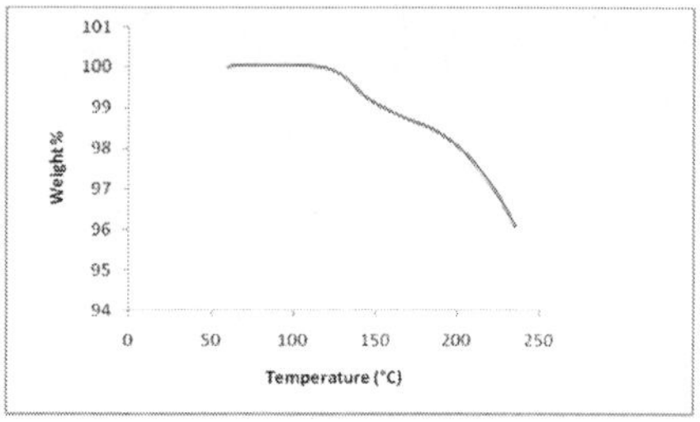

(Succinic modified)

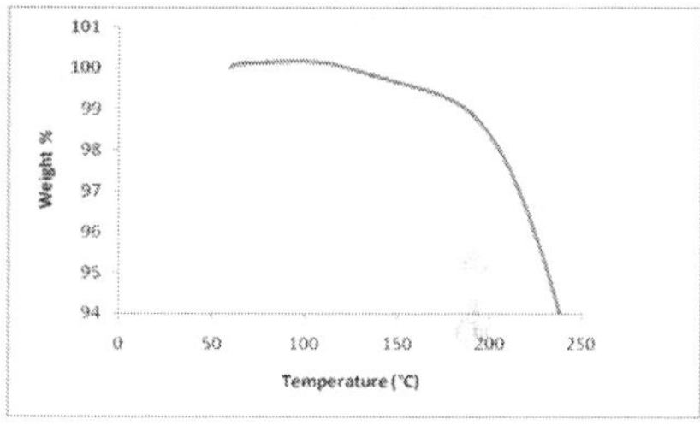

(Acetic modified)

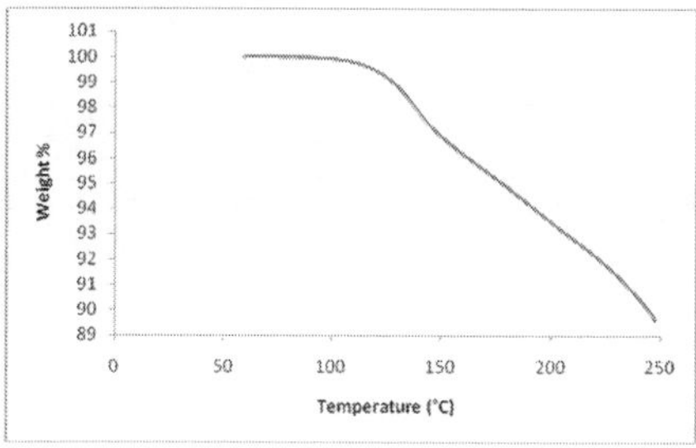

(Maleic modified)

Figure 6. Thermogravimetric analyses (TGA) plots of unmodified and modified lignin under oxidizing conditions. Reproduced from Ref. 33. Copyright (2014, The Royal Society of Chemistry) (33).

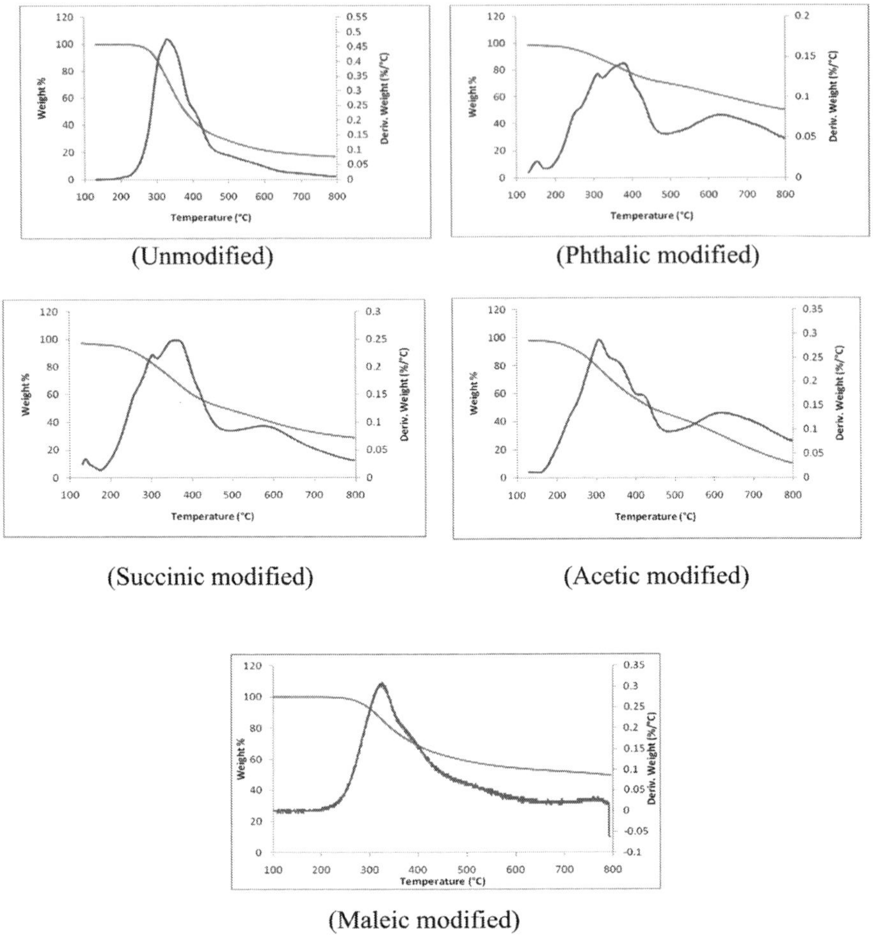

Figure 7. Thermogravimetric analyses (TGA) plots of unmodified and modified lignins under carbonizing conditions. Reproduced from Ref. 33. Copyright (2014, The Royal Society of Chemistry) (33).

As expected, carbon fibers derived from most of the modified lignins exhibited higher BET surface area than that of the unmodified lignin derived fiber (Table 1). Interestingly, in TGA analysis; corresponding lignin precursors also showed relatively uniform weight loss trends compared to the unmodified lignin. All surface areas were found to be in the range of commercial graphitic carbon materials used in anode.

Table 1. BET surface areas pore volumes of carbonized oxidized and carbonized unmodified and modified lignins[a]. Adapted from Ref. 33. Copyright (2014, The Royal Society of Chemistry) (*33*).

Carbon type	*BET Surface area (m^2/g)*	*Adsorbed pore volume (cm^3/g)*
Phthalic Anhydride modified	32.7230 (1.5566)	0.0133 (0.0009)
Acetic anhydride modified	37.4293 (1.2133)	0.0151 (0.0006)
Succinic anhydride modified	15.3236 (1.8268)	0.0077 (0.0008)
Maleic Anhydride modified	6.3865 (1.2363)	0.0018 (0.0007)
Unmodified	8.1976 (0.7089)	0.0042 (0.003)

[a] Standard deviation in each case is listed inside bracket

Scanning electron microscopy (SEM) images of carbon fibers derived from modified lignins (Figure 8) indicated that chemical modification of the lignin precursor produced a heightened degree of sub-micron porosity in lignin carbon fibers. Most fiber cross sections displayed a cylindrical hollow void, coincident with the fiber's axial direction with a diameter of 10-25 μm. The increased microscale porosity increases battery capacity and ion diffusion rates while these voids decrease the tensile and shear strengths of the fiber. Most importantly, the porous network of stable graphitic structures enhances resistance to exfoliation and facilitates rapid exchange of lithium ions between the electrolyte and the graphitic carbon of the anode.

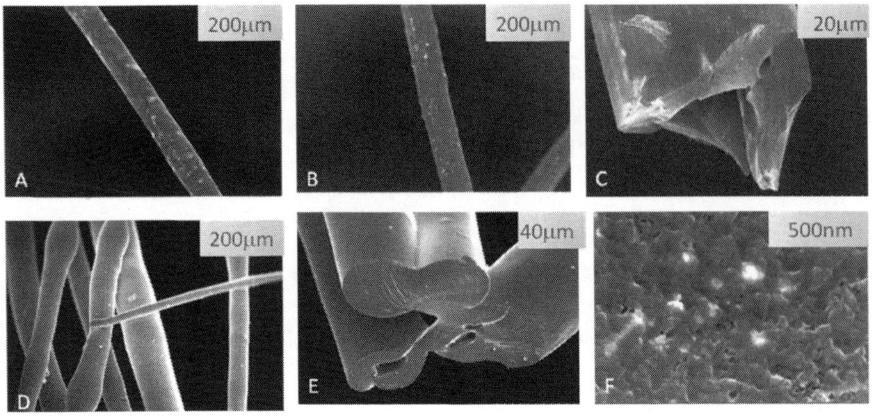

Figure 8. Scanning electron microscope (SEM) images of lignin carbon fibers. Where, A= unmodified Alcell hardwood lignin (AHL); B = phthalic anhydride modified AHL; C = phthalic anhydride modified AHL, showing cylindrical hollow; D = acetic anhydride modified AHL; E = fiber-fiber fusion and F = high resolution image of phthalic anhydride modified AHL, showing submicron porosity. Reproduced from Ref. 32. Copyright (2014, American Chemical Society) (32).

Finally, the resistivity measurement of modified carbon fibers showed resistivity in the range of 10^{-5}-10^{-4} Ωm, which was comparable to typical values for graphite (10^{-6}-10^{-5} Ωm) (*30*). Thus, all lignin carbon fiber samples exhibited sufficient conductivity for use as anode materials.

In another study, Tenhaeff at al. studied the electrochemical performance of organosolv lignin derived carbon fibers as anode materials (*34*). This work involved unmodified lignin. In this work, lignin was converted to three dimensional, electrically interconnected networks by air driven melt processing technique. The lignin fibers were then thermally stabilized in air. By controlling the temperature and rheological properties during stabilization process, fused, self standing porous mats (Figure 9) were obtained which were subsequently converted to a carbon fiber-based monolithic structure by pyrolyzing in an inert atmosphere at 600°C. To compare the electrochemical performance of these materials; two electrode designs were selected. The first design used monolithic fused lignin carbon fiber mats.

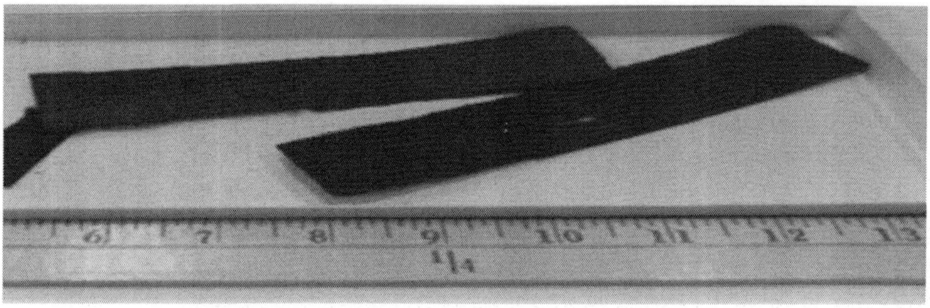

(a)

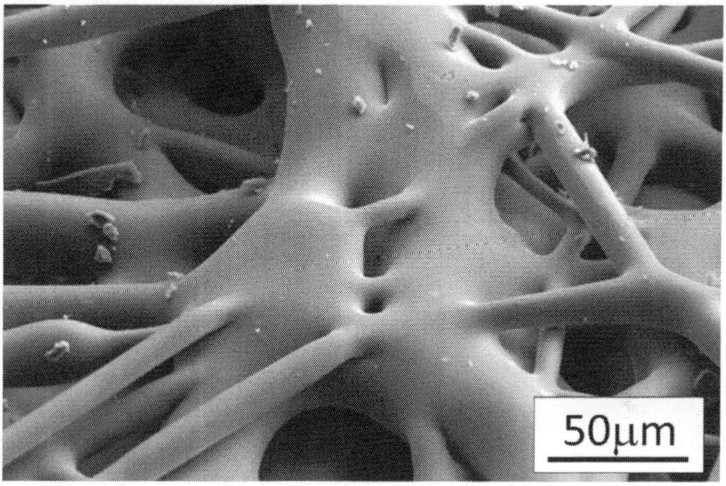

(b)

Figure 9. a) digital image of lignin carbon fiber fused mats after carbonization, b) SEM image of fiber fiber fusion in lignin carbon fiber mats. Reproduced from Ref. 34. Copyright (2014, John Wiley and Sons) (34).

These mats were interconnected and thus, acted as both active material and current collector. In the second design, powdered lignin carbon fibers were added to a conductive additive and a polymeric binder and the resulting slurry was coated on the copper current collector. Both electrode designs involving lignin carbon fiber showed high columbic efficiency. The second design (conventional electrode design consisting lignin carbon fibers) carbonized at 1000°C showed high reversible charge capacity (approximately 350 mAhg^{-1}) (Figure 10) comparable to the capacity of industry standard graphite (372 mAhg^{-1}). Lignin carbon fiber mats, on the other hand showed low capacity. Interestingly, lignin carbon fibers carbonized at 2000°C, showed good cyclic stability over 40 cycles in solution of $LiPF_6$ in polyethylene carbonate, which is a preferred electrolyte constitutes for low temperature application. Graphite, in contrast, rapidly exfoliates due to solvent intercalation in polyethylene carbonate. These results were obtained without tuning the battery design or material processing parameters. Thus, with proper optimization of the whole process and tuning of the battery design, excellent performance could be obtained. For example, the use of chemically modified lignin would facilitate optimization.

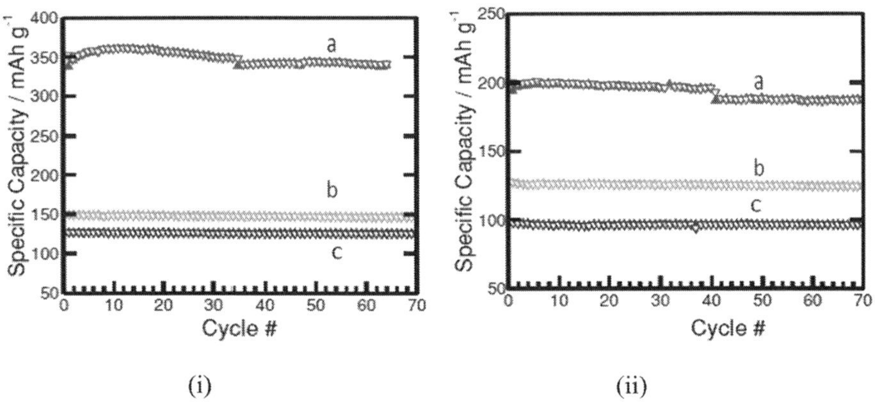

Figure 10. (i) Charge capacities as a function of cycling and carbonization temperatures (1000 (a), 1500 (b), and 2000°C (c) in a) slurry-coated electrode and ii) fused mats. Reproduced from Ref. 34. Copyright (2014, John Wiley and Sons) (34).

Conclusions

Technologies that incorporate abundant and environmentally sustainable lignin as a precursor are rapidly developing. Recent insights into the chemistry of converting lignin to carbon fibers have led to the development of a new class of anode materials with significant promise for low cost energy storage solutions with energy densities. Li-ion batteries are currently considered the most efficient and economically viable energy storage system enabling new environmentally friendly technologies, such as ultra mobile electronic devices, power tools, electric vehicles, etc. The unique characteristics of lignin carbon fiber-based anode materials and their superior efficiency in high density applications combined with lower production costs make it a viable alternative to graphite anodes; in particular, for lowering the cost of energy storage systems is the automotive industry. In recent years, a fundamental understanding of lignin precursor chemistry and the structure of lignin carbon fibers has been obtained. However, several challenges remain, which are associated with quality and consistency of lignin precursor chemistry. In analogy to the petrochemical industry, standardized methods for extraction and analysis of lignin from lignocellulosic biomass are required to ensure a continuous supply of lignin precursors with consistent purity and chemical composition. The precursor chemistry determines the outcome of the conversion process and the degree of porosity and turbostratic disorder in lignin carbon fibers. There is also significant potential for further improvements in charge capacities and cycle stability by tuning the manufacturing process and lignin precursor composition.

References

1. Baker, D. A.; Rials, T. G. *J. Appl. Polym. Sci.* **2013**, *130*, 713–728.
2. Dhillon, G. S.; Gassara, F.; Brar, S. K.; Verma, M. In *Lignin properties and applications in biotechnology and bioenergy*, 1st ed.; Paterson, R. J., Ed.; Nova Science Publishers Inc.: New York, 2012; pp 1−34.
3. Saito, T.; Brown, R. H.; Hunt, M. A.; Pickel, D. L.; Pickel, J. M.; Messman, J. M.; Baker, F. S.; Keller, M.; Naskar, A. K. *Green Chem.* **2012**, *14*, 3295–3303.
4. Zakzeski, J.; Bruijnincx, P. C. A.; Jongerius, A. L.; Weckhuysen, B. M. *Chem. Rev.* **2010**, *110*, 3552–3599.
5. Bozell, J. J.; O'Lenick, C. J.; Warwick, S. *J. Agric. Food Chem.* **2011**, *59*, 9232–9242.
6. Harton, S. E.; Pingali, S. V.; Nunnery, G. A.; Baker, D. A.; Walker, S. H.; Muddiman, D. C.; Koga, T.; Rials, T. G.; Urban, V. S.; Langan, P. *ACS Macro Lett.* **2012**, *1*, 568–573.
7. Ragauskas, A. J.; Beckham, G. T.; Biddy, M. J.; Chandra, R.; Chen, F.; Davis, M. F.; Davison, B. H.; Dixon, R. A.; Gilna, P.; Keller, M.; Langan, P.; Naskar, A. K.; Saddler, J. N.; Tschaplinski, T. J.; Tuskan, G. A.; Wyman, C. E. *Science* **2014**, *344*, 709–719.
8. Baker, D. A.; Gallego, N. C.; Baker, F. S. *J. Appl. Polym. Sci.* **2012**, *124*, 227–234.
9. Baker, F. S.; Griffith, W. L.; Compere, A. L. In *Process Report*; Oak Ridge National Laboratory: 2005.
10. Sudo, K.; Shimizu, K. *J. Appl. Polym. Sci.* **1992**, *44*, 127–134.
11. Luo, J.; Genco, J.; Cole, B.; Fort, R. *BioResources* **2011**, *6*, 4566–4593.
12. Braun, J. L.; Holtman, K. M.; Kadla, J. F. *Carbon* **2005**, *43*, 385–394.
13. Chung, D. L. *Carbon Fiber Composites*; Butterworth-Heinemann: New York, 1994.
14. Matsumoto, T.; Mochida, I. *Carbon* **1992**, *30*, 1041–1046.
15. Yang, C. Q.; Simms, J. R. *Carbon* **1993**, *31*, 451–459.

16. Drbohlav, J.; Stevenson, W. T. K. *Carbon* **1995**, *33*, 693–611.
17. Saito, T.; Perkins, J. H.; Vautard, F.; Meyer, H. M.; Messman, J. M.; Tolnai, B.; Naskar, A. K. *ChemSusChem* **2014**, *7*, 221–228.
18. Sarkanen, S.; Teller, D. C.; Hall, J.; McCarthy, J. L. *Macromolecules* **1981**, *14*, 426–434.
19. Xiao, B.; Sun, X. F.; Sun, R. *Polym. Degrad. Stab.* **2001**, *71*, 223–231.
20. Sudo, K.; Shimizu, K.; Nakashima, N.; Yokoyama, A. *J. Appl. Polym. Sci.* **1993**, *48*, 1485–1491.
21. Rozman, H. D.; Tan, K. W.; Kumar, R. N.; Abubakar, A. *Polym. Int.* **2001**, *50*, 561–567.
22. Liu, Y.; Xue, J. S.; Zheng, T.; Dahn, J. R. *Carbon* **1996**, *34*, 193.
23. Isaev, I.; Salitra, G.; Soffer, A.; Cohen, Y. S.; Aurbach, D.; Fischer, J. *J. Power Sources* **2003**, *119–121*, 28.
24. Broussely, M.; Biensan, P.; Simon, B. *Electrochim. Acta* **1999**, *45*, 3–22.
25. Dahn, J. R.; Zheng, T.; Liu, Y.; Xue, J. S. *Science* **1995**, *270*, 590–593.
26. Isaev, I.; Salitra, G.; Soffer, A.; Cohen, Y. S.; Aurbach, D.; Fischer, J. *J. Power Sources* **2003**, *119*, 28–33.
27. Han, Y. S.; Yu, J. S.; Park, G. S.; Lee, J. Y. *J. Electrochem. Soc.* **1999**, *146*, 3999–4004.
28. Takami, N.; Satoh, A.; Hara, M.; Ohsaki, T. *J. Electrochem. Soc.* **1995**, *142*, 371–379.
29. Fey, G. T. K.; Cho, Y.D.; Chen, C.L.; Huang, K.P.; Lin, Y.C.; Kumar, T.P.; Chan, S.H. *Int. J. Chem. Eng. Appl.* **2011**, *2*, 20–25.
30. Milczarek, G.; Inganäs, O. *Science* **2012**, *335*, 1468–1471.
31. Wang, S.-X.; Yang, L.; Stubbs, L. P.; Li, X.; He, C. *ACS App. Mater. Interfaces* **2013**, *5*, 12275–12282.
32. Chatterjee, S.; Jones, E. B.; Clingenpeel, A. C.; McKenna, A. M.; Rios, O.; McNutt, N. W.; Keffer, D. J.; Johs, A. *ACS Sustainable Chem. Eng.* **2014**, *2*, 2002–2010.
33. Chatterjee, S.; Clingenpeel, A.; McKenna, A.; Rios, O.; Johs, A. *RSC Adv.* **2014**, *4*, 4743–4753.
34. Tenhaeff, W. E.; Rios, O.; More, K.; McGuire, M. A. *Adv. Funct. Mater.* **2014**, *24*, 86–94.

Chapter 12

Synthesis of Nanoparticles for Biomass Conversion Processes

Jared T. Wabeke,[1] Hazim Al-Zubaidi,[1] Clara P. Adams,[1] Liyana A. Wajira Ariyadasa,[1] Setare Tahmasebi Nick,[1] Ali Bolandi,[2] Robert Y. Ofoli,[2,*] and Sherine O. Obare[1,*]

[1]Department of Chemistry and the Nanotechnology Research and Computation Center, Western Michigan University, Kalamazoo, Michigan 49008, United States
[2]Department of Chemical Engineering and Materials Science, Michigan State University, East Lansing, Michigan 48824, United States
[*]E-mail: ofoli@egr.msu.edu; sherine.obare@wmich.edu

The design and synthesis of well-defined metallic and bimetallic nanoparticles (NPs), their characterization, and the assessment of how their size and shape-dependent properties influence their applications are important areas of investigation for advancing green technologies that protect the environment. This chapter reviews recent advances in the design and synthesis of metallic, bimetallic and semiconductor nanoparticles and their emerging applications in the production of energy and chemicals from biorenewable materials. This contribution focuses on nanoparticle-mediated processes for biomass transformation that include hydrogenation, hydrogenolysis, decarboxylation, small molecule oxidation, conversion of cellulosic materials, hydrocarbon formation, and production of fuel cells. These processes all have significant potential for development of green technologies.

Introduction

The search for viable alternatives to reduce the dependence on fossil fuels for both energy and chemicals production is a significant 21st century challenge. Biorenewables have the potential to replace fossil fuels as a significant source of raw materials for fuels and chemical production, and has therefore become an increasingly important and active area of research globally. Biomass materials are a viable alternative to fossil fuels due to their renewability, sustainability and availability. However, the chemistry involved in obtaining products from biomass relative to crude oil presents a number of challenges as well as a variety of opportunities. The challenges lie in the physical properties of crude oil relative to biomass. While the former is volatile, non-polar and consists mostly of unfunctionalized hydrocarbons, the latter is non-volatile, polar and consists of molecules that are highly functionalized. As a result of these differences in physical properties, well-established techniques used for petroleum processing are unsuitable for biomass conversion, creating an urgent need to develop alternative methods.

Biomass can be categorized into two types of materials: (1) oleaginous feedstock and (2) carbohydrates (*1*). The former constitutes a minor portion (less than 5%) of biomass, and includes triglycerides, free fatty acids (FFA), and oils (*1*). Triglycerides make up the largest portion of oleaginous feedstock and have some of the highest hydrogen to carbon (H/C) ratios, combined with relatively low amounts of oxygen (*2*). Free fatty acids (FFA) are present mainly in low-grade, processed oleaginous feedstock after the hydrolysis of triglycerides. Sources of oils and fats include vegetable oils (including soybean, rapeseed and palm oil), food industry waste streams, and algae (*3–5*). Carbohydrates make up the largest portion (~75%) of the biomass available for conversion. This group of compounds includes mono- and polysaccharides. Monosaccharides are less abundant in nature, but can be found in plants such as sugarcane and sugar beet. Generally, monosaccharides are used for the production of ethanol via fermentation.

Polysaccharides include a large list of molecules including starch, cellulose and hemicellulose. Starch is very useful in the production of monosaccharides since it is easily broken down into glucose or other sugars. Cellulose is similar in nature to starch; however, cellulose has a linear structure and thus forms long strands and rigid bundles of plant material. This makes cellulose much more difficult to break down, thus requiring more advanced techniques for its processing. Biomass comprising polysaccharides is generally found in a mixture of starches, cellulose and hemicellulose, and falls under the classification of lignocelluloses (*1*).

The energy density of fuels from biomass must be increased to make biomaterials acceptable substitutes for crude oil. Crude oil contains a much higher quantity of carbon-hydrogen bonds and has very low oxygen content when compared to accessible biomass (*1–6*). Crude oil also has a higher density than accessible biomass. Biomass densification, procedures such as compaction or chemical modifications, are used to increase the energy density of lignocellulosic biomass (*6*, *7*). Chemical modifications can be considered pretreatment steps that increase the density and, more importantly, the energy density of biomass feedstock. The focus of most pretreatments is the reduction of the moisture content of biomass, which generally increases the energy density of the raw materials (*8–11*). Other methods used for the pretreatment of lignocellulose include pyrolysis, the direct thermal decomposition of biomass feedstock (*12*), and the controlled synthesis of useful molecules referred to as platform molecules (*13*). Pyrolysis treatment leads to the production of bio-oils comprising 15-30 wt% aqueous phase and 35-50 wt% non-aqueous phase, with the remainder comprising aromatic compounds such as benzene and toluene (*14*). The energy content of bio-oils depends on the composition of the biomass initially used in the thermal decomposition. The production of platform molecules includes processes such as dehydration of sugars (*15–17*), depolymerization of lignocellulosic materials into simple sugars (*18*), and the production of levulinic acid and γ-valerolactone (*19*, *20*). Neither of these processes yields a direct replacement for conventional fuels, but each produces biomass with a high energy density for use in refinery processes.

Oleaginous materials can be processed using a few different methods, and studies on converting them to conventional fuels are an ongoing area of active research (*21–31*). For example, fluid catalytic cracking can be used to produce fuels from vegetable oils, animal fats and waste oils or fats. There are several reactions that occur in the cracking process, including hydrogen transfer, coking reactions, isomerization and cracking of alkanes, alkenes, naphthenes, and alkyl aromatics (*32*). Typically, the catalytic cracking of oleaginous materials is carried out over acid catalysts such as zeolite (*21–24*), aluminum-containing materials (*25–29*), and amorphous materials (*30*, *31*). The choice of catalyst is important in determining the resulting products. A zeolite catalyst tends to produce more of the desired organic liquid products, which comprise hydrocarbons similar in boiling point to gasoline, kerosene and diesel; on the other hand amorphous catalysts tend to produce higher amounts of gases (*33–35*). Typically, conversion reactions involving catalytic cracking are performed over temperature ranges of 300 to 500°C, with conversions over 80% (*33–35*). However, other studies have been conducted under milder reaction conditions (*36*, *37*).

Hydrotreatment is used in the refining process to remove sulfur, nitrogen, metals and oxygen from feedstock. Reaction conditions typically require temperatures of 300 to 450°C and pressures of 35 to 170 bar. In addition, hydrogen is necessary for the process and can be costly, due to the high amounts of oxygen present in biomass. Hydrotreatment yields reaction products that are indistinguishable from those of petroleum-based diesel fuels, thus making this process very valuable (*38*, *39*). However, side reactions decrease the production of green diesel, the fuel produced through biomass conversion (*40–42*). Unfortunately, most biomass comprises sugars and polymers, which require more complex processes for their conversion due to the low energy densities of these types of biomass (*1*).

Catalytic deoxygenation of free fatty acids (FFA) is also used for the production of biofuels. This process is currently of research interest because it requires much less hydrogen than hydrotreatment (*43–45*). Catalysts used for the deoxygenation process are generally palladium (Pd), which leads to decarboxylation; and platinum (Pt), which leads to decarbonylation, usually utilizing low hydrogen pressures (*43–48*). The refinement of these catalysts is important to increasing the efficiency of both processes. Most of the hydrogen in this process is used to saturate the fatty acids not consumed for deoxygenation, which increases the energy density of the material (*49*).

Lignocellulosic feedstock can be refined using pyrolysis techniques into bio-oil for use in refineries. Bio-oils are commonly chemically transformed into hydrocarbons using zeolite catalysts (*50*). Currently, FCC processes used on bio-oils cause a separation as they approach critical working temperatures. The processes that have been shown to occur include cracking, polymerization, coke formation, oligomerization, cyclization, and isomerization. The reactions that occur can be moderated using different types of catalysts (*51–53*). One way to increase the energy density of bio-oil is hydrotreating prior to its use in refineries. The hydrotreating process decreases the oxygen content of the bio-oil, which decreases the side products formed in the FCC process. Lignocellulose has also been refined to sugars for use in fuel production. Sugars are generally used in the fermentation process to produce ethanol, a fuel with low energy density. However, sugars are also being used to produce platform molecules, which are then used to synthesize target fuel molecules (*54*, *55*). Sugars can also be directly converted into hydrocarbon fuels. These fuels are directly compatible with current gasoline and diesel fuels and have high energy densities. To form hydrocarbons, sugars need to be dehydrated and then reduced to increase the degree of saturation in the compounds. Most techniques require large amounts of hydrogen in order to produce the desired products (*56–58*). Aqueous phase reforming (APR) is an alternative to procedures that require high use of hydrogen (*59–62*). This process allows for the production of hydrogen gas in the liquid phase, which is much cleaner than techniques using hydrocarbons.

Recently, nanoparticles have played a significant role in the conversion of biorenewables to high value-added chemicals and fuels. Nanoparticles offer a high surface area to volume ratio, robustness, high catalytic activity, tunable morphology, unique chemical and physical characteristics relative to their bulk counterparts, and the ability to modify their surfaces, providing unique opportunities for the catalytic conversion of biomass. Many types of nanoparticles, including metals, metal oxides and semiconductors, are used in the conversion of biorenewables for the production of fuels and other usable chemicals. This chapter presents key examples of the use of nanoparticles in the areas of hydrogenation, cellulose conversion, small molecule oxidation, production of liquid fuels, and fatty acid decarboxylation. It also discusses the role of nanoparticles in the design of fuel cells.

Synthesis of Nanoscale Materials Relevant for Biomass Conversion

The last two decades have seen tremendous progress toward development of synthetic procedures for various particles with nanoscale dimensions. Metallic, bimetallic and semiconductor nanoparticles have each played a significant role in the conversion of biomass into both chemicals and fuels. The shape, size, morphology and dispersity of nanoparticles govern their chemical and physical properties as well as their reactivity. Therefore, it is essential that synthetic procedures be designed to produce well-defined nanoparticles. The synthesis of nanoparticles involves control of the nucleation and growth processes (*63*). Here we provide a general overview of synthetic procedures for metallic, bimetallic, and semiconductor nanoparticles that are relevant for biomass conversion.

Synthesis of Metallic Nanoparticles

Wet-chemical synthesis procedures are the most commonly used methods for the preparation of metallic nanoparticles. The following are examples of wet-chemical methods designed to produce the most widely used metal nanoparticles for biomass conversion. The synthesis of zerovalent metal nanoparticles requires the selection of an appropriate metal precursor, a reducing agent or reduction process, and a stabilizer to coat the particle surface (*64*). The most common precursors for synthesis of metallic nanoparticles are inorganic salts or inorganic complexes. The stabilizer is used to control particle growth, size, and shape, and to reduce the tendency for particle aggregation. The reducing agent, whether it is strong or mild toward a specific reaction, must also be carefully chosen as it governs the rate of the nucleation and growth processes (*65–67*). The synthesis of metallic nanoparticles can be carried out in solution or in a water-in-oil emulsion (*68–71*). The selection of the method is based on the precursors used as well as the nature of the desired nanoscale particles.

Wet-chemical processes rely on the control of the nucleation and growth processes. The pioneering work of Ostwald (*72*) and Turkevich (*73*) provided an improved understanding of the processes that lead to the formation of metal nanoparticles. Bönnemann's important work also showed that there is a need for a metal salt precursor, a stabilizing agent, a reducing agent, and a suitable solvent for the preparation of metal nanoparticles (*74*). Based on Bönnemann's work, the formation of metallic nanoparticles can be represented by the equation:

$$MX_n(NR_4)_m + (n\text{-}m)Red \rightarrow M_{NP} + (n\text{-}m)(Red^+X^-) + m(NR_4^+X^-) \qquad [1]$$

where M represents a metal, X is the anion from the metal precursor, R is an alkyl group, and Red is the reducing agent. The metal atoms are reduced to the zerovalent state, leading to the nucleation and growth of nanoparticles.

The most common metals for biomass conversion include palladium (Pd), platinum (Pt) and ruthenium (Ru). There are several methods for the synthesis of Pd nanoparticles (*75–82*), and various supports have been used to immobilize Pd NPs such as carbon, mesocellular foams, carbon nanotubes, and silica. Particles can be prepared in the presence of the supports, or *ex-situ* and then immobilized on supports. Inductive coupled plasma atomic emission spectroscopy (ICP-AES) is then usually used to determine the metal content of particles immobilized on the supports. Sonochemical methods have also been used in the synthesis of NPs, including Pd (*77–79*). A notable example was reported by Dhas and Gedanken, where they used a sonochemical process to prepare Pd NPs with a diameter of 10-20 nm (*77*). In this reaction, palladium acetate was combined with a stabilizer, myristyltrimethylammonium bromide, in a 1:2 mole ratio and sonicated for 30 minutes to produce the Pd NPs. The reaction was carried out in tetrahydrofuran or methanol solvent (*77*).

Our group developed a facile procedure for the synthesis of well-defined Pd NPs that were stabilized by thioether ligands (*80–82*). The synthetic procedure included the pyrolysis of palladium acetate [$Pd_3(OAc)_6$] in toluene, using *n*-dodecyl sulfide as the stabilizing agent. The pyrolysis process did not require the addition of a chemical reducing agent. Monodisperse Pd NPs with an average diameter of 1.7 ± 0.2 nm, 1.9 ± 0.2 nm, 2.5 ± 0.1 nm, 3.5 ± 0.1 nm, and 4.1 ± 0.1 nm were obtained by controlling the reaction parameters, in this case, reaction time and temperature. Figure 1 shows the transmission electron microscopy (TEM) images of Pd NPs obtained by this procedure. Other thioether ligands, including biotin, have proven to be highly effective in the synthesis of Pd NPs using the straightforward pyrolysis procedure, leading to the formation of monodisperse particles (*81*). Furthermore, the reaction was not limited to using [$Pd_3(OAc)_6$]. In fact, using [$PdCl_4$]Na_2 as a metal precursor was also effective in the presence of polar solvents and using hydrazine as the reducing agents (*80*).

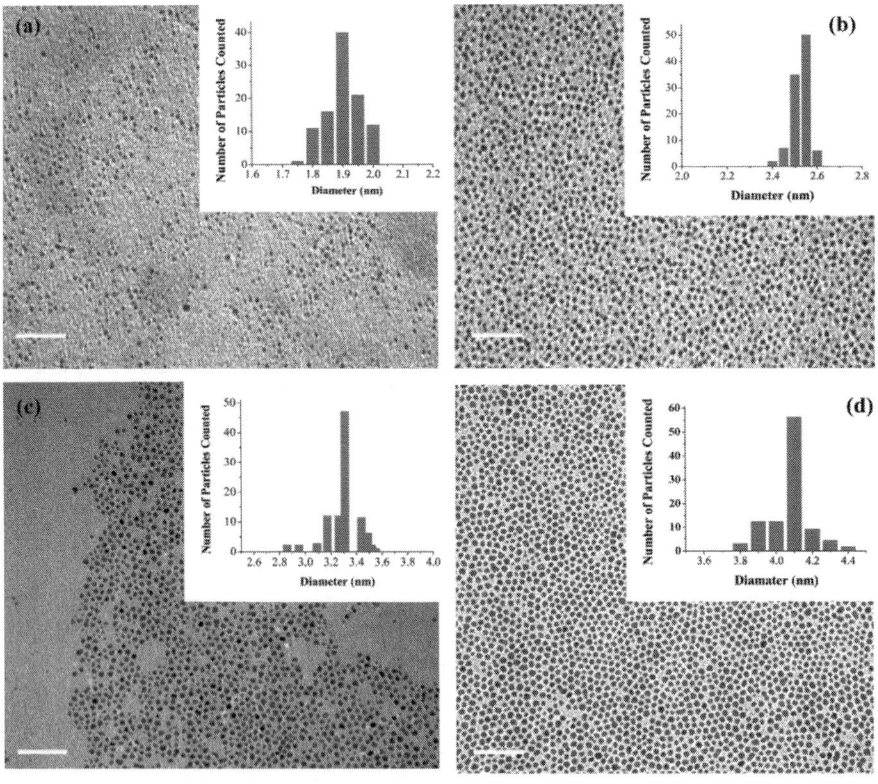

Figure 1. TEM images for Pd NPs with size of (a) 1.9 ± 0.2 nm (b) 2.5 ± 0.1 nm (c) 3.3 ± 0.1 nm, and (d) 4.1 ± 0.1 nm. Reproduced from reference listing (80). Copyright 2007, ACS.

Oftentimes, however, when nanoparticles are synthesized in an organic solvent, their use in green processes may be limited. Thus, if the preparation of metal NPs in an aqueous solvent does not lead to well-defined particles, NPs could be prepared in different medium and transferred to the aqueous phase without compromising the stability of the particles. In our work, we showed that a ligand exchange process was effective in removing the Pd NPs from an organic solvent and transferring them into an aqueous solution. The process was carried out combining *n*-dodecyl sulfide-stabilized Pd NPs that were prepared in toluene with the sodium salt of bis(11-carboxyundecyl) sulfide that was dissolved in water. This combination led to the formation of two visible layers after 5 minutes of stirring. The bottom aqueous layer was black, indicating that the Pd NPs were in the aqueous phase, while the top organic (toluene) layer became colorless.

In a similar process, we prepared well-defined Ru NPs stabilized by polyvinylpyrrolidone (PVP) (*83*). Ruthenium is an excellent catalyst for hydrogenation and deoxygenation reactions, which are essential processes in biomass conversion. Typically, Ru NP catalysts are synthesized and supported on a surface to create a heterogeneous catalyst. The support can be in the form of polymers, silica, carbon, carbon nanofibers (*84*) or carbon nanotubes (*85–87*). Metallic NPs can be synthesized in several ways by utilizing different metal salts, reducing agents and stabilizers.

Synthesis of Bimetallic Nanoparticles

Bimetallic NPs, for example, platinum-gold (Pt-Au) and platinum-bismuth (Pt-Bi), have been used for biomass conversion. Significant interest has been devoted toward synthesizing bimetallic NPs mainly because the resulting particles have been found to display improved reactivity in comparison to their single-metal counterparts (*88–92*). The synthesis methods and distribution of each element affect the crystal structure of the bimetallic NPs, which can be completely different from the bulk alloy. Successive reduction, co-reduction and thermal decomposition are the main methods used for preparing bimetallic NPs. The method selected can be used to control the particle morphology.

Successive reduction of metal ions is a procedure used to synthesize bimetallic NPs. The metal ion precursors must have a large difference in reduction potential. The metal ion with the more negative reduction potential is reduced first, followed by the second metal, which has a lower negative reduction potential. In the co-reduction method, two corresponding metal ions are simultaneously reduced in the presence of a stabilizing ligand. In the process of thermal decomposition, bimetallic NPs are synthesized by heating the precursors in different ratios, in the presence of the stabilizers. This method is based on the change in the nucleation and growth processes upon increasing the temperature (*88–92*).

Reduction of the metal salts over a support has been used to produce bimetallic clusters. Meitzner *et al.* (*91*) described the synthesis of rhodium-copper clusters dispersed on silica. A solution was created containing rhodium trichloride, copper nitrate and silica. The material was dried and then exposed to hydrogen at high temperatures to facilitate the reduction of the metal salts. The material was cooled under helium and was handled to reduce air exposure. This procedure illustrates a co-impregnation synthesis method in which silica is introduced to an aqueous solution of metal salts.

Synthesis of Semiconductor Nanoparticles

Semiconductor NPs are a class of materials that constitute important catalysts for a variety of applications in industry due to their unique size- and shape-dependent properties and reactivity. The band gap of semiconductor NPs is highly dependent on their size; the smaller the NP, the larger the band gap and vice versa. Thus, controlling particle size in semiconductor NP synthesis is a crucial step. A well-defined synthesis method should ideally result in the formation of pure, monodisperse crystalline particles (*93–96*). Until recently, the most common method for synthesis of semiconductor NPs was by precipitation from a colloidal solution (*96*). More recent methods include those in which a host material, such as a sol-gel system or a polymer, is used (*93*). Anaerobic methods and high temperature organometallic synthesis have also been used (*96*).

In the case of synthesis of metal oxide NPs, an important type of semiconductor NP, many different approaches have been reported (*97–109*). The sol-gel and hydrothermal methods are two routes often used in metal oxide NP preparation. The sol-gel method, which involves the hydrolysis and polymerization of the precursors to form a colloidal suspension, has been used in the synthesis of a variety of semiconductor NPs (*97–100*). The hydrothermal method involves placing a solution of the precursor in a pressure autoclave and heating the solution above the boiling point of water (*101*). The process results in a saturated vapor solution. Vijayalakshmi and Rajendran (*102*) reported the preparation of titanium dioxide (TiO_2) NPs under ambient conditions. They showed that the hydrothermal process produced TiO_2 NPs that were highly crystalline compared to particles obtained by the sol-gel methods. The sol-gel method (*103*), solvothermal method (*104*), micelles and inverse micelles method (*105*), chemical vapor deposition (*106*), direct oxidation (*107*), electrodeposition (*108*), and microwave methods (*109*) are also effective for the synthesis of semiconductor NPs.

Applications of Nanoparticles in Biorenewables Transformation

Metallic NPs, including bimetallic and semiconductor NPs, are used to catalyze many reactions related to biomass conversion. Their increased catalytic activity, attributed to their size-dependent properties, when compared to conventional catalysts, provide a significant impact on green technologies. These particles have been used to catalyze several types of reactions, such as conversion of sugars, decarboxylation of fatty acids, hydrolysis of cellulosic materials, and production of biodiesel. They have been used for advances in fuel cell technology. These types of NPs are also important in other biorenewable transformations, such as hydrogenation (*110*) and small molecule oxidation (*111–130*).

Conversion of Sugars

Ruthenium NPs have been effective as catalysts in the conversion of sugars. For example, Zhao *et al.* developed a Ru catalyst supported on carbon nanofibers (CNF) for the hydrogenolysis of sorbitol to ethylene glycol and propylene glycol (*131*, *132*). A proposed adopted mechanism is shown in Figure 2 (*132*). The wet impregnation, calcination and reduction methods were employed to prepare the Ru/CNF catalyst. The investigations were performed to assess the effect of calcination on catalyst properties. Calcination of the as-prepared Ru/CNF catalyst were conducted at 180, 240, and 300°C for 5 hours, leading to catalysts designated Ru/CNF-180°C, Ru/CNF-240°C, Ru/CNF-300°C, respectively. TEM images depicted 1.0 nm as-prepared Ru/CNF, Ru/CNF-180°C and Ru/CNF-240°C particles, while Ru/CNF-300°C gave a particle size of 10 nm.

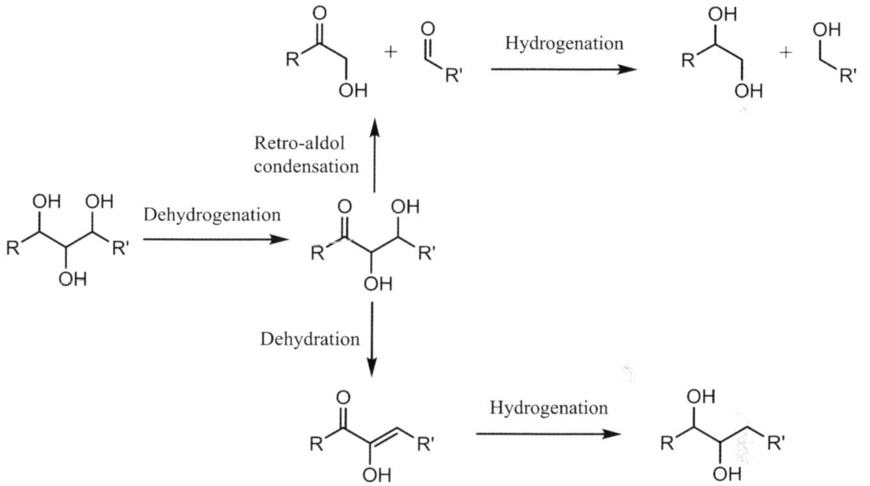

Figure 2. Proposed mechanism of polyol hydrogenolysis. Reproduced from reference listing (132). Copyright 2010, Elsevier.

Treatment of the catalyst by calcination resulted in a decrease in sorbitol conversion but an increase in glycol selectivity when compared to previous studies (*133*, *134*). Ru/CNF-240°C was the better catalyst for the sorbitol conversion as it displayed the highest selectivity with suitable yields for ethylene glycol and propylene glycol. The larger NPs (10 nm) or lower Ru dispersion, along with temperature increase (Ru/CNF-300°C catalyst), led to a decrease in sorbitol conversion and glycol selectivity, which in turn resulted in fewer Ru active sites for sorbitol hydrogenolysis.

Base promoters, such as calcium oxide and sodium hydroxide, were used in sorbitol hydrogenolysis. A base promoter is often required for efficient sorbitol conversion; however, it is possible that product selectivity may be reduced. The results indicated that when sodium hydroxide was used, the sorbitol conversion was higher but the selectivity was lower than that of calcium hydroxide, in which the opposite occurred. If the amount of catalyst is increased, sorbitol conversion also increases. Unfortunately, this decreases the selectivity for glycerol and ethylene glycol. Others (*133–137*) have suggested that the primary use of a base promoter is to prevent leaching of metal particles from the catalyst. Zhao *et al.* confirmed that the base promoter could improve sorbitol hydrogenolysis by neutralizing the organic acids derived from side-reactions (*131*, *132*). It is apparent from these studies that NPs are emerging as central catalysts for important reactions relevant to biorenewables transformation.

Production of Hydrocarbons

Obtaining hydrocarbons from crude oil is a mature process. However, due to the significant structural differences between biomass and crude oil, converting biomass to hydrocarbons is a difficult yet essential process in obtaining useful materials from renewables. Dumesic's group studied the conversion of sugars and polyols to monofunctional hydrocarbons such as hydrophobic alcohols, ketones, carboxylic acids, and heterocyclic compounds (*138*), and liquid fuels using Pt-Re catalysts. While the catalysts were not on the nanoscale, the work done by Dumesic's group opened avenues for using NPs for comparative studies and understanding catalytic processes on the nanoscale. In the study carried out by Dumesic's group, the Pt-Re catalyst was developed by controlling the rate of C-C and C-O bond cleavage, which resulted in the production of different monofunctional hydrocarbons. To convert biomass-derived carbohydrates, such as glucose and sorbitol, into liquid fuels, most of the oxygen atoms in the starting material should be removed. This was achieved using a C-C coupling process and isomerization, which also increases the molecular weight of diesel and jet fuels by producing branched hydrocarbons.

Polyol/sugar supplies enough hydrogen to partially deoxygenate the remainder of the feed. More than 90% of the energy content will be retained in the reaction products by reformation of part of the polyol/sugar on the Pt/Re catalyst during the production of hydrocarbons. The reformation can lead to the production of CO_2, an important factor in the conversion of polyols to monofunctional hydrocarbons since the hydrogen required for the deoxygenation reactions will be generated as a byproduct. The reformation process can also lead to the water gas shift reaction, during which CO is adsorbed onto the catalyst surface and then reacted with water to produce H_2 and CO_2. The catalyst surface was covered with strongly adsorbed CO and highly oxygenated reaction intermediates. Therefore, cleavage of C-O is favored compared to cleavage of C-C on Pt (*139*). The rate of C-O cleavage in the hydrogenolysis reaction is promoted by oxygenated hydrocarbons on Re since the oxygen and hydroxyl groups bind strongly to Re (*140*). Thus, the presence of oxygen and/or hydroxyl groups can mediate the effect of Re in the Pt-Re-C catalyst (*141*).

Biomass can be converted to chemicals and fuels at temperatures above 770K (*142*). Chemicals will form liquid products categorized as bio oil. It is a complex mixture containing more than 300 highly oxygenated compounds. The catalytic conversion of glucose or sorbitol using the Pt-Re/C leads to liquid products such as alkanes, olefins, and aromatics, with suitable molecular weights for transportation fuels.

Nanoparticle-Fostered Transformation of Cellulose into Commodity Chemicals

Fukuoka *et al.* successfully demonstrated the conversion of cellulose into sugar alcohols, sorbitol and mannitol, using supported Ru (Ru/HUSY (H form of ultrastable Y zeolite)) and Pt (Pt/γ-Al_2O_3) catalysts (*143*). This investigation utilized a variety of support materials and metal precursors that included Rh, Pd, Ir, and Ni. However, only Ru and Pt led to the high yields of the sugar alcohols that were of interest. The production of sorbitol and mannitol by the hydrogenation of glucose is shown in Figure 3. Relative to other supports, γ-Al_2O_3, HUSY(40), SiO_2-Al_2O_3, and HUSY(20) were chosen because of their high yields. A novel 'green' catalytic method was employed, and it was found that Pt/γ-Al_2O_3 gave sugar alcohols in 31% yield with sorbitol being the main product (sorbitol: 25%, mannitol: 6%) with a molar ratio of 4:1 (sorbitol:mannitol).

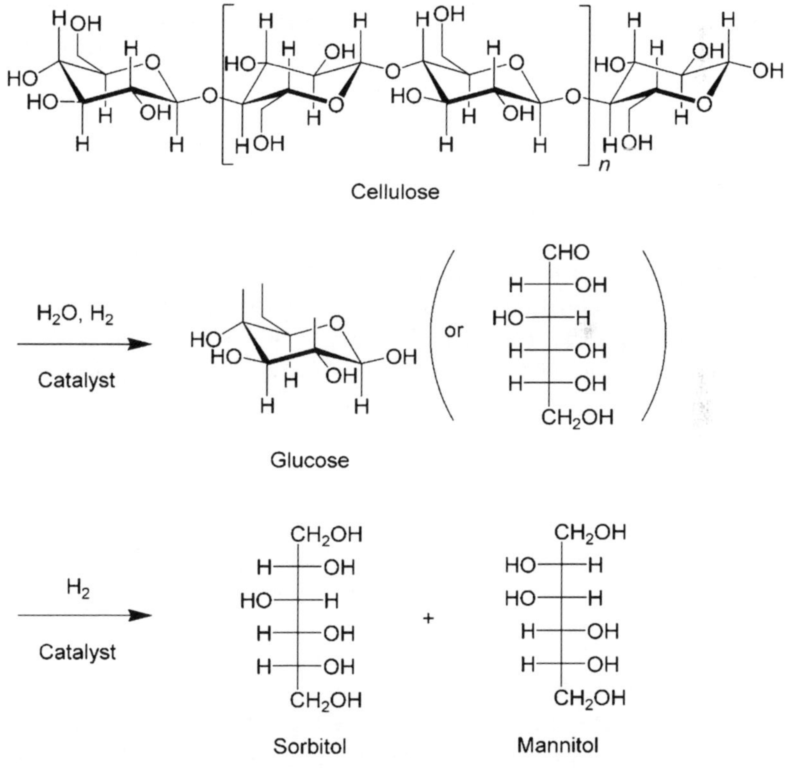

Figure 3. Hydrolysis of cellulose to glucose and the reduction of glucose to sorbitol and mannitol. Reproduced from reference listing (143). Copyright 2006, Wiley.

The reaction temperature range was 443-473K, with the highest yields observed over the Pt/γ-Al_2O_3 catalyst at 463 K and the Ru/HUSY(20) catalyst for the conversion of cellulose gave ~26% yield. The catalysts displayed environmentally friendly properties by being reusable for up to three catalytic runs, with no deactivation occurring. Additional reactions in this study consisted of the hydrolysis of cellulose into glucose by the supported catalysts under H_2 pressure. The mechanistic pathway shows H_2 being adsorbed onto the metal surface in a dissociative behavior, and the hydrogen species reversibly deposited on the support surface. The glucose yield was less than 4%, indicating that the metal drives the hydrolysis of cellulose. This study is in agreement with Zhang *et al.*, who reported that crystallinity, degree of polymerization, availability of chain ends, and fraction of accessible bonds are important factors that contribute to the conversion of cellulose (*144*).

A Ru carbon nanotube (CNT) catalyst was developed by Deng *et al.* to convert cellulose with a crystallinity of 33% into sorbitol in an aqueous medium in the presence of hydrogen (*145*). SiO_2, CeO_2, MgO, Al_2O_3, and CNTs were used as support for the Ru catalyst. Each was prepared by the impregnation process. After the drying step, the materials were calcined in air at 350°C and reduced by H_2 at 350°C. Cellulose samples were also prepared, which consisted of treating commercial grade cellulose (Alfa Aesar, crystallinity of 85%) with H_3PO_4 under various time and temperature conditions.

Additional studies included various catalysts (Fe, Co, Ni, Pd, Pt, Rh, Ru, Ir, Ag, and Au) supported on CNTs for the conversion of the treated cellulose with a crystallinity of 33% (*145*). Among all the catalysts investigated, Ru/CNT gave the highest yield of sorbitol. SiO_2, CeO_2, MgO were found to be inadequate supports, giving yields of 7%, 5%, and 0%, respectively, of sorbitol. Although the acidic Al_2O_3-supported Ru gave a better catalytic performance with a 22% yield of sorbitol, the CNT-supported Ru provided the highest yield of 69% of sorbitol. It was confirmed that CNT was the most effective support for the Ru catalyst for the conversion of cellulose into sorbitol, demonstrating its stability with only ~8% decrease in the sorbitol yield after four recycles.

The hydrolysis of cellulose to glucose can be catalyzed by solid acids (*146–152*), which is considered a possible intermediate for sorbitol formation and one of the more important factors for the conversion of cellulose into sorbitol. Analyses were conducted to pre-treat the CNT with concentrated HNO_3 to generate acidic groups, such as carboxylic acid groups, on their surfaces. It was found that acidic functional groups generated on CNT surfaces contributed to the conversion of cellulose into sorbitol, resulting in increased sorbitol production from the Ru/CNT catalyst

Fujita and co-workers investigated the bioelectrocatalytic process for cellulosic materials (fructose and cellobiose) in an ionic liquid (IL) medium such as choline dihydrogen phosphate ([ch][dhp]) (*153*). ILs are unique materials with physicochemical properties by which they display thermal and chemical stability. ILs are expected to act as a matrix for applications in protein preservation, drug delivery, biosensor development and biofuel cells.

Several steps are essential for energy conversion in an IL medium: (1) extraction of cellulose from biomass, (2) depolymerization of cellulose to produce mono-, di-, or oligosaccharides, and (3) collection of electrons from the saccharides. This investigation analyzed cellobiose dehydrogenase (CDH) and fructose dehydrogenase (FDH) in hydrated [ch][dhp]. Gold nanoparticles

(Au NPs) were used to develop the FDH-immobilized Au NP electrode and were prepared according to the literature (*154*). In the presence of D-fructose, catalytic oxidation currents based on the direct electron transfer (DET) reaction of FDH in hydrated [ch][dhp] were observed; however, there were no catalytic currents in the absence of D-fructose. Also, no current was observed with the FDH-unmodified electrode, even in the presence of D-fructose, indicating that the bioelectrocatalysis of FDH successfully took place in the hydrated [ch][dhp].

Due to the large surface area-to-volume ratio of the Au NPs, the FDH-Au NP electrode was effective in hydrated [ch][dhp] also enabling a high level of protein loading. Fujita *et al.* also did additional analysis regarding the DET reaction of CDH immobilized on the electrode in an aqueous solution without a mediator, using a carbon NP modified carbon paper electrode (*153*). Catalytic currents were observed when D-cellobiose was added to the hydrated [ch][dhp]; however, no current of CDH was detected in the absence of D-cellobiose, nor was a current detected with the CDH-unmodified electrode, even in the presence of D-cellobiose. According to these results, CDH catalyzes the oxidation of cellobiose through DET in hydrated [ch][dhp] and the CDH-immobilized electrode stored in hydrated [ch][dhp] at room temperature was found to be stable for more than 3 weeks.

Continued studies on the conversion of cellulose are paramount for advancements in green technologies and biorenewables transformation. While nanoparticles have proven to be effective catalysts, there remain many opportunities for advancements in this area.

Decarboxylation of Fatty Acids

The full reduction of biofuels to hydrocarbons is very attractive because it enables biorenewables to serve as direct replacements for fossil fuels; in particular, it creates a resource that is compatible with our current infrastructure (*155*, *156*). Oxygenated biofuels, which contain alkyl esters of long chain fatty acids, have several disadvantages, including high pour point, high cloud point, low energy density, and the potential for incompatibility with current delivery systems (*157*). Deoxygenation of biorenewable fuels is an important process for producing direct substitutes of fuels from crude oil.

Deoxygenation of fatty acids (FA) can be achieved by decarboxylation, decarbonylation or full reduction with hydrogen, as shown in Figure 4. In each case, the process produces an alkane (*158*).

$$C_{17}H_{35}(CO)OH \longrightarrow CO_2 + H_3C{-}C_{16}H_{33}$$

$$C_{17}H_{35}(CO)OH \longrightarrow CO + H_2O + H_2C{=}C_{16}H_{32}$$

$$C_{17}H_{35}(CO)OH \xrightarrow{3H_2} 2H_2O + H_5C_2{-}C_{16}H_{33}$$

Figure 4. Three different pathways for deoxygenation of fatty acids.

Three main types of catalysts are normally used to carry out deoxygenation reactions: supported sulfided metal catalysts such as $NiMo/Al_2O_3$ and $CoMo/Al_2O_3$ (*158–160*); porous acidic or basic catalysts such as zeolites; acidic-H-MCM-41, H-ZSM-S, and basic-porous $SnO-ZnO-Al_2O_3$ and supported metal catalysts. The first two types of catalysts are costly, generally inefficient thus leading to impure products, and require high temperature and pressure conditions.

The third type of catalyst holds promise, as reflected by several studies (*158*). Murzin and coworkers reported that catalytic decarboxylation is a low energy technique, and that zerovalent Pd has the highest activity (*158*). Jones and coworkers synthesized highly dispersed Pd NPs on ultra-porous silica mesocellular foam (MCF) and effectively catalyzed the deoxygenation of fatty acids (*161*). The silica support was functionalized with an amine, urea or mercapto silanes and characterized to be Pd-MCF-X, where X denotes the functionalized group. The mesocellular support minimizes the effect of mass transfer limitations and prevents pore blockage (*161*). The functionalization of MCF with silanes leads to monodispersity of the Pd precursor inside the support. It causes the formation of small Pd NPs with an average size of 2 nm after reduction of the precursor with evenly distributed particles inside the support. Low cost FFA feedstock taken from a waste stream contained C_{18} acid as the main constituent. Decarboxylation of FFA was done in a batch reaction over the Pd-MCF-X at 300°C for 6 hours (Figure 5). Conversion of stearic acid to *n*-heptadecene was 89.1%, 37%, and 84.8% for the Pd catalyst containing urea, mercapto and amine, respectively, with no side products. The group used Pd-MCF for decarboxylation of stearic acid to *n*-heptadecane to test the deactivation process and to develop a procedure to regenerate the catalytic activity. Pd-MCF-U and Pd-MCF-N showed the best catalytic activity with having sufficient Pd NPs inside the support. The proposed reason for the low catalytic activity of Pd-MCF-S is that it has a much lower number of active sites due to the formation of PdS during the high temperature treatments used for the preparation of the catalyst. These results are consistent with the previous work done by Murzin and co-workers (*162*, *163*). The deactivation of supported Pd NP catalysts during decarboxylation of fatty acids was also investigated.

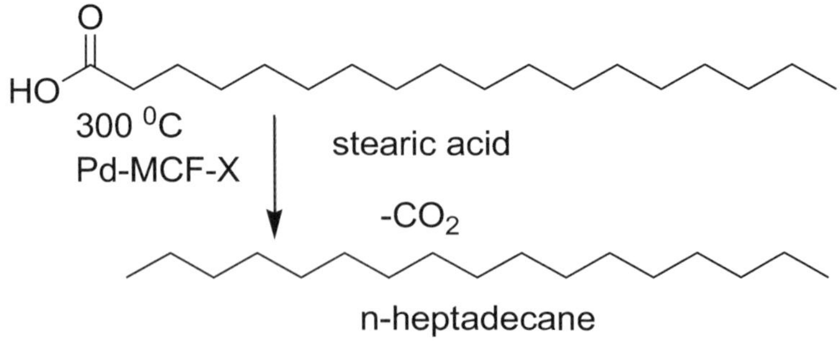

Figure 5. Decarboxylation of stearic acid to n-heptadecane. Reproduced from reference listing (161). Copyright 2010, Elsevier.

Han *et al.* developed a one-step synthesis method for Pd NPs on SBA-15 (highly ordered mesoporous silicate) by a sol-gel reaction (*162*). Murzin's group used similar nanoparticles in a semi batch reactor to perform the deoxygenation reaction, which could control the flow rate of the carrier gas and inlet/outlet pressures (*163*). The catalyst oxidation was prevented by introducing hydrogen gas into the system at 60°C before the reaction. To perform the reaction, Pd-SBA-15 catalyst was added to the reaction mixture of stearic acid and dodecene in an inert atmosphere. Once the reaction was complete the catalyst was recovered by filtering and washing with acetone. The recovered catalyst was characterized for actual Pd content, metal dispersion and average crystalline size. The physical properties of the catalysts with various metal loading percentages of Pd were studied. It was reported that the Pd dispersion decreased while the size of the particle increased with higher metal loading (*164*). The study also showed that the large surface area and porosity of the supporting material containing Pd NPs made it more suitable for the catalytic reaction of large fatty acid molecules to proceed. The kinetics of the deoxygenation of stearic acid was also studied (Figure 6). To compare the physical properties of SBA-15 and the Pd-SBA-15, N_2 adsorption-desorption isotherms of a Pd-loaded catalyst was used, and it confirmed that the mesoporous channels were not disturbed by the deposition of Pd.

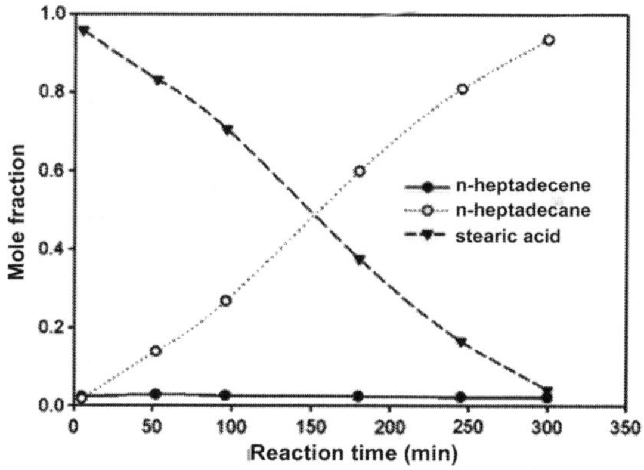

Figure 6. Kinetic study of the deoxygenation of stearic acid over a palladium catalyst. Reproduced from reference listing (163). Copyright 2010, Springer.

Murzin and coworkers used Pd NPs supported on carbon for the deoxygenation of stearic acid (*164*). A metal prior oxide support was used for deposition of the metal because of improved metal dispersion (37%), and a smoother surface that increased reactivity. The most important observation of this study was that the composition of the product mixture depended on the type of support and the surface groups present. For instance, there were equal amounts of *n*-pentadecane and *n*-heptadecane in the liquid phase; however there were a range of compounds such as CO, CO_2, methane, propane and compounds with C_1, C_3 in the gas phase, which is different from the case of Pd-MCF-X and Pd-SBA-15. They reported that an increased number of acid groups on the catalyst surface plays a significant role in giving a range of deoxygenated products. Acidic groups on the surface of the catalyst facilitated better transportation of reactants to the catalyst and easy removal of product from the catalyst.

Production of Biodiesel

Production of biodiesel has been one of the most extensively researched areas in biomass conversion. Recently, Deng *et al.* showed that Mg-Al (hydrotalcite) NPs were effective in the production of biodiesel from *Jatropha* oil (*165*). *Jatropha curcas L.* trees grow abundantly in Central and South America, Africa, India and South China, and are an excellent source of biomass. Even though *Jatropha* oil is non-edible, it is similar to edible oils (rapeseed, sunflower, palm and soybean) (*166*) that are commonly used for biodiesel production. The hydrotalcite NPs were used in the transesterification of *Jatropha* oil. Several experiments were conducted under various reaction conditions that included the methanol/*Jatropha* oil molar ratio, catalyst concentration, temperature and ultrasonic power to assess biodiesel yield.

A two-step transesterification process was used for biodiesel production. Previous investigations show that the first step is acid-esterification of the *Jatropha* oil (*167, 168*). The second step is solid base catalytic transesterification. The conversion of *Jatropha* oil yields diglyceride, monoglyceride, glycerin and fatty acid methyl esters (FAMEs) (*169, 170*). Palmitic acid, palmitoleic acid, stearic acid, oleic acid, linoleic acid and linolenic acid with a biodiesel yield of 13.79%, 0.95%, 6.33%, 42.61%, 26.34%, and 0.06%, respectively, consist of the FAMEs used in this study. Increasing the molar ratio from 3:1 to 4:1 ultimately resulted in an increase in the methyl esters. As the ratio increased beyond 4:1, no effect was observed on the methyl ester yield. A high activity was detected due to the large surface area-to-volume ratio of the calcined nanoparticles and the strong basic sites. The ester yield increased significantly from ~53% to ~94% when the catalyst concentration increased from 0.5 to 1.0 wt%. A weight percent above 1.0 would cause a reduction in the biodiesel yield. The reaction temperature also played an important role in biodiesel production. At lower temperature (303 K), the biodiesel yield was only ~52%; however, at 318 K, biodiesel yield was ~94%. Temperatures above 318 K resulted in a decrease in biodiesel yield. Ultrasonic radiation, along with catalytically active nanoparticles synthesized by using urea and MHT, facilitates the mixing of reactants with catalysts and is the possible reason for the high conversion rate. The best ultrasonic power was 210 W, which resulted in a biodiesel yield of 95.2%. Deactivation studies were conducted on the catalyst to ensure stability and ultimately the Mg-Al catalyst could be used 8 times, yielding 89.1% biodiesel.

Crossley *et al.* investigated the use of NPs to catalyze biofuel upgrade reactions at the oil/water interface (*171*). Pd nanohybrids (5 wt% Pd/SWNT/SiO_2) were found to stabilize water-oil emulsions and determine the catalytic reactivity at the liquid/liquid interface. Oxide NPs are also of interest because they have been previously shown to stabilize oil-in-water emulsions due to their hydrophilicity (*172–174*). On the other hand, carbon nanotubes are attractive because they produce emulsions due to their hydrophobicity (*175*). The hybrid

NPs were prepared using a previous method (*176*) and the group was able to tune the composition so as to modify the hydrophilic-hydrophobic balance to obtain water-in-oil or oil-in-water emulsions (*177*). The objectives of the investigation were to: (1) extend the utility of the nanohybrids by incorporating a transition metal (Pd) to make them catalytically active for hydrogenation and (2) add a solid base to catalyze condensation reactions. By depositing Pd on the hydrophilic surface, the aqueous reaction would be catalyzed; in contrast, the deposition of Pd on the hydrophobic face would support an organic solvent reaction. Two types of preparations with nanotubes were conducted: (A) incipient wetness impregnation to deposit Pd on single-walled carbon nanotubes (SWNT), and (B) MgO used as a support instead of SiO_2.

Vanillin (4-hydroxy-3-methoxybenzaldehyde) was an ideal compound used to investigate the catalytic activity of Pd-nanohybrids to phenolic hydrodeoxygenation in a water-in-oil (decalin) emulsion because it contains three different types of oxygenated functional groups (hydroxyl, aldehyde, and ether) and it is partially soluble in both aqueous and organic phases. Three different temperatures (100, 200, and 250°C) were used for the reactions. The nanohybrid particles were filtered, which allowed the emulsion to be broken and the two liquid phases were separated by GC-MS and GS-FID to monitor the migration of the products when going from aqueous to organic phase. Changes in the chemoselectivity were observed when the temperature increased. At 100°C, vanillin alcohol was produced by hydrogenation of the aldehyde and further converted into 2-methoxy-4-methylphenol (*p*-creosol) by means of hydrogenolysis, which was termed the dominant pathway at 200°C. Upon further increase of the temperature (250ºC), decarbonylation of the aldehyde group became the leading pathway (conversion to *o*-methoxyphenol (guaiacol)).

Furthermore, the reactivity of molecules that are soluble in either the organic or the aqueous phase was explored. Octanal, which is soluble in oil, and glutaraldehyde, which is soluble in water, were chosen for this investigation. They carried out reactions whereby octanal and glutaraldehyde were (i) separately placed in equimolar solutions of water and decalin, (ii) separately dissolved in pure decalin, and (iii) separately dissolved in pure water. Since the Pd-nanohybrids were placed along the hydrophilic side for catalyzing purposes, conversion of glutaraldehyde was expected. Only 58% conversion of glutaraldehyde was obtained in the aqueous phase, while 98% was converted in the emulsion under the same conditions (3 h at 100°C). Cyclic hemiacetal and valerolactol were the major products from this glutaraldehyde conversion. The results showed that selectivity could be modified by controlling the rates of reaction and migration out of a phase where the catalyst is located (Pd on the hydrophilic side). When Pd was placed on the hydrophobic side (octanal reaction), the opposite results were observed, where 9.1% of 1-octanol was the only dominant product. Finally, a tandem reaction catalyzed by MgO instead of SiO_2 was investigated. In this case, Pd-catalyzed hydrogenation was followed by aldol condensation of 5-methylfurfural and acetone. MgO imparts basicity to the nanohybrids, which in turn helped stabilize the Pd particles on both the hydrophilic and hydrophobic sides. Ultimately, this allowed the reactions to take place at the liquid/liquid interface by using a bifunctional catalyst that contains both metal and basic sites.

Design of Fuel Cells

Biofuel cells are an important alternative energy source and are classified into two major categories: microbial-based fuel cells and enzymatic fuel cells, depending on whether the enzymes are inside microorganisms or outside of the living cells (*178–187*). One of the advantages of these fuel cells is the reduction of O_2 without production of hydrogen peroxide as a side-product.

Basu *et al.* reported the application of platinum-based bimetallic NPs in the design of a direct glucose alkaline fuel cell (Figure 7) (*188*). Pt-Au and Pt-Bi bimetallic catalysts were synthesized by reduction of their corresponding metal precursors on carbon support. Sodium borohydride was used as the reducing agent and the reaction was carried out in an aqueous solution. In a typical batch direct glucose fuel cell (DGFC), activated charcoal was used as the cathode in a wide range of glucose and electrolyte concentrations and at ambient temperature. The bimetallic NPs were utilized as the anode material in the DGFC. Chronoamperometry (CA) and cyclic voltammetry (CV) were used to characterize glucose oxidation in the presence of PtAu/C and PtBi/C. The cyclic voltammetry data showed a reduction in the anodic peak height for Pt-Bi/C sample. This reduction is related to catalyst poisoning due to strong adsorption of intermediate products. In an alkaline medium, the Pt-Au/C catalyst is more active than the Pt-Bi because the former resists catalytic poisoning. This was confirmed by comparing the long-term poisoning rate of each catalyst via chronoamperometric measurements at constant potential. In another experiment, the open current voltage was obtained on different anodes. Using similar reaction conditions, the Pt-Bi/C anode potential was measured to be 0.8 V, while both Pt-Au/C and commercial Pt-Ru/C anode potentials were measured to be 0.9 V. Investigation of the power densities at varying concentrations of glucose confirmed that the Pt-Au/C anode is a more effective catalyst for the oxidation of glucose than both Pt-Bi/C and Pt-Ru/C anodes.

In other work, Liu Deng and coworkers developed a biofuel cell that was powered by ethanol and alcoholic beverages (*189*). Alcohol dehydrogenase (ADH) can utilize the electrons from different alcohols and is considered an important enzyme for designing biofuel cells. However, this enzyme has a high potential for over oxidizing NADH at the electrode surface. The ethanol bioanode is constructed of a film of Au nanoparticles, ADH and Meldola's blue (MDB). The oxygen biocathode is prepared by immobilization of Au nanoparticles and laccase on a sulfonated (3-mercaptopropyl)-trimethoxysilane sol-gel (PSSG)-Chitosan (CHI) composite membrane. The performance of the biofuel cell can be increased by removing the high potential for NADH oxidation, which is achieved by combination of Au NPs and MDB in the bioanode. The higher loading of the enzyme on the electrode, which is due to the 3D structure of carbon that resulted in increased surface area of the electrode, combined with the advantageous combination of Au NPs and MDB, increased the power density of this ethanol/oxygen biofuel cell roughly five times higher than previous reports (*190*).

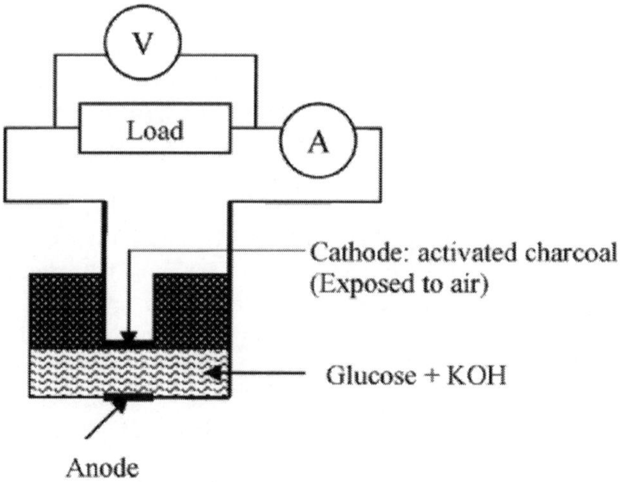

Figure 7. Schematic of a direct glucose fuel cell. Reproduced from reference listing (188). Copyright 2011, Elsevier.

The oxidation of glucose was also investigated using Pt-Au alloy nanoparticles (*191*). Various NPs including Pt, Au, Pt_{50}-Au_{50} alloy and Pt_{20}-Au_{80} were chosen as the anodic material, and the oxidation potential of each anode was investigated using electrochemical methods. In the cathode of a typical cell, oxygen was reduced on laccase co-immobilized with 2,2′-azinobis (3-ethylbenzothiazoline-6-sulfonate) diammonium salt. Based on the results of this investigation, three different potential regions can be observed in which the electrooxidation of glucose occurs. Two of these three regions are related to the pure Pt and pure Au potentials, while the third relates to the potential range between the formation of Pt-OH and Au-OH. The current densities in such a cell were reported to increase in the presence of Au while, in the same work, Pt proved to be a better choice for dehydrogenation of glucose (*192*). The performance of the biofuel cell can be improved by keeping the concentration of glucose below 10 mM, for medical purposes, and optimizing the nominal surface composition of Pt-Au.

Habrioux and coworkers have reported their investigation on the development of a new biocathode and its application in a glucose/oxygen biofuel cell (*193*). The biocathode was constructed from bilirubin oxidase (BOD) (*194–196*) adsorbed on Vulcan XC 72 R and immobilized on a nafion matrix. ABTS was utilized as an electrochemical mediator in the composition of the biocathode due to improvements of the kinetic parameters for the biocathode. Furthermore, $Au_{70}Pt_{30}$ bimetallic nanoparticles were used as an anode as reported previously in the literature (*197*). After oxidation, Au increases the current density, and dehydrogenation of the anomeric carbon of glucose occurs at lower potentials on the Pt side. It was concluded that the cooperative effect between Pt and Au makes this material active. The designed biofuel cell displayed higher performance in comparison to a membrane-less concentric system (*198*, *199*).

Hoa *et al.* introduced a novel biohydrogen fuel cell (*200*). In this study, the *Escherichia coli* strain MC13-4 was used as a source of hydrogen production. The system consists of a bioreactor for hydrogen production accompanied by a filter to remove unwanted gases and a fuel cell responsible for producing electricity. A consecutive polymer nanocomposite was used in the assembly of the anode in the fuel cell. They offer two different configurations to prepare the anode with the composition of Pt immobilized on functionalized multi-walled carbon nanotubes (Pt/fMWCNTs) and polyaniline (PANI). The former consists of core-shell structure with each Pt/fMWCNT bound to PANI, and the latter consists of a thin film of PANI coated on the surface of a Pt/fMWCNT layer. The advantage of utilizing PANI in this system is its ability to improve electron and proton transfer by supporting the catalyst. The conformational differences between the two anode composites result in different power densities. The maximum power density can be obtained from the system consisting of a thin film of PANI nanofibers coating a Pt/fMWCNTs/carbon paper as the anode.

Ryu *et al.* reported the construction of a glucose biofuel cell using carbon nanotubes (CNTs) with Pt nano-islands (*201*). The study was conducted to achieve a maximum glucose oxidation while using a constant amount of Pt. The morphology of the catalyst was modified utilizing intense pulse light (IPL). Cyclic voltammetry was used to monitor the effect of catalyst morphology on glucose oxidation. The results show an improvement of glucose oxidation with IPL irradiation on Pt-CNT electrode in comparison to unmodified Pt-CNT electrode. The power densities of biofuel cells containing the modified Pt–CNT electrode is 4.3 times higher than the unmodified electrode.

Miu and coworkers have investigated the potential application of metallic-semiconductor nanosystem (Pt-SiO_2) as a proton exchange membrane/electro-catalyst assembled in miniaturized micro fuel cells (*202*). In this work, SiO_2 NPs and a nanostructured silicon layer were studied as two different catalyst substrates, and it was shown that the structure of the support can cause different crystalline structures to form and alter the size of the Pt NPs during the deposition process. It was shown that porous silicon nanostructures have advantages over SiO_2 NPs, such as providing the possibility that Pt might grow inside the nanostructured layer as well as on the surface of the substrate.

Conclusions

In the last decade, tremendous efforts have been devoted toward developing new and robust materials for the transformation of biorenewables to chemicals and fuels. Nanoparticles have emerged as viable materials for biomass conversion due to their increased surface reactivity and the ability to tune their catalytic activity by controlled synthesis. Reactions that are central to green technologies, including hydrogenation, alcohol oxidation, hydrocarbon production, conversion of sugars, biodiesel production and design of fuel cells have been impacted positively by the use of NPs. It is important to note that an understanding of the specific catalytic mechanisms, the stability of the catalyst, and the surface chemistry of the NPs during these reactions are areas that are still not well understood. Improvements in analytical techniques for *in situ* catalyst characterization accompanied by computational studies are needed to provide further insights into catalysis at the nanoscale. Nonetheless, the field is promising and is an area of rapid growth.

Acknowledgments

R.Y.O. and S.O.O. are grateful for support from the Michigan Economic Development Corporation. S.O.O. gratefully appreciates support of this work through the National Science Foundation under awards CHE 1005456 and DMR 0963678. S.O.O. further thanks the Western Michigan University Office for the Vice President for Research for funding under the Faculty Research and Creative Activities Award (FRACAA) Program.

References

1. Melero, J. A.; Iglesias, J.; Garcia, A. *Energy Environ. Sci.* **2012**, *5*, 7393–7420.
2. Crocker, M.; Andrews, R. The Rationale for Biofuels. *Thermochemical Conversion of Biomass to Liquid Fuels and Chemicals*; Crocker, M., Ed.; Royal Society of Chemistry: Cambridge, U.K., 2010; Vol. 1, pp 1−25.
3. Greenwell, H. C.; Laurens, L. M. L.; Shields, R. J.; Lovitt, R. W.; Flinn, J. K. *J. R. Soc., Interface* **2010**, *7*, 703–726.
4. Wijffels, R. H.; Barbosa, M. J.; Eppink, M. H. M. *Biofuels, Bioprod. Biorefin.* **2010**, *4*, 287–295.
5. Tran, N. H.; Bartlett, J. R.; Kannangara, G. S. K.; Milev, A. S.; Volk, H.; Wilson, M. S. *Fuel* **2010**, *89*, 265–274.
6. Bhattacharya, S. C.; Sett, S.; Shrestha, R. M. *Energy Sources* **1989**, *11*, 161–182.
7. Tumuluru, J. S.; Wright, C. T.; Hess, J. R.; Kenney, K. L. *Biofuels, Bioprod. Biorefin.* **2011**, *5*, 683–707.
8. Demirbas, A. *J. Anal. Appl. Pyrolysis* **2004**, *71*, 803–815.
9. Lam, P. S.; Sokhansanj, S.; Bi, X.; Lim, C. J.; Melin, S. *Energy Fuels* **2011**, *25*, 1521–1528.
10. van der Stelt, M. J. C.; Gerhauser, H.; Kiel, J. H. A.; Ptasinski, K. J. *Biomass Bioenergy* **2011**, *35*, 3748–3762.
11. Chew, J. J.; Doshi, V. *Renewable Sustainable Energy Rev.* **2011**, *15*, 4212–4222.
12. Chew, T. L.; Bhatia, S. *Bioresour. Technol.* **2008**, *99*, 7911–7922.
13. Huang, H. -J; Ramaswamy, S.; Tschirner, U. W.; Ramarao, B. V. *Sep. Purif. Technol.* **2008**, *62*, 1–21.
14. Yaman, S. *Energy Convers. Manage.* **2004**, *45*, 651–671.
15. Nakamura, Y.; Morikawa, S. *Bull. Chem. Soc. Jpn.* **1980**, *53*, 3705–3706.
16. Takagaki, A.; Ohara, M.; Nishimura, S.; Ebitani, K. *Chem. Commun.* **2009**, *41*, 6276–6278.
17. Ohara, M.; Takagaki, A.; Nishimura, S.; Ebitani, K. *Appl. Catal., A* **2010**, *383*, 149–155.

18. Horvath, I. T.; Mehdi, H.; Fábos, V.; Boda, L.; Mika, L. T. *Green Chem.* **2008**, *10*, 238–242.
19. Fitzpatrick, S. W. U.S. Patent, No. 4,897,497, 1990.
20. Fitzpatrick, S. W. U.S. Patent, No. 5,608,105, 1997.
21. Prasad, Y. S.; Bakhshi, N. N.; Mathews, J. F.; Eager, R. L. *Can. J. Chem. Eng.* **1986**, *64*, 278–284.
22. Milne, T. A.; Evans, R. J.; Nagle, N. *Biomass* **1990**, *21*, 219–232.
23. Leng, T. Y.; Mohamed, A. R.; Bhatia, S. *Can. J. Chem. Eng.* **1999**, *77*, 156–162.
24. Twaiq, F. A.; Mohamed, A. R.; Bhatia, S. *Microporous Mesoporous Mater.* **2003**, *64*, 95–107.
25. Twaiq, F. A.; Mohamed, A. R.; Bhatia, S. *Fuel Process. Technol.* **2003**, *84*, 105–120.
26. Ooi, Y. S.; Zakaria, R.; Mohamed A. R., A. R.; Bhatia, S. *Appl. Catal., A* **2004**, *274*, 15–23.
27. Ooi, Y. S.; Bhatia, S. *Microporous Mesoporous Mater.* **2007**, *102*, 310–317.
28. Demirbas, A. *Energy Convers. Manage.* **2009**, *50*, 2782–2801.
29. Boocock, D. G. B.; Konar, S. K.; Mackay, A.; Cheung, P. T. C.; Liu, J. *Fuel* **1992**, *71*, 1291–1297.
30. Vonghia, E.; Boocock, D. G. B.; Konar, S. K.; Leung, A. *Energy Fuels* **1995**, *9*, 1090–1096.
31. Corma, A.; Miguel, P. J.; Orchilles, A. V. *J. Catal.* **1994**, *145*, 171–180.
32. Katikaneni, S. P. R.; Adjaye, J. D.; Bakhshi, N. N. *Can. J. Chem. Eng.* **1995**, *73*, 484–497.
33. Idem, R. O.; Katikaneni, S. P. R.; Bakhshi, N. N. *Fuel Process. Technol.* **1997**, *51*, 101–125.
34. Pioch, D.; Lozano, P.; Rasoanantoandro, M. C.; Graille, J.; Geneste, P.; Guida, A. *Oleagineux* **1993**, *48*, 289–292.
35. Tian, H.; Li, C.; Yang, C.; Shan, H. *Chin. J. Chem. Eng.* **2008**, *16*, 394–400.
36. Chew, T. L.; Bhatia, S. *Bioresour. Technol.* **2009**, *100*, 2540–2545.
37. Stumborg, M.; Wong, A.; Hogan, E. *Bioresour. Technol.* **1996**, *56*, 13–18.
38. Aatola, H.; Larmi, M.; Sarjovaara, T.; Mikkonen, S. *SAE Int. J. Engines* **2009**, *1*, 1251–1262.
39. Gusmao, J.; Brodzki, D.; Djega-Mariadassou, G.; Frety, R. *Catal. Today* **1989**, *5*, 533–544.
40. da Rocha Filho, G. N.; Brodzki, D.; Djega-Mariadassou, G. *Fuel* **1993**, *72*, 543–549.
41. Simacek, P.; Kubicka, D.; Sebor, G.; Pospisil, M. *Fuel* **2009**, *88*, 456–460.
42. Simakova, I.; Simakova, O.; Maki-Arvela, P.; Simakov, A.; Estrada, M.; Murzin, D. Y. *Appl. Catal., A* **2009**, *355*, 100–108.
43. Snare, M.; Maki-Arvela, P.; Simakova, I. L.; Myllyoja, J.; Murzin, D. Y. *Russ. J. Phys. Chem. B* **2009**, *3*, 1035–1043.
44. Snare, M.; Kubickova, I.; Maki-Arvela, P.; Eranen, K.; Murzin, D. Y. *Ind. Eng. Chem. Res.* **2006**, *45*, 5708–5715.
45. Kubickova, I.; Snare, M.; Eranen, K.; Maki-Arvela, P.; Murzin, D. Y. *Catal. Today* **2005**, *106*, 197–200.

46. Snare, M.; Kubickova, I.; Maki-Arvela, P.; Chichova, D.; Eranen, K.; Murzin, D. Y. *Fuel* **2008**, *87*, 933–945.
47. Maki-Arvela, P.; Kubickova, I.; Snare, M.; Eranen, K.; Murzin, D. Y. *Energy Fuels* **2007**, *21*, 30–41.
48. Roberts, W. L.; Lamb, H. H.; Stikeleather, L. F.; Turner, T. L. U.S. Patent No. 105,813, 2011.
49. Bridgwater, A. V. *Appl. Catal., A* **1994**, *116*, 5–47.
50. Adjaye, J. D.; Bakhshi, N. N. *Fuel Process. Technol.* **1995**, *45*, 161–183.
51. Katikaneni, S. P. R.; Adjaye, J. D.; Bakhshi, N. N. *Energy Fuels* **1995**, *9*, 1065–1078.
52. Sharma, R. K.; Bakhshi, N. N. *Energy Fuels* **1993**, *7*, 306–314.
53. Gruter, G. J. M.; Dautzenberg, F. Eur. Patent No. 1,834,950, 2007.
54. Gruter, G. J. M.; Dautzenberg, F. Eur. Patent No. 2,050,742, 2007.
55. Huber, G. W.; Shabaker, J. W.; Dumesic, J. A. *Science* **2003**, *300*, 2075–2077.
56. Huber, G. W.; Cortright, R. D.; Dumesic, J. A. *Angew. Chem., Int. Ed.* **2004**, *43*, 1549–1551.
57. Huber, G. W.; Chheda, J. N.; Barrett, C. J.; Dumesic, J. A. *Science* **2005**, *308*, 1446–1450.
58. Cortright, R. D.; Davda, R. R.; Dumesic, J. A. *Nature* **2002**, *418*, 964–967.
59. Shabaker, J. W.; Huber, G. W.; Dumcsic, J. A. *J. Catal.* **2004**, *222*, 180–191.
60. Davda, R. R.; Dumesic, J. A. *Angew. Chem., Int. Ed.* **2003**, *42*, 4068–4071.
61. Chheda, J. N.; Huber, G. W.; Dumesic, J. A. *Angew. Chem., Int. Ed.* **2007**, *46*, 7164–7183.
62. Murray, C. B.; Norris, D. J.; Bawendi, M. G. *J. Am. Chem. Soc.* **1993**, *115*, 8706–8715.
63. Ahlrichs, R.; Fenske, D.; Oesen, H.; Schneider, U. *Angew. Chem., Int. Ed. Engl.* **1992**, *31*, 323–326.
64. Xu, J.; Bhattacharyya, D. *Ind. Eng. Chem. Res.* **2007**, *46*, 2348–2359.
65. Meng, Z. H.; Liu, H. L.; Liu, Y.; Zhang, J.; Yu, S.; Cui, F. Y.; Ren, N. Q.; Ma, J. *J. Membr. Sci.* **2011**, *372*, 165–171.
66. Freemantle, R. G.; Liu, M.; Guo, W.; Obare, S. O. Approaches to Synthesis and Characterization of Spherical & Anisotropic Palladium Nanomaterials. In *Metallic Nanomaterials for Life Sciences*; Kumar, C. S. S. R., Ed.; Wiley-VCH: Weinheim, 2009, Vol. 1; pp 305−355.
67. Wang, C. B.; Zhang, W. X. *Environ. Sci. Technol.* **1997**, *31*, 2154–2156.
68. Martino, A.; Stoker, M.; Hicks, M.; Bartholomew, C. H.; Sault, A. G.; Kawola, J. S. *Appl. Catal., A* **1997**, *161*, 235–248.
69. Li, F.; Vipulanandan, C.; Mohanty, K. K. *Colloids Surf., A* **2003**, *223*, 103–112.
70. Shamsuzzoha, L. W. M.; Ritchie, S. M. C. *J. Nanopart. Res.* **2005**, *7*, 469–476.
71. Wu, L.; Ritchie, S. M. C. *Chemosphere* **2006**, *63*, 285–292.
72. Ostwald, W. *Lehrbuck der Allgemeinen Chemie*; W. Engelmann Publ.: Leipzig, Germany, 1896; Vol. 2, Part 1.
73. Turkevitch, J. *Gold Bull.* **1985**, *18*, 86–91.
74. Bönneman, H.; Brijoux, W.; Brinkmann, R.; Dinjus, E.; Fretzen, T.; Joussen, B.; Korall, J. *Angew. Chem., Int. Ed.* **1992**, *31*, 323.

75. Niu, Y.; Yeung, L. K.; Crooks, R. M. *J. Am. Chem. Soc.* **2001**, *123*, 6840–6846.
76. Garcia-Martinez, J. C.; Scott, R. W. J.; Crooks, R. M. *J. Am. Chem. Soc.* **2003**, *125*, 11190–11191.
77. Arul Dhas, N.; Gedanken, A. *J. Mater. Chem.* **1998**, *8*, 445–450.
78. Chen, W.; Cai, W.; Lei, Y.; Zhang, L. *Mater. Lett* **2001**, *50*, 53–56.
79. Kan, C.; Cai, W.; Li, C.; Zhang, L.; Hofmeister, H. *J. Phys. D: Appl. Phys.* **2003**, *36*, 1609–1614.
80. Ganesan, M.; Freemantle, R.; Obare, S. O. *Chem. Mater.* **2007**, *19*, 3464–3471.
81. Lin, R.; Freemantle, R. G.; Kelley, N. M.; Fielitz, T. R.; Obare, S. O.; Ofoli, R. Y. *Nanotechnology* **2010**, *21*, 325605.
82. Lin, R.; Ma, X.; Fielitz, T. R.; Obare, S. O.; Ofoli, R. Y. *Catal. Commun.* **2012**, *18*, 168–175.
83. Ma, X.; Lin, R.; Beuerle, C.; Jackson, J. E.; Obare, S. O.; Ofoli, R. Y. *Nanotechnology* **2014**, *25*, 045701.
84. Zhao, L.; Zhou, J.; Chen, H.; Zhang, M.; Sui, Z.; Zhou, X. *Korean J. Chem. Eng.* **2010**, *27*, 1412–1418.
85. Deng, W.; Tan, X.; Fang, W.; Zhang, Q.; Wang, Y. *Catal. Lett.* **2009**, *133*, 167–174.
86. Zhao, L.; Zhou, J. H.; Sui, Z. J.; Zhou, X. G. *Chem. Eng. Sci.* **2010**, *65*, 30–35.
87. Chen, P.; Zhang, H.; Lin, G.; Hong, Q.; Tsai, K. R. *Carbon* **1997**, *35*, 1495–1501.
88. Sinfelt, J. H. *J. Catal.* **1973**, *29*, 308–315.
89. Sinfelt, J. H. *Acc. Chem. Res.* **1987**, *20*, 134–139.
90. Meitzner, G.; Via, G. H.; Lytle, F. W.; Sinfelt, J. H. *J. Chem. Phys.* **1983**, *78*, 2533–2541.
91. Meitzner, G.; Via, G. H.; Lytle, F. W.; Sinfelt, J. H. *J. Chem. Phys.* **1983**, *78*, 882–889.
92. Meitzner, G.; Via, G. H.; Lytle, F. W.; Sinfelt, J. H. *J. Chem. Phys.* **1985**, *83*, 4793–4799.
93. Pickett, N. L.; O'Brien, P. *Chem. Rec.* **2001**, *1*, 467–479.
94. Skaff, H.; Emrick, T. Semiconductor nanoparticles: Synthesis, Properties, and Integration into Polymers for the Generation of Novel Composite Materials. In *Nanoparticles: Building blocks for Nanotechnology*; Rotello, V., Ed.; Springer Science + Business Media, Inc.: New York, NY, U.S.A., 2004; Chapter 2.
95. Weller, H. *Angew. Chem., Int. Ed. Engl.* **1993**, *32*, 41–53.
96. Rogach, A. L.; Talapin, D. V.; Weller, H. Semiconductor Nanoparticles. In *Colloids and Colloid Assemblies: Synthesis, Modification, Organization and Utilization of Colloid Particles*; Caruso, F., Ed.; Wiley-VCH Verlag GmbH & Co. KGaA: Weinheim, FRG, 2003; Chapter 2.
97. Schwartz, J. A.; Contescu, C.; Contescu, A. *Chem. Rev.* **1995**, *95*, 477–510.
98. Wright, A. P.; Davis, M. E. *Chem. Rev.* **2002**, *102*, 3589–3614.
99. Lu, Z.-L.; Lindner, E.; Mayer, H. A. *Chem. Rev.* **2002**, *102*, 3543–3578.
100. Chen, X.; Mao, S. S. *Chem. Rev.* **2007**, *107*, 2891–2959.

101. Chae, S. Y.; Park, M. K.; Lee, S. K.; Kim, T. Y.; Kim, S. K.; Lee, W. *Chem. Mater.* **2003**, *15*, 3326–3331.
102. Vijayalakshmi, R.; Rajendran, V. *Arch. Appl. Sci. Res.* **2012**, *4*, 1183–1190.
103. Niederberger, M.; Bartl, M. H.; Stucky, G. D. *Chem. Mater.* **2002**, *14*, 4364–4370.
104. Xu, J.; Ge, J.-P.; Li, Y.-D. Li. *J. Phys. Chem. B* **2006**, *110*, 2497–2501.
105. Lim, K. T.; Hwang, H. S. *Langmuir* **2004**, *20*, 2466–2471.
106. Seifried, S.; Winterer, M.; Hahn, H. *Chem. Vap. Deposition* **2000**, *6*, 239–244.
107. Wu, J.-M. *J. Cryst. Growth* **2004**, *269*, 347–355.
108. Lei, Y.; Zhang, L. D.; Fan, J. C. *Chem. Phys. Lett.* **2001**, *338*, 231–236.
109. Wu, X.; Jiang, Q.-Z.; Ma, Z.-F.; Fu; Shangguan, W.-F. *Solid State Commun.* **2005**, *136*, 513–517.
110. Dauthal, P.; Mukhopadhyay, M. *Ind. Eng. Chem. Res.* **2013**, *52*, 18131–18139.
111. Zamani, F.; Hosseini, S. M. *Catal. Commun.* **2014**, *43*, 164–168.
112. Zhang, L.; Li, P.; Yang, J.; Wang, M.; Wang, L. *ChemPlusChem* **2014**, *79*, 217–222.
113. Polshettiwar, V.; Varma, R. S. *Org. Biomol. Chem.* **2009**, *7*, 37–40.
114. Hou, W.; Dehm, N. A.; Scott, R. W. J. *J. Catal.* **2008**, *253*, 22–27.
115. Frank, A. J.; Rawski, J.; Maly, K. E.; Kitaev, V. *Green Chem.* **2010**, *12*, 1615–1622.
116. Deplanche, K.; Merroun, M. L.; Casadesus, M.; Tran, D. T.; Mikheenko, I. P.; Bennett, J. A.; Zhu, J.; Jones, I. P.; Attard, G. A.; Wood, J.; Selenska-Pobell, S.; Macaski, L. E. *J. R. Soc., Interfaces* **2012**, *9*, 1705–1712.
117. Liang, D.; Gao, J.; Wang, J.; Chen, P.; Wei, Y.; Hou, Z. *Catal. Commun.* **2011**, *12*, 1059–1062.
118. Taarning, E.; Madsen, A. T.; Marchetti, J. M.; Egeblad, K.; Christensen, C. H. *Green Chem.* **2008**, *10*, 408–414.
119. Ketchie, W. C.; Fang, Y.; Wong, M. S.; Murayama, M.; Davis, R. J. *J. Catal.* **2007**, *250*, 94–101.
120. Ketchie, W. C.; Murayama, M.; Davis, R. J. *J. Catal.* **2007**, *250*, 264–273.
121. Bianchi, C.; Porta, F.; Prati, L.; Rossi, M. *Top. Catal.* **2000**, *13*, 231–236.
122. Porta, F.; Prati, L.; Rossi, M.; Coluccia, S.; Marta, G. *Catal. Today* **2000**, *61*, 165–172.
123. Demirel, S.; Kern, P.; Lucas, M.; Claus, P. *Catal. Today* **2007**, *122*, 292–300.
124. Demirel, S.; Lehnert, K.; Lucas, M.; Claus, P. *Appl. Catal., B* **2007**, *70*, 637–643.
125. Hayashi, T.; Inagaki, T.; Itayama, N.; Baba, H. *Catal. Today* **2006**, *117*, 210–213.
126. Dimitratos, N.; Lopez-Sanchez, J. A.; Anthonykutty, J. M.; Brett, G.; Carley, A. F.; Tiruvalam, R. C.; Herzing, A. A.; Kiely, C. J.; Knight, D. W.; Hutchings, G. J. *Phys. Chem. Chem. Phys.* **2009**, *11*, 4952–4961.
127. Yan, Y. M.; Tel-Vered, R.; Yehezheli, O.; Cheglakov, Z.; Willner, I. *Adv. Mater.* **2008**, *20*, 2365–2370.
128. Katz, E.; Willner, I. *Electroanalysis* **2003**, *15*, 913–947.

129. Alfonta, L.; Bardea, A.; Khersonsky, O.; Katz, E.; Willner, I. *Biosens. Bioelectron.* **2001**, *16*, 675–687.
130. Katz, E.; Alfonta, L.; Willner, I. *Sens. Actuators, B* **2001**, *76*, 134–141.
131. Zhao, L.; Zhou, J.; Chen, H.; Zhang, M.; Sui, Z.; Zhou, X. *Korean J. Chem. Eng.* **2010**, *27*, 1412–1418.
132. Zhao, L; Zhou, J. H.; Sui, Z. J.; Zhou, X. G. *Chem. Eng. Sci.* **2010**, *65*, 30–35.
133. Wang, K. Y.; Hawley, M. C.; Furney, T. D. *Ind. Eng. Chem. Res.* **1995**, *34*, 3766–3770.
134. Andrews, M. A.; Klaeren, S. A. *J. Am. Chem. Soc.* **1989**, *111*, 4131–4133.
135. Ye, L. M.; Duan, X. P.; Lin, H. Q.; Yuan, Y. Z. U.S. Patent No. 4366332, 1982.
136. Chopade, S. P.; Miller, D. J.; Jackson, J. E.; Werpy, T. A.; Frye, J. G.; Zacher, A. H. U.S. Patent No. 6291725, 2001.
137. Sirkar, A. K. U.S. No Patent 4338472, 1982.
138. Simonetti, D. A.; Rass-Hansen, J.; Kunkes, E. L.; Soares, R. R.; Dumesic, J. A. *Green Chem.* **2007**, *9*, 1073–1083.
139. Alcala, R.; Mavrikakis, M.; Dumesic, J. A. *J. Catal.* **2003**, *218*, 178–190.
140. Pallassana, V.; Neurock, M. *J. Catal.* **2002**, *209*, 289–305.
141. Zhang, J.; Vukmirovic, M. B.; Sasaki, K.; Nilekar, A. U.; Mavrikakis, M.; Adzic, R. R. *J. Am. Chem. Soc.* **2005**, *127*, 12480–12481.
142. Czernik, S.; Bridgwater, A. V. *Energy Fuels* **2004**, *18*, 590–598.
143. Fukuoka, A.; Dhepe, P. L. *Angew. Chem., Int. Ed.* **2006**, *45*, 5161–5163.
144. Zhang, Y. P.; Lynd, L. R. *Biotechnol. Bioeng.* **2004**, *88*, 797–824.
145. Deng, W.; Tan, X.; Fang, W.; Zhang, Q.; Wang, Y. *Catal. Lett.* **2009**, *133*, 167–174.
146. Klemm, D.; Heublein, B.; Fink, H. P.; Bohn, A. *Angew. Chem., Int. Ed.* **2005**, *44*, 3358–3393.
147. Gallezot, P. *Chem. Soc. Rev.* **2012**, *41*, 1538–1558.
148. Huber, G. W.; Iborra, S.; Corma, A. *Chem. Rev.* **2006**, *106*, 4044–4098.
149. Dhepe, P. L.; Fukuoka, A. *Catal. Surv. Asia* **2007**, *11*, 186–191.
150. Dhepe, P. L.; Fukuoka, A. *ChemSusChem.* **2008**, *1*, 969–975.
151. Yan, N.; Zhao, C.; Luo, C.; Dyson, P. J.; Liu, H.; Kou, Y. *J. Am. Chem. Soc.* **2006**, *128*, 8714–8715.
152. Suganuma, S.; Nakajima, K.; Kitano, M.; Yamguchi, D.; Kato, H.; Hayahsi, S.; Hara, M. *J. Am. Chem. Soc.* **2008**, *130*, 12787–12793.
153. Fujita, K.; Nakamura, N.; Murata, K.; Igarashi, K.; Samejima, M.; Ohno, H. *Electrochim. Acta* **2011**, *56*, 7224–7227.
154. Frens, G. *Nat. Phys. Sci.* **1973**, *241*, 20–22.
155. Ford, J. P.; Immer, J. G.; Lamb, H. H. *Top. Catal.* **2012**, *55*, 175–184.
156. Ash, M.; Dohlman, E. *Oil Crops Situation and Outlook Year Book: Electronic Outlook Report from the Economic Research Service*; United States Department of Agriculture: 2007. Available online at http://usda.mannlib.cornell.edu/usda/ers/OCS//2000s/2007/OCS-01-16-2007.pdf [accessed September 14th, 2014].
157. Lotero, E.; Liu, Y.; Lopez, D. E.; Suwannakarn, K.; Bruce, D. A.; Goodwin, J. G. *Ind. Eng. Chem. Res.* **2005**, *44*, 5353–5363.

158. Madsen, A. T.; Rozmysłowicz, B.; Simakova, I. L.; Kipilő, T.; Leino, A. R.; Kordás, K.; Eränen, K.; Mäki-Arvela, P.; Murzin, D. Y. *Ind. Eng. Chem. Res.* **2011**, *50*, 11049–11058.
159. Senol, O. I.; Viljava, T.-R.; Krause, A. O. I. *Appl. Catal., A* **2007**, *326*, 236–244.
160. Ryymin, E.-M.; Honkela, M. L.; Viljava, T.-R.; Krause, A. O. I. *Appl. Catal., A* **2009**, *358*, 42–48.
161. Ping, E. W.; Wallace, R.; Pierson, J.; Fuller, T. F.; Jones, C. W. *Microporous Mesoporous Mater.* **2010**, *132*, 174–180.
162. Han, P.; Wang, X.; Qui, X.; Ji, X.; Gao, L. *J. Mol. Catal. A: Chem.* **2007**, *272*, 136–141.
163. Lestari, S.; Maki-Arvela, P.; Eranen, K.; Beltramini, J.; Max-Lu, G. Q.; Murzin, D. Y. *Catal. Lett.* **2010**, *134*, 250–257.
164. Lestari, S.; Simakova, I.; Tokarev, A.; Maki-Arvela, P.; Eranen, K.; Murzin, D. Y. *Catal. Lett.* **2008**, *122*, 247–251.
165. Deng, X.; Fang, Z.; Liu, Y. H.; Yu, C. L. *Energy* **2011**, *36*, 777–784.
166. Gui, M. M.; Lee, K. T.; Bhatia, S. *Energy* **2008**, *33*, 1646–1653.
167. Aramendía, M. A.; Borau, V.; Jiménez, C.; Marinas, J. M.; Ruiz, J. R.; Urbano, F. J. *Appl. Catal., A* **2003**, *255*, 301–308.
168. Deng, X.; Fang, Z.; Liu, Y. H. *Energy Convers. Manage.* **2010**, *51*, 2802–2807.
169. Xie, W. L.; Li, H. *J. Mol. Catal. A: Chem.* **2006**, *255*, 1–9.
170. Barakos, N.; Pasias, S.; Papyannakos, N. *Bioresour. Technol.* **2008**, *99*, 5037–5042.
171. Crossley, S.; Faria, J.; Shen, M.; Resaco, R. E. *Science* **2010**, *327*, 68–72.
172. Binks, B. P. *Curr. Opin. Colloid Interface Sci.* **2002**, *7*, 21–41.
173. Binks, B. P.; Whitby, C. P. *Langmuir* **2004**, *20*, 1130–1137.
174. Binks, B. P.; Philip, J.; Rodrigues, J. A. *Langmuir* **2005**, *21*, 3296–3302.
175. Wang, R. K.; Park, H. O.; Chen, W. C.; Silvera-Batista, C.; Reeves, R. D.; Butler, J. E.; Ziegler, K. J. *J. Am. Chem. Soc.* **2008**, *130*, 14721–14728.
176. Resasco, D. E.; Alvarez, W. E.; Pompeo, F.; Balzano, L.; Herrera, J. E.; Kitiyanan, B.; Borgna, A. *J. Nanopart. Res.* **2002**, *4*, 131–136.
177. Shen, M.; Resasco, D. E. *Langmuir* **2009**, *25*, 10843–10851.
178. Heller, A. *Phys. Chem. Chem. Phys* **2004**, *6*, 209–216.
179. Katz, E.; Willner, I.; Kotlyar, A. B. *J. Electroanal. Chem.* **1999**, *479*, 64–68.
180. Tayhas, G.; Palmore, R.; Kim, H.-H. *J. Electroanal. Chem.* **1999**, *464*, 110–117.
181. Ikeda, T.; Kano, K. *J. Biosci. Bioeng.* **2001**, *92*, 9–18.
182. Chen, T.; Barton, S. C.; Binyamin, G.; Gao, Z.; Zhang, Y.; Kim, H.-H.; Heller, A. *J. Am. Chem. Soc.* **2001**, *123*, 8630–8631.
183. Tsujimura, S.; Tatsumi, H.; Ogawa, J.; Shimizu, S.; Kano, K.; Ikeda, T. *J. Electroanal. Chem.* **2001**, *496*, 69–75.
184. Kim, H. H.; Mano, N.; Zhang, Y.; Heller, A. *J. Electroanal. Chem.* **2003**, *150*, A209–A213.
185. Mano, N.; Mao, F.; Heller, A. *J. Am. Chem. Soc.* **2003**, *125*, 6588–6594.
186. Mano, N.; Mao, F.; Heller, A. *J. Electroanal. Chem.* **2005**, *574*, 347–357.
187. Farneth, E.; D'Amore, M. B. *J. Electroanal. Chem.* **2005**, *581*, 197–205.

188. Basu, D.; Basu, S. *Electrochim. Acta* **2011**, *56*, 6106–6113.
189. Deng, L.; Shang, L.; Wen, D.; Zhai, J.; Dong, S. *Biosens. Bioelectron.* **2010**, *26*, 70–73.
190. Deng, L.; Wang, F. A.; Chen, H. J.; Shang, L.; Wang, L.; Wang, T.; Dong, S. J. *Biosens. Bioelectron.* **2008**, *24*, 329–333.
191. Habrioux, A.; Sibert, E.; Servat, K.; Vogel, W.; Kokoh, K. B.; Alonso-Vante, N. *J. Phys. Chem. B* **2007**, *111*, 10329–10333.
192. Möller, H.; Pistorius, P. C. *J. Electroanal. Chem.* **2004**, *570*, 243–255.
193. Habriouxa, A.; Napporna, T.; Servata, K.; Tingryb, S.; Kokoh, K. B. *Electrochim. Acta* **2010**, *55*, 7701–7705.
194. Yan, Y.-M.; Baravik, I.; Tel-Vered, R.; Willner, I. *Adv. Mater.* **2009**, *21*, 4275–4279.
195. Murata, K.; Kajiya, K; Nakamura, N.; Ohno, H. *Energy Environ. Sci.* **2009**, *2*, 1280–1285.
196. Hussein, L.; Feng, Y. L.; Alonso-Vante, N.; Urban, G.; Krüge, M. *Electrochim. Acta* **2011**, *56*, 7659–7665.
197. Habrioux, A.; Vogel, W.; Guinel, M.; Guetaz, L.; Servat, K.; Kokoh, B.; Alonso-Vante, N. *Phys. Chem. Chem. Phys.* **2009**, *11*, 3573–3579.
198. Habrioux, A.; Servat, K.; Tingry, S.; Kokoh, K. B. *Electrochem. Commun.* **2009**, *11*, 111–113.
199. Kadirgan, F.; Kannan, A. M.; Atilan, T.; Beyhan, S.; Ozenler, S.; Suzer, S.; Yörür, A. *Int. J. Hydrogen Energy* **2009**, *34*, 9450–9460.
200. Hoa, L. Q.; Sugano, Y.; Yoshikawa, H.; Saito, M.; Tamiya, E. *Biosens. Bioelectron.* **2010**, *25*, 2509–2514.
201. Ryu, J.; Kim, H. S.; Hahn, H. T.; Lashmore, D. *Biosens. Bioelectron.* **2010**, *25*, 1603–1608.
202. Miu, M.; Danila, M.; Ignat, T.; Craciunoiu, F.; Kleps, I.; Simion, M.; Bragaru, A.; Dinescu, A. *Superlattices Microstruct.* **2009**, *46*, 291–296.

Chapter 13

Agricultural Residues as Feedstocks for Lactic Acid Fermentation

Daniel Pleissner and Joachim Venus*

Department of Bioengineering, Leibniz-Institute for Agricultural Engineering Potsdam-Bornim, Max-Eyth-Allee 100, 14469 Potsdam, Germany
***E-mail: jvenus@atb-potsdam.de**

Enormous amounts of agricultural residues are produced globally every year from crops and mostly burned or disposed. Alternatively, the use of agricultural residues as feedstocks in biotechnological processes provides an innovative way to convert *'no-value'* material into *'value-added'* products. This chapter introduces to opportunities of using agricultural feedstocks in fermentative lactic acid production. Examples of fermentations using straw, rapeseed residues and grass press juice as substrates are shown. Furthermore, pre-treatment methods and future perspectives of agricultural residues as substrates for the fermentative production of lactic acid are presented.

Introduction

Biotechnological processes and bio-based products are nowadays in focus of various studies as promising alternatives to petrochemical routes and products, and an option to overcome the challenges of resource depletion (*1*). The strategy of '*white biotechnology*' allows the utilization of carbon, nitrogen and phosphorus compounds from hydrolyzed organic material by assimilation in biomass and metabolites of microorganism, to be used for the production of chemicals, materials and energy. Lactic acid (2-hydroxypropionic acid), for instance, is a well-known metabolic product. It serves as a building block for a variety of key chemicals and bulk polymers, and has an enormous industrial potential.

In 1780, lactic acid was first discovered by Carl Wilhelm Scheele in sour milk and first considered to be a milk component, but later identified as a metabolic product of certain microorganisms. In 1856, Louise Pasteur discovered the potential of the genus *Lactobacillus* for the fermentative lactic acid formation. The first industrial lactic acid production was started in the United States in 1881. Nowadays, lactic acid is produced biotechnologically in technical reactors on an industrial scale using suitable bacterial and fungal strains (*2*, *3*). An advantage of microbial lactic acid production to the production from fossil oil is the selective formation of one of the two optical isomers (D(-)- and L(+)-lactic acid, Figure 1), by selection of an appropriate microbial strain (*4*).

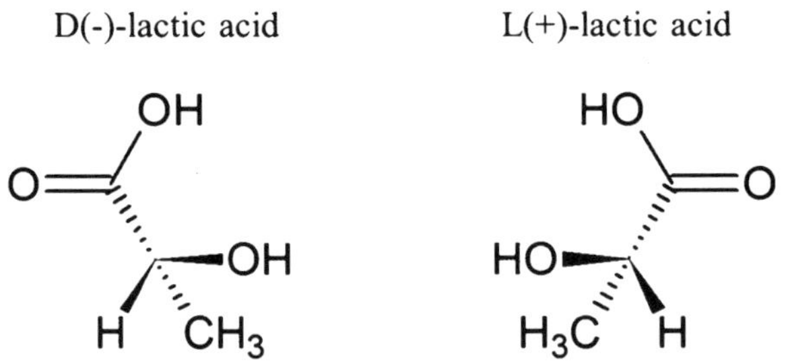

Figure 1. Chemical structure of D(-)- and L(+)-lactic acid.

Due to its properties, lactic acid is used for pH regulation, flavoring, improving of microbial quality and mineral fortification by the food industry, and for skin lightening, skin hydration and as an antimicrobial agent by the cosmetic industry. The pharmaceutical industry is also interested in lactic acid and uses it in tablets, prostheses, surgical sutures and drug delivery systems. The chemical industry uses lactic acid for example, as green solvent and cleaning agent (*5*). Lactic acid is considered the most potential feedstock monomer for chemical conversions (*5–8*). It has been receiving attention as a monomer for the production of poly(lactic acid), a biobased and potentially biodegradable polymer (bio-plastic) with a melting temperature in the range from 130°C to 180°C. For application purposes of poly(lactic acid), it is important to recognize that properties like melting temperature and degree of crystallinity depend on the sequence of L(+)- and D(-)-lactic acid enantiomers in the polymer. Thus, by alternating both enantiomers, poly(lactic acid) polymers with various properties regarding thermostability can be synthesized (*9*).

The global lactic acid market was estimated to be 714.2 kilo tons in 2013, which is expected to reach 1,960.1 kilo tons by 2020 (*10*). Growth in demand for lactic acid and its salts and esters in industrial applications will be driven mainly by lactic acid-based biodegradable polymers and to a lesser degree by lactate solvents (*3*, *8*).

As mentioned above, pure L(+)- and D(-)-lactic acid can be produced biotechnologically. The microbial conversion of substrates into lactic acid requires the presence of carbon, complex nitrogen and phosphorus sources. In conventional fermentations, glucose, protein hydrolysates (e. g. yeast extract and peptone) and an inorganic phosphorus salt are used as carbon, nitrogen and phosphorus sources, respectively. Particularly the nitrogen source is essential for the conversion of glucose into lactic acid. For instance, yeast extract, was found to contain all necessary amino acids and other organic nitrogen compounds needed to maintain cell development and lactic acid formation. Yeast extract, however, is too expensive and processes based on it are cost-inefficient (*11*). Due to the low price of lactic acid, its biotechnological production depends on the use of inexpensive substrates. In order to further decrease the costs of lactic acid fermentation processes, alternative nutrient sources, like by-products and residues from agricultural materials (e. g. grass, straw and rapeseed) are of considerable interest.

The present chapter introduces to opportunities of using agricultural residues as feedstocks in fermentative lactic acid production. Examples of fermentations using straw, rapeseed residues and grass press juice are presented. Furthermore, the chapter introduces to pre-treatment methods and future perspectives of agricultural residues as substrates for fermentative lactic acid production.

Agricultural Residue

The term agricultural residue used here is defined as biomass derived from sugar, starchy, cellulosic and oil crops. Bentsen *et al.* (*12*) estimated a global crop residue production from barley, maize, rice, soybean, sugar cane and wheat at around 3.7 Pg dry matter per year between 2006 and 2008. This amount can theoretically be used as a source of nutrients in biotechnological processes for the production of *'value-added'* compounds. The value of agricultural residues as nutrient source depends on their specific contents of cellulose, hemicellulose, lignin, starch, protein and minerals. The contents can vary between different crops and a screening prior to fermentations is recommended. Table 1 gives an overview of different residues and their compositions.

Most of the agricultural residues contain cellulose, which consists of glucose units linked by beta(1.6)-glycosidic bonds. Glucose is the carbon source of choice for many lactic acid fermentations, and thus, hydrolyzed cellulose a potential source of carbon. Residues from grain production, corn cob, cornstalks, corn stover, wheat and barley straws, can contain more than 40 % (w/w) cellulose. Similar amounts were found in switchgrass, while lupine contains 20 % (w/w) cellulose. The cellulose content of lucerne can vary between 8 % and 35 % (w/w) depending on the botanical fraction analyzed (Table 1). Under consideration of a global production of 1,376 Tg per year, cellulose is certainly an interesting and promising source of carbon for biotechnological processes (*12*).

Table 1. Comparison of composition of agricultural residues

Agricultural residues	*Cellulose [%, w/w]*	*Hemicellulose [%, w/w]*	*Lignin [%, w/w]*	*Total protein[a] [%, w/w]*	*Ash [%, w/w]*	*Ref.*
Corn cob	45	35	15	2	3	(*19–21*)
Corn stalks	39-47	26-31	3-5	5	7.3	(*20–22*)
Corn stover	35-40	21-25	11-19	6	11	(*18, 23, 24*)
Straw						
-Wheat[b]	32-41	25-29	23-24	3-5	4-10	(*23, 25*)
-Barley[b]	33-43	24-33	6-10	2-4	2-10	(*25–27*)
Rapeseed	4-5	3	N/A	35.9	6.5	(*28, 29*)
Grass						
-Lucerne[b]	8-35	N/A	2-11	2-5	N/A	(*30*)
-Lupine	20[c]	71[c]	1[c]	38	4	(*31*)
-Switchgrass	45	31-32	22-23	1-5	3-4	(*20, 32, 33*)

[a] Total nitrogen × 6.25 [b] Variations in composition result from the analysis of different botanical fractions [c] Quantified in hull

Another compound that occurs in large quantities in agricultural residues is hemicellulose. Hemicellulose is a heterogeneous polymer containing various amounts of pentose and hexose sugars (e. g. xylose, arabinose, glucose, galactose, mannose, fucose and mannuronic acid), with an global production of 848 Tg per year (*12*, *13*). Lupines can contain more than 71 % (w/w) hemicellulose, while residues from grain production contain around 30 % (w/w) (Table 1). Pentose and hexose sugars in hemicellulose are potential carbon sources, however, not all lactic acid producing microbial strains grow on these sugars and produce lactic acid at a high yield (*14–16*).

Lignin, with its highly complex structure of aromatic alcohols, is an additional major carbon component of agricultural residues. Wheat straw and switchgrass can contain more than 20 % (w/w, Table 1) and the global production was estimated at 666 Tg per year (*12*). However, due to the different interunit linkages, the affinity to form a more condensed structure and poor product sensitivity, lignin is difficult to treat and to use as carbon source in fermentations (*17*).

Starch is present in only small quantities. Corn stover and switchgras, for instance, can contain 0.4 and 1.0 % (w/w) starch, respectively (*18*). This amount might be too little in order to play a considerable role as carbon source for lactic acid fermentations.

In order to supply sufficient nitrogen to microbes and to maintain the microbial conversion of carbon into lactic acid, the nitrogen/protein content of agricultural residues needs to be high. However, organic materials with high nitrogen/protein contents (e. g. lupine and rapeseed, Table 1) are mainly used in food and feed productions, and not available in large amounts for biotechnological processes. Agricultural residues are usually poor in proteins and contents of 6 % (w/w) and less were quantified in corn stover, straw, lucerne and switchgrass (Table 1).

Beside cellulose, hemicellulose, lignin, starch and proteins, agricultural residues contain minerals like phosphorus. In plant tissue, phosphorus is present in phytic acid, polyphosphates, phospholipids, nucleotide phosphates and nucleic acids (*34*, *35*). Phosphorus can be quantified in organic material after oxidation of the organic matter and analysis of the remaining ash. For instance, ash of *Miscanthus x giganteus* contains up to 10 % (w/w) phosphorus pentoxide (*36*). Yusiharni and Gilkes (*37*) quantified 0.2 % to 1.3 % (w/w) phosphorus in ash of Australien native plants. As shown by the two examples, the phosphorus content can be highly variable in different plant materials. Therefore, preliminary experiments are needed in order to investigate if the amount of phosphorus in feedstocks is available to microorganisms and sufficiently high to support an efficient lactic acid production.

Pre-Treatment of Agricultural Residues

Cellulose, hemicellulose, lignin, proteins and phosphorus containing compounds in agricultural residues are mostly arranged in recalcitrant structures. That structure needs to be broken-up by a pre-treatment in order to make carbon, nitrogen and phosphorus compounds accessible to microbes. The pre-treatment includes the hydrolytic conversion of cellulose, hemicelluloses, proteins and phosphorus compounds into utilizable nutrients (e. g. pentose and hexose sugars, amino acids and phosphate) by chemical (acidic), biological and/or enzymatical methods (Figure 2). Pre-treatment methods are energy-intensive and the selection of an efficient method is crucial for the overall economy of a biotechnological process.

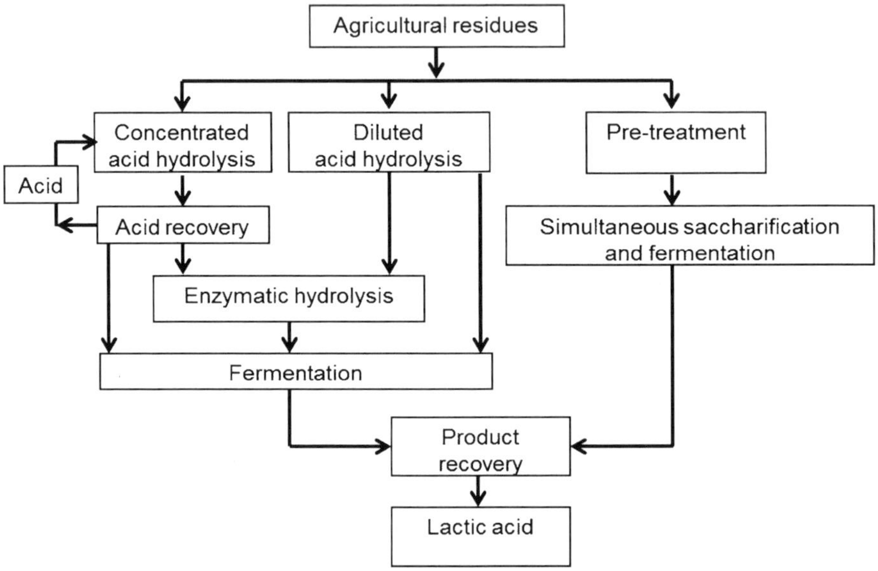

Figure 2. Process flow for the utilization of agricultural residues.

For the chemical hydrolysis of residues, concentrated or diluted acids (e. g. sulfuric acid and hydrochloric acid) are used. Hydrolysis using concentrated acids results in a high yield (around 80 %) of reducing sugars, however, the acid causes corrosion of equipment and needs to be recovered or neutralized afterwards (*38*). It should be admitted here that concentrated acids do not only hydrolyze cellulose and hemicellulose, but also proteins and phosphorus compounds. Thus, a nutrient rich hydrolysate can be produced in one batch. Contrarily, diluted acids have the advantage that they do not harm the equipment and an acid recovery step can be skipped. However, relatively low yields of around 50 % of reducing sugars and the need of high temperatures are serious disadvantages (*38*). For instance, Ji *et al.* (*39*) reported the hydrolysis of wheat straw by a continuous approach using diluted H_2SO_4. The hydrolysis was carried out at 215°C and a relatively low yield of 60.8 % of reducing sugars was obtained.

Zhou *et al.* (*40*) recently reported another chemical approach for the hydrolysis of cellulosic biomass using acidic ionic catalysts like 1-propyl sulfonic acid-2-phenyl imidazoline hydrogensulfate. The hydrolysis was carried out at a relative low temperature of 100°C for 60 min and a relatively high yield of reducing sugars of 85.1 % was obtained. This approach might be a promising alternative to acid hydrolysis.

A pre-treatment of agricultural residues is required when enzymes are used for hydrolysis in an enzymatic approach. The pre-treatment in form of an acidic pre-digestion or physicochemical treatment (e. g. steam explosion) ensures that the recalcitrant structure is accessible to enzymes (*41*). Cellulases, hemicellulases, proteases and phosphatases needed for the digestion of cellulose, hemicellulose, proteins and polyphosphate, respectively, can be added and the process performed at low temperatures. However, enzymes need to be recovered after hydrolysis and reused in order to keep costs of enzymes low.

Another opportunity to hydrolyze agricultural residues is a biological approach. By this approach, hydrolytic enzymes are secreted by an organism (bacteria or fungi, or a combination of both), being cultured in presence of the residues in solid state or submerged fermentations (*42*, *43*). This method has the advantage that no pure and expensive enzyme formulations are required. Biological hydrolysis can also be used directly in combination with lactic acid fermentation in order to reduce the number of process steps. A combination of hydrolysis and fermentation is called simultaneous saccharification and fermentation (SSF), and has been shown for the production of lactic acid from e. g. cassava starch, corn stover and sweet potatoes (*44–48*). SSF is one of the most used processes to produce lactic acid from lignocellulosic materials and an opportunity to prevent enzyme inhibition by the product (*47*, *49–51*). An example of SSF of rapeseed residues is given below.

Fermentation

Agricultural residues can be used to obtain sugar solutions that may be usefully exploited for the production of lactic acid (Figure 2) through the following steps: a) pre-treatment, to break down the lignocellulosic structure, b) enzymatic hydrolysis, to depolymerize lignocellulose to fermentative sugars, c) sugar fermentation to lactic acid by lactic acid bacteria and d) separation and purification of lactic acid (*20*, *52*). Because of the relatively low price of lactic acid, one of the major challenges in its large-scale fermentative production is the cost of the nutrient source. Lactic acid can be produced from a wide spectrum of carbon sources including starchy materials, many food industry by-products (e. g. molasses and whey), agricultural residues and by-products (e. g. hydrolysates of straw, cottonseed hulls, corn cob, corn stalks, rapeseed residues, wheat bran and brewer's spent grains) and various other renewable resources like green biomass. Together with the need of low-cost carbon, there is an additional demand of suitable supplements, which should not cause additional costs and problems in view of impurities. Therefore, the kind of nutrients as well as the optimization of their concentration is essential. It is likely, that one of the future trends in lactic acid production will end up in mixtures of different low-cost raw materials in order to avoid the use of expensive complex supplements (*8*, *53*). In the following, a couple of examples are given for the use of different agricultural residues for lactic acid production via several process regimes and/or microorganisms.

Straw

The industrial production of fermentation products (e. g. ethanol and lactic acid) is facing the challenge of redirecting the production process from fermentation of relatively easily convertible but expensive starchy materials to complex but inexpensive lignocellulosic materials such as wood and residues from crops. Unlike starch, which contains homogenous and easily hydrolysable polymers, lignocellulosic material is often rich in polysaccharides such as cellulose and hemicellulose, which are embedded in a matrix of lignin (Table 1). The resulting hydrolysates contain a mixture of hexoses and pentoses, which all are potential substrates for lactic acid producing microbes (*54*).

Besides the species *Lactobacilli*, also *Bacillus* strains and the fungus *Rhizopus* have been widely investigated for the production of lactic acid from wheat straw, by-products from corn processing and rice straw (*55–60*). The pre-treatment of lignocellulosic material can be followed by enzymatic hydrolysis to complete the release of sugar monomers. As already mentioned above, the pre-treatment results in the hydrolysis of cellulose into hexoses, while hemicellulose is converted into pentoses. Thus, in contrast to starch, the hydrolysis of lignocellulosic material results in the release of pentose sugars in addition to hexose sugars. This implies, that appropriate fermenting organisms need to be selected, which are able to convert both hexose and pentose sugars into desired fermentation products. However, microorganisms are mostly unable to efficiently metabolize pentoses and thus, extensive metabolic engineering is necessary to improve their performance on lignocellulosic materials.

Furthermore, hydrolysates from lignocellulosic materials may contain inhibitors such as furfural, phenols and carboxylic acids, which can potentially inhibit the fermenting organism. Therefore, the organism must also be tolerant to these inhibitors. The inhibitory effect of the hydrolysates can be reduced by applying a detoxification process prior to fermentation (*61*, *62*). However, the inclusion of this extra process step increases significantly the total cost of the fermentation product. Therefore, it would be economically beneficial that the microorganisms used are capable of producing fermentation products from undetoxified hydrolysates.

With regard to the above mentioned issues, processes were developed for producing L(+)- and D(-)-lactic acid with high enantiopurity from lignocellulosic material (*63*, *64*). Advantageously, a bacterium was found to efficiently metabolize hexoses and pentoses and to produce enantiomerically pure lactic acid. In Figure 3, fermentations of *Bacillus coagulans* and *Lactobacillus coryniformis* subsp. *torquens* on lignocellulosic hydrolysate are shown. *Bacillus coagulans* efficiently converted all sugars (pentoses and hexoses, 60 g L^{-1}), and produced more than 60 g L^{-1} L(+)-lactic acid (Figure 3A). This extent of conversion is higher than previously described yields in the relevant prior art (*65*). *Lactobacillus coryniformis* subsp. *torquens* formed more than 40 g L^{-1} D(-)-lactic acid with an enantiopurity of >99 % from 40 g L^{-1} of lignocellulosic sugars (Figure 3B). Using a suitable downstream processing system including ultrafiltration, softening, electrodialysis, ion-exchange, decolorization and evaporation, several grades of L(+)- and D(-)-lactic acid, up to pharmaceutical grade, can be produced.

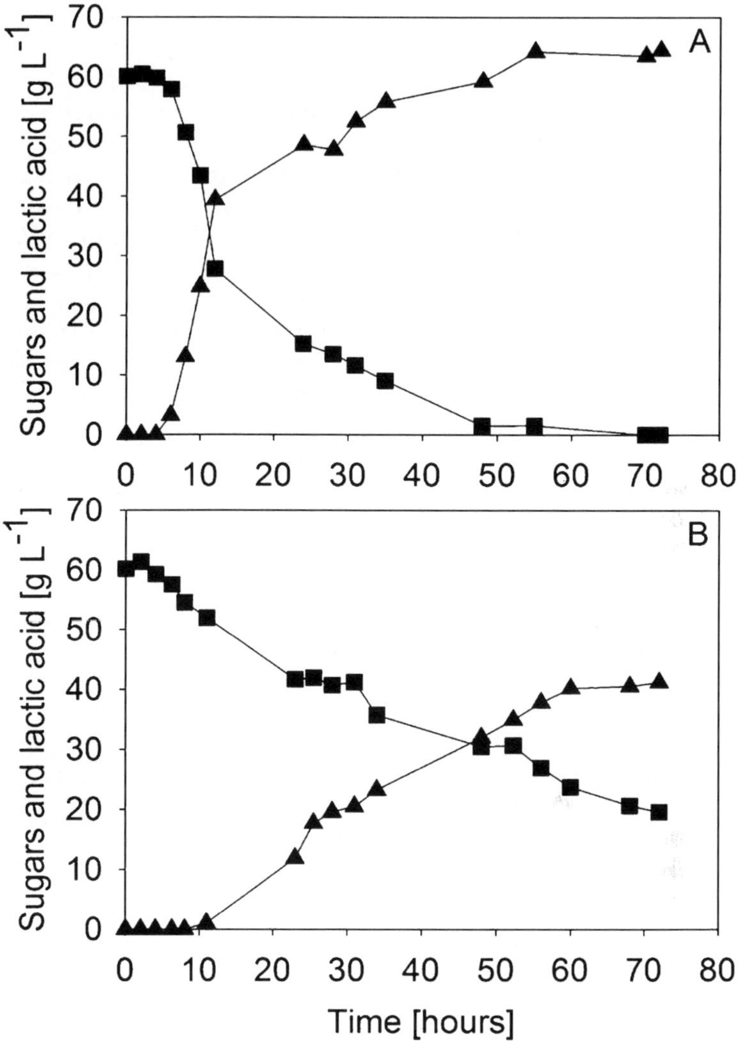

Figure 3. Performance in sugar (▪, hexoses and pentoses) consumption and lactic acid (▴) formation of the L(+)-lactic acid producer Bacillus coagulans (A) and the D(-)-lactic acid producer Lactobacillus coryniformis subsp. torquens (B) when cultured in lignocellulosic hydrolysate.

Rapeseed

In contrast to other related materials, there are only a few studies on the use of rapeseed residues (e. g. straw, meal and cake) as substrates for the fermentative production of bio-based products (*66–68*). There are two main types of rape: winter rape, which occurs mainly in maritime locations in Western, Central and Eastern Europe and hand canola, which is to be found in Canada, China, India and Europe. Rapeseed is the second most abundant rapeseed crop after soybean, with a worldwide cultivation of 65 million tons in 2012. Use of the oil, obtained by different pressing techniques, as a fuel was of growing interest in recent years. From the procedures of oil pressing/extraction, high amounts of crop material remain. Two of these residual materials are rapeseed meal or cake with high solid contents. Rapeseed cake contains predominantly proteins, carbohydrates, lignin, oil and ash. Its nutrient composition and proteins are affected by the process used in rapeseed oil production. Most of the industrially produced oil is obtained by hexane extraction combined with heat pressing. A gentle alternative is cold pressing at a temperature of 50-60°C. Rapeseed hulls, which are rich in fiber, lignin and other polyphenols, can be partially removed before oil pressing (*69*). Because of the before mentioned composition (Table 1), the residual material is partly used in feed production. However, the high protein content makes it a promising nitrogen source in lactic acid production and extensive studies have been performed at the Leibniz-Institute for Agricultural Engineering Potsdam-Bornim. The studies dealt with two different aims. First, the use of rapeseed press residues as nitrogen source (e. g. amino acids, proteins and minerals) for bacteria in combination with other fermentable sugars (*70, 71*). Second, the use of rapeseed meal as a lignocellulosic source of carbohydrates, which can be pre-treated followed by enzymatic hydrolysis in order to release sugars for subsequent lactic acid fermentation. Depending on the yield of the different upstream processes, the hydrolysates can be mixed with other substrate components and/or used solely.

Several samples (press cake/meal, high oleic rapeseed meal (HO), eruca rapeseed meal and protein extracted meal (protex)) were analyzed for some basic parameters like dry matter content (DM) and sugar, nitrogen and phosphorus contents (Table 2). The analysis revealed differences between samples, which are expected to influence the performance of fermentative processes.

Table 2. Analytical figures of different rapeseed residues

Sample	*DM*[a] *[%]*	*Sugars [%, w/w]*	*Lignin [%, w/w]*	*NH_4-N*[b] *[mg kg^{-1}]*	*N_{tot}*[c] *[g kg^{-1}]*	*P_{tot}*[d] *[g kg^{-1}]*
Cake	98.1	8.1	11.6	87.9	47.5	10.5
Cake	91.6	5.6	10.2	152.6	59.3	13.0
Meal	96.9	4.2	10.3	220.1	54.4	11.2
HO	94.3	0.1	13.0	-	54.7	13.3
Eruca	95.4	0.7	9.7	187.2	64.0	17.6
Protex	97.3	0.5	12.0	100.3	24.6	5.0

[a] Dry matter determined at 105°C [b] Ammonium nitrogen [c] Total nitrogen [d] Total phosphorus

Figure 4 shows a comparison of eight different rapeseed residues (RSR 1-8) regarding final lactic acid concentration. After the pre-treatment of the tested rapeseed materials and mixing with other carbon sources (e. g. glucose and molasses) the lactic acid concentration reached nearly 80 % of the concentration obtained in a conventional nutrient broth supplemented with yeast extract. However, using RSR 8 the lactic acid concentration was considerably lower. This result underlines the importance of a preliminary test prior to fermentations at larger-scale.

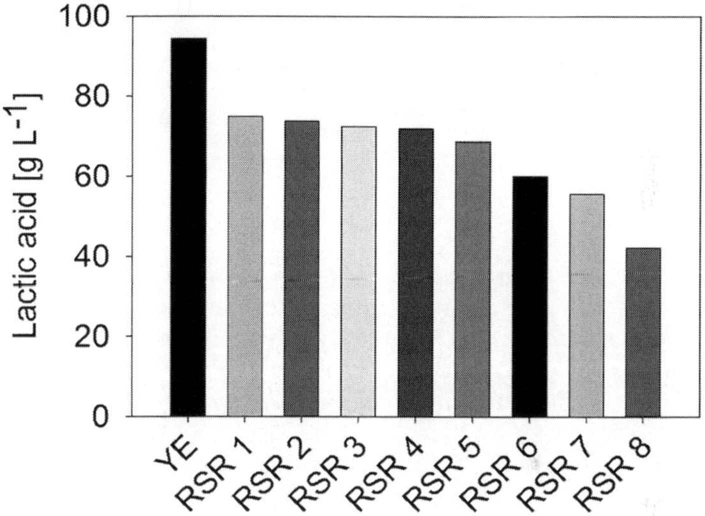

Figure 4. Comparison of lactic acid concentrations obtained by fermenting different rapeseed residues (RSR 1-8) or yeast extract (YE) as a conventional nitrogen source.

For the direct use of rapeseed residues as carbon and nutrient source, different pre-treatment procedures followed by the application of different enzymes were carried out. In Figure 5 a comparison of a separated hydrolysis and fermentation process and a SSF approach is shown. Irrespective the type of process, the direct approach did only result in a lactic acid concentration of 30-40 g L^{-1}, which is far below the concentrations shown in Figure 4 and most-likely caused by a limitation in carbon. The additional use of different hydrolytic enzymes did not result in higher lactic acid concentrations.

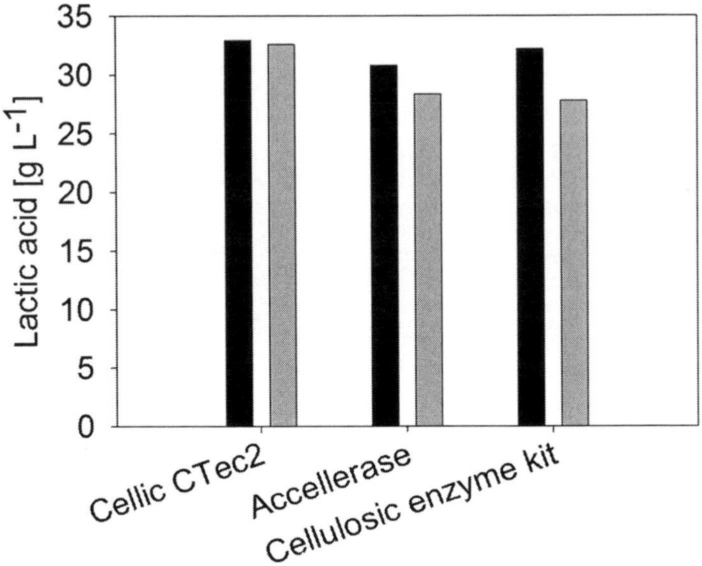

Figure 5. Comparison of lactic acid concentrations obtained by fermenting rapeseed residues as carbon and nitrogen sources in a direct approach. Pre-treatment and fermentation were carried out separately (black column) and simultaneously (gray column). Furthermore, different hydrolytic enzymes (Cellic® CTec2, accellerase and a cellulosic enzyme kit) were investigated.

Grass

'*Green Biorefinery*' concepts facilitate the multiple use of fresh green biomass (*72*). Press cakes, for instance, can be used as solid fuel or fibrous composite materials (*73*). A useful combination of green biomass processing for the production of fodder pellets and the utilization of the green press juices in lactic acid fermentation was described by Andersen and Kiel (*74*), and Vodnar *et al.* (*75*).

Originally, grass press juice was a residue from green fodder pellet production, but after identification of its valuable components, a number of approaches for its industrial utilization were developed. Thomsen *et al.* (*76*), for instance, investigated the production of a feed concentrate from residual juices that is even suitable for non-ruminants.

Generally, the grass press juices contain a variety of nitrogen compounds and inorganic salts, and can act as a substitute for synthetic nutrients in already existing processes. For instance, the microbial polyhydroxyalkanoates production, can even become more efficient when grass press juices are used as nutrient sources (*77–79*). Therefore, numerous studies were carried out in order to further investigate the utilization of green press juices (*80–82*).

Due to its low carbohydrate content, green press juice cannot serve as a source of carbon for lactic acid production. Therefore, an external carbon source is required. Numerous strains were tested for their suitability to grow and to produce lactic acid in presence of green press juice. By now, a strain of *Lactobacillus paracasei* was identified to be the most appropriate one for lactate production from grass press juice (*78*). The optimum values of cell growth and lactic acid production were determined in discontinuous fermentation experiments with pH control. At near optimum conditions, *Lactobacillus paracasei* is able to accumulate more than 100 g lactic acid L^{-1} in the medium. The yield of lactic acid can be higher than 90 %, and the duration of fermentation varies between 23 and 54 hours. The results correspond well with the outcomes of other studies carried out at similar conditions with other *Lactobacillus* strains (*83*). After changing to a continuous fermentation mode with cell retention and control of glucose concentration, the best results were obtained by feeding of alfalfa press juice instead of synthetic nutrients (*84*). Together with high biomass concentrations and the number of living cells, the productivity increased considerably compared to fermentations performed using well adapted synthetic nutrients (Table 3). Performance of *Lactobacillus paracasei* on barley hydrolysate in combination with green press juice was in the same range like the performance of another strain of the same species on pure glucose, MRS minerals, peptone and yeast extract (*85*). A similar performance, in a membrane cell-recycle bioreactor, was documented by Kwon *et al.* (*86*). They observed the parallel appearance of greatly increased cell density and high lactic acid productivity of about 21.6 g L^{-1} h^{-1} when grass press juice was used.

Table 3. Main results of lab-scale continuous lactic acid fermentations

Test conditions[a]	*Biomass [g L^{-1}]*	*Lactic acid [g L^{-1}]*	*Productivity [g L^{-1} h^{-1}]*
Synthetic salts & nitrogen sources	9.2	80	7.5
Synthetic salts & nitrogen sources	18.0	50	9.1
Change to green press juice	110.0	60	21.5
Green press juice	110.0	65	25.0

[a] Nutrients added to a basic barley hydrolysate broth

Another set of experiments performed with grass press juice from different harvesting sites in Germany revealed a maximum volumetric lactic acid productivity of 2.5 g L^{-1} h^{-1} and a yield of 0.66. It could further be shown that the composition of the fermentation broth (i. e. nitrogen content) has a strong impact on the product formation and sugar consumption (*87*). Generally, the higher the concentration of nitrogen compounds in green press juice, the better the microbial performance.

Future Perspectives and Concluding Remarks

The aim of this chapter is to raise the awareness on the potential and opportunities of utilizing by-products and residues from agricultural materials in lactic acid fermentation. The utilization of agricultural feedstocks enables the conversion of '*no value waste*' into '*value-added products*'. Furthermore, it contributes to solutions to overcome the depletion in resources in the future and to the establishment of a green society. By the presented studies, the possibility of substituting of synthetic nutrients by renewable resources like renewable carbon sources and green biomass without any loss of productivity could be shown. However, technological challenges need to be met by intensive research activities in order to use the complete organic matter, particularly the lignin fraction, as substrate in fermentations. Also concepts need to be developed ensuring a utilization of the globally produced agricultural residues.

Acknowledgments

The authors would like to acknowledge the financial support from the German Agency of Renewable Resources (SynRg, FKZ 22023008/Raps). The authors further acknowledge Giovanna Rehde, Kerstin Mehlmann, Sigrid Quilitz and Roland Schneider from the Leibniz-Institute for Agricultural Engineering Potsdam-Bornim for carrying out the experimental and analytical work.

References

1. Pfaltzgraff, L. A.; De bruyn, M.; Cooper, E. C.; Budarin, V.; Clark, J. H. *Green Chem* **2013**, *15*, 307–314.
2. Litchfield, J. H. *Adv. Appl. Microbiol.* **1996**, *42*, 45–95.
3. Abdel-Rahman, M. A.; Tashiro, Y.; Sonomoto, K. *Biotechnol. Adv.* **2013**, *31*, 877–902.
4. Hofvendahl, K.; Hahn-Hägerdal, B. *Enzyme Microb. Technol.* **2000**, *26*, 87–107.
5. Wee, Y.-J.; Kim, J.-N.; Ryu, H.-W. *Food Technol. Biotechnol.* **2006**, *44*, 163–172.
6. Jim Jem, K.; van der Pol, J.; de Vos, S. In *Plastics from Bacteria - Natural Functions and Applications*; Chen, G. G.-Q., Ed.; Microbiology Monographs; Springer-Verlag: Berlin, Heidelberg, 2010; Vol. 14, pp 323–346.
7. Castillo Martinez, F. A.; Balciunas, E. M.; Salgado, J. M.; Domínguez González, J. M.; Converti, A.; Pinheiro de Souza Oliveira, R. *Trends Food Sci. Technol.* **2013**, *30*, 70–83.
8. Taskila, S.; Ojamo, H. *Lactic Acid Bacteria - R & D for Food, Health and Livestock Purposes*; Kongo, M., Ed.; InTech: 2013; DOI: 10.5772/51282, pp 615-632.
9. Jamshidi, K.; Hyon, S. K.; Ikada, Y. *Polymers* **1988**, *29*, 2229–2234.
10. Grandviewresearch. http://www.grandviewresearch.com/industry-analysis/lactic-acid-and-poly-lactic-acid-market, October 28th, 2014.
11. Teleyadi, S.; Cheryan, M. *Appl. Microbiol. Biotechnol.* **1995**, *43*, 242–248.
12. Bentsen, N. S.; Felby, C.; Thorsen, B. J. *Prog. Energy Combust. Sci.* **2014**, *40*, 59–73.

13. Popa, V. I. *Polysaccharides in Medical Applications*; Dumitri, S., Ed.; Marcel Dekker: New York, 1996; p 107.
14. Neureiter, M.; Danner, H.; Madzingaidzo, L.; Miyafuji, H.; Thomasser, C.; Bvochora, J.; Bamusi, S.; Braun, R. *Chem. Biochem. Eng. Q.* **2004**, *18*, 55–63.
15. Doran-Peterson, J.; Cook, D. M.; Brandon, S. K. *Plant J.* **2008**, *54*, 582–592.
16. Boguta, A. M.; Bringel, F.; Martinussen, J.; Jensen, P. R. *Microb. Cell Fact.* **2014**, *13*, 97, DOI: 10.1186/s12934-014-0097-0.
17. Azadi, P.; Inderwildi, O. R.; Farnood, R.; King, D. A. *Renewable Sustainable Energy Rev.* **2013**, *21*, 506–523.
18. Kaliyan, N.; Morey, R. V. *Am. Soc. Agric. Biol. Eng.* **2009**, *52*, 907–920.
19. Brunner, T.; Kanzian, W.; Obernberger, I.; Theissl, A. http://bios-bioenergy.at/uploads/media/Paper-Brunner-Combustion-properties-of-maize-cobs-2011-06.pdf, October 28th, 2014.
20. Abdel-Rahman, M. A.; Tashiro, Y.; Sonomoto, K. *J. Biotechnol.* **2011**, *156*, 286–301.
21. Avila-Segura, M.; Barak, P.; Hedtcke, J. L.; Posner, J. L. *Biomass Bioeng.* **2011**, *35*, 1190–1195.
22. Huang, Y.; Dong, H.; Shang, B.; Xin, H.; Zhu, Z. *Appl. Energy* **2011**, *88*, 947–952.
23. Mosier, N.; Wyman, C.; Dale, B.; Elander, R.; Lee, Y. Y.; Holtzapple, M.; Ladisch, M. *Bioresour. Technol.* **2005**, *96*, 673–686.
24. Islamiyati, R.; Rasjid, S.; Ismartoyo, A. N. *Int. J. Sci. Technol.* **2013**, *2*, 149–152.
25. Aman, P.; Nordkvist, E. *Swedish J. Agric. Res.* **1983**, *13*, 61–67.
26. Garda-Aparicio, M. A. P.; Ballesteros, I.; Gonzales, A.; Oliva, J. M.; Ballesteros, M.; Negro, M. A. J. *Appl. Biochem. Biotechnol.* **2006**, *129−132*, 278–288.
27. Rowell, M. R. *ACS Symp. Series* **1992**, *476*, 12–27.
28. Bell, J. M.; Jeffers, H. F. *Can. J. Anim. Sci.* **1976**, *56*, 269–271.
29. Naczk, M.; Shahidi, F. In *Canola and Rapeseed -Production, Chemistry, Nutrition, and Processing Technology*; Shahidi, F., Ed.; Springer-Verlag: Berlin Heidelberg, 1990; pp 211−220.
30. Bailey, R. W.; Allison, R. M.; O'Connor, K. F. *Proc. N. Z. Grassl. Assoc.* **1970**, *32*, 127–136.
31. Mohamed, A. A.; Rayas-Duarte, P. *Cereal Chem.* **1995**, *72*, 643–647.
32. Lindsey, K.; Johnson, A.; Kim, P.; Jackson, S.; Labbé, N. *Biomass Bioenergy* **2013**, *56*, 29–37.
33. Kim, Y.; Mosier, N. S.; Ladisch , M. R.; Pallapolu, V. R.; Lee, Y. Y.; Garlock, R.; Balan, V.; Dale, B. E.; Donohoe, B. S.; Vinzant, T. B.; Elander, R. T.; Falls, M.; Sierra, R.; Holtzapple, M. T.; Shi, J.; Ebrik, M. A.; Redmond, T.; Yang, B.; Wyman, C. E.; Warner, R. E. *Bioresour. Technol.* **2011**, *102*, 11089–11096.
34. Williams, S. G. *Plant Physiol.* **1970**, *45*, 376–381.
35. Schachtman, D. P.; Reid, R. J.; Ayling, S. M. *Plant Physiol.* **1998**, *116*, 447–453.

36. Baxter, X. C.; Darvell, L. I.; Jones, J. M.; Barraclough, T.; Yates, N. E.; Shield, I. *Fuel* **2012**, *95*, 50–62.
37. Yusiharni, E.; Gilkes, R. *Geoderma* **2012**, *189-190*, 369–380.
38. Galbe, M.; Zacchi, G. *Appl. Microbiol. Biotechnol.* **2002**, *59*, 618–628.
39. Ji, W.; Shen, Z.; Wen, Y. *Chem. Eng. J.* **2015**, *260*, 20–27.
40. Zhuo, K.; Du, Q.; Bai, G.; Wang, C.; Chen, Y.; Wang, J. *Carbohydr. Polym.* **2015**, *115*, 49–53.
41. Kumar, P.; Barrett, D. M.; Delwiche, M. J.; Stroeve, P. *Ind. Eng. Chem. Res.* **2009**, *48*, 3713–3729.
42. Rodriguez, A.; Perestelo, F.; Carnicero, A.; Regalado, V.; Perez, R.; De la Fuente, G.; Falcon, M. A. *FEMS Microbiol. Ecol.* **1996**, *21*, 213–219.
43. Pleissner, D.; Kwan, T. H.; Lin, C. S. K. *Bioresour. Technol.* **2014**, *158*, 48–54.
44. Wyman, C. E.; Spindler, D. D.; Grohmann, K. *Biomass Bioenergy* **1992**, *3*, 301–307.
45. Kádár, Z.; Szengyel, Z.; Réczey, K. *Ind. Crops Prod.* **2004**, *20*, 103–110.
46. Nguyen, C. M.; Choi, G. J.; Choi, Y. H.; Jang, K. S.; Kim, J.-C. *Biochem. Eng. J.* **2013**, *81*, 40–46.
47. Zhao, K.; Qiao, Q.; Chu, D.; Gu, H.; Dao, T. H.; Zhang, J.; Bao, J. *Bioresour. Technol.* **2013**, *135*, 481–489.
48. Chookietwattana, K. *APCBEE Procedia* **2014**, *8*, 156–160.
49. John, R. P.; Anisha, G. S.; Madhavan Nampoothiri, K.; Pandey, A. *Biotechnol. Adv.* **2009**, *27*, 145–152.
50. Qi, X.; Tang, Y.; Jian, H.; Li, X.; Jiang, J. *Adv. Mater. Res.* **2011**, *152-153*, 1404–1411.
51. Gullon, B.; Garrote, G.; Alonso, J. L.; Parajo, J. C. *J. Agric. Food Chem.* **2007**, *55*, 5580–5587.
52. Moldes, A. B.; Torrado, A.; Converti, A.; Dominguez, J. M. *Appl. Biochem. Biotechnol.* **2006**, *135*, 219–227.
53. Koutinas, A. A.; Vlysidis, A.; Pleissner, D.; Kopsahelis, N.; Lopez Garcia, I.; Kookos, I. K.; Papanikolaou, S.; Kwan, T. H.; Lin, C. S. *Chem. Soc. Rev.* **2014**, *43*, 2587–2627.
54. Maas, R. H. W.; Bakker, R. R.; Jansen, M. L. A.; Visser, D.; de Jong, E.; Eggink, G.; Weusthuis, R. A. *Appl. Microbiol. Biotechnol.* **2008**, *78*, 751–758.
55. Garde, A.; Jonsson, G.; Schmidt, A. S.; Ahring, B. K. *Bioresour. Technol.* **2002**, *81*, 217–223.
56. Yao, R. S.; Qi, B. K.; Deng, S. S.; Liu, N.; Peng, S. C.; Cui, Q. F. *Bioresources* **2007**, *2*, 389–398.
57. Gao, L.; Yang, H.; Wang, X.; Huang, Z.; Ishii, M.; Igarashi, Y.; Cui, Z. *Bioresour. Technol.* **2008**, *99*, 2742–2748.
58. Wang, Q.; Zou, D.; Ma, H.; Ji, Y.; Wang, X. *Chem. Biochem. Eng. Q.* **2010**, *24*, 371–376.
59. Saito, K.; Hasa, Y.; Abe, H. *J. Biosci. Bioeng.* **2012**, *114*, 166–169.
60. Zhang, Y.; Chen, X.; Luo, J.; Qi, B.; Wan, Y. *Bioresour. Technol.* **2014**, *158*, 396–399.

61. Ye, L.; Hudari, M. S. B.; Zhi, L. I.; Chuan, W. U. J. *Biochem. Eng. J.* **2013**, *83*, 16–21.
62. Jeong, S.-Y.; Trinh, L. T. P.; Lee, H.-J.; Lee, J.-W. *Bioresour. Technol.* **2014**, *152*, 444–449.
63. Walsh, P.; Venus, J. EP 2013/059184 [WO/2013/164423], 2013.
64. Walsh, P.; Venus, J. EP 2013/059186 [WO/2013/164425], 2013.
65. Li, Y.; Cui, F. *Sustainable Biotechnology - Sources of Renewable Energy*; Singh, O. V., Harvey, S. P., Eds.; Springer-Verlag: Berlin, Heidelberg, 2010; pp 211−228.
66. Castro, E.; Díaz, M. J.; Cara, C.; Ruiz, E.; Romero, I.; Moya, M. *Bioresour. Technol.* **2011**, *102*, 1270–1276.
67. Chen, K.; Zhang, H.; Miao, Y.; Wei, P.; Chen, J. *Enzyme Microb. Technol.* **2011**, *48*, 339–344.
68. Lomascolo, A.; Uzan-Boukhris, E.; Sigoillot, J.-C.; Fine, F. *Appl. Microbiol. Biotechnol.* **2012**, *95*, 1105–1114.
69. Rommi, K.; Hakala, T. K.; Holopainen, U.; Nordlund, E.; Poutanen, K.; Lantto, R. *J. Agric. Food Chem.* **2014**, *62*, 7989–7997.
70. Bell, J. M. *J. Anim. Sci.* **1984**, *58*, 996–1010.
71. Zhang, S. B.; Wang, Z.; Xu, S. Y. *J. Am. Oil Chem. Soc.* **2007**, *84*, 97–105.
72. Kamm, B.; Hille, C.; Schonicke, P.; Dautzenberg, G. *Biofpr* **2010**, *4*, 253–262.
73. Biowert Industrie GmbH, BIOWERT-bio based industry. http://www.biowert.de, October 28th, 2014.
74. Andersen, M.; Kiel, P. *Ind. Crop Prod.* **2000**, *11*, 129–137.
75. Vodnar, D. C.; Venus, J.; Schneider, R.; Socaciu, C. *Chem. Eng. Technol.* **2010**, *3*, 468–474.
76. Thomsen, M. H.; Bech, D.; Kiel, P. *Chem. Biochem. Eng. Q.* **2004**, *18*, 37–46.
77. Koller, M.; Bona, R.; Hermann, C.; Horvat, P.; Martinz, J.; Neto, J.; Pereira, L.; Varila, P.; Braunegg, G. *Biocatal. Biotransform.* **2005**, *23*, 329–337.
78. Venus, J. *Biotechnol. J.* **2006**, *1*, 1428–1432.
79. Leiss, S.; Venus, J.; Kamm, B. *Chem. Eng. Technol.* **2010**, *33*, 2102–2105.
80. Ajibola, O. O. *Food Chem.* **1984**, *13*, 181–192.
81. Balcerek, M.; Pielech-Przybylska, K.; Patelski, P. *Biomass Bioenergy* **2011**, *35*, 4841–4848.
82. Martel, C. M.; Parker, J. E.; Jackson, C. J.; Warrilow, A. G. S.; Rolley, N.; Greig, C.; Morris, S. M.; Donnison, I. S.; Kelly, D. E.; Kelly, S. L. *Bioresour. Technol.* **2011**, *102*, 1503–1508.
83. Berry, A. R.; Franco, C. M. M.; Zhang, W.; Middelberg, A. P. J. *Biotechnol. Lett.* **1999**, *21*, 163–167.
84. Kamm, B.; Kamm, M.; Venus, J. *Trends in Biotechnology Research*; Hearns, E. C., Ed.; Nova Science Publishers: New York, 2006; pp 199−223.
85. Richter, K.; Nottelmann, S. *Eng. Life Sci.* **2004**, *4*, 426–432.
86. Kwon, S.; Yoo, I. K.; Lee, W. G.; Chang, H. N.; Chang, Y. K. *Biotechnol. Bioeng.* **2001**, *73*, 25–34.
87. Papendiek, F.; Venus, J. *Chem. Biochem. Eng. Q.* **2014**, *28*, 375–382.

Editors' Biographies

Sherine O. Obare

Sherine O. Obare is a Professor of Inorganic Chemistry and also the Associate Chair and Graduate Director at the Department of Chemistry at Western Michigan University. She received her undergraduate degree in Chemistry at West Virginia State University, and she received her Ph.D. at the University of South Carolina with Professor Catherine J. Murphy. She completed a two-year Dreyfus postdoctoral fellowship at Johns Hopkins University with Professor Gerald. J. Meyer. In 2004, Dr. Obare joined Western Michigan University as an Assistant Professor, was promoted to the rank of Associate Professor with tenure in 2009, and was promoted to full Professor in 2014. Her research interests lie in the area of designing nanoscale materials for environmental remediation, improved healthcare, and alternative energy. Her work has been featured in over 100 publications in the form of journal articles, review articles, book chapters, and conference presentations. Dr. Obare's research program has been funded by the National Science Foundation, the Department of Defense, the Army Research Office, the National Institutes of Health, the Department of Education, and the Michigan Economic Development Corporation.

Dr. Obare is the recipient of several awards including the 2009 International Union of Pure and Applied Chemistry (IUPAC) Young Observer Award, the National Science Foundation CAREER award, and the American Competitiveness and Innovation (ACI) Fellowship. She is also an Associate Editor for *The Journal of Nanomaterials*. At Western Michigan University, Dr. Obare has been the recipient of the 2012 Emerging Scholar Award and the 2012 Faculty Achievement Award in Professional and Community Service. Dr. Obare has served as the Director of the NIH-sponsored Bridges to the Baccalaureate Program at Western Michigan University — a program that recruits underrepresented minority students from community colleges in Michigan and supports them in their pursuits for advanced degrees in biomedical and behavioral sciences. Dr. Obare served on the NOBCChE Executive Board for three years and chaired the National NOBCChE Secondary Education Science Bowl and Science Fair for seven years.

Dr. Obare has organized symposia for the Division of Environmental Chemistry and the Colloids and Surface Chemistry Division of the American Chemical Society. She has also served on the International Conference Planning Committee for the Annual Meetings for Advanced Oxidation Technologies since 2009. In 2009, she served as a Scientific Committee Member for the International Conference on Nanoscale Science and Technology in Tunisia that was sponsored by the Alexander von Humboldt Foundation in 2012. In 2013 Obare was named as one of the top 25 female professors in Michigan.

Rafael Luque

Professor Rafael Luque received a Ph.D. in Chemistry from the Universidad de Cordoba, Spain in 2005. His research program and expertise focus on (nano)materials science, heterogeneous (nano)catalysis, microwave and flow chemistry, biofuels, and green chemical methods in synthetic organic chemistry. He has published over 220 research articles, filed 3 patent applications, and edited seven books; he has also written several book chapters. He has given Invited, Keynote, and Plenary lectures, along with keynote addresses, at several international conferences. Professor Luque is also heavily involved in chemical education and promoting science in developing countries. He is a member of the Editorial Advisory Boards of several prestigious journals including *Chemical Society Reviews* (RSC), *Catalysis Communications* (Elsevier), *Current Organic Synthesis* (Bentham Publishers), *Sustainable Chemical Processes* (Chemistry Central), and *Current Green Chemistry* (Bentham Publishers). He is also Editor-in-Chief of the porous section of the journal *Materials* and Series Editor of *Topics in Current Chemistry* (Springer).

Professor Luque's recent awards include the Marie Curie Prize from Instituto Andaluz de Quimica Fina in Spain (2011), the Green Talents award from the Federal Ministry of Education and Research in Germany (2011), the TR35 Spain from Technology Review and MIT as one of the top 10 young entrepreneurs in Spain (2012), and, most recently, the RSC Environment, Sustainability and Energy Early Career Award (2013) from the Royal Society of Chemistry UK. He has also been recently honored as 2013 Distinguished Engineering Fellow and Visiting Professor from CBME at Hong Kong University of Science and Technology. He currently serves as Chinese Academy of Sciences Visiting Professor at the State Key Laboratory of Electroanalytical Chemistry of the Changchun Institute of Applied Chemistry.

Prof. Luque combines his academic duties with his activities as a young entrepreneur; he co-founded the spin-off companies Starbon® Technologies at York, UK (2011, http://www.starbon-technologies.com/) — which markets novel biomass derived carbonaceous materials — along with Green Applied Solutions S.L. (GAS, http://greenappliedsolutions.com/) in Cordoba, Spain (2012), which is an R&D and consultancy technological company providing waste valorization methods for marketable products. More recently, he has taken part in the founding of Posidonia Oceanica S.L. (2013, www.posidoniaoceanica.com), in Spain.

Indexes

Author Index

Adams, C., 219
Al-Zubaidi, H., 219
Ariyadasa, L., 219
Bolandi, A., 219
Chatterjee, S., 203
Esposito, D., 53
Friščić, T., 161
García-Álvarez, J., 37
Gonzalez, M., 69
Hunsen, M., 117
Jin, R., 139
Johs, A., 203
Julien, P., 161
Kirchhecker, S., 53
Lam, W., 187
Leahy, K., 129
Li, G., 139
Lin, C., 187
Luque, R., 1
Mack, A., 129
Mack, J., 129
Mottillo, C., 161
Nick, S., 219
Obare, S., 1, 219
Ofoli, R., 219
Pleissner, D., 247
Rios, O., 203
Saito, T., 203
Soh, L., 9
Vaddula, B., 69
Venus, J., 247
Wabeke, J., 219
Yalla, S., 69
Yu, C., 139
Zhou, W., 139

Subject Index

A

Amino acid-derived imidazolium zwitterions
bio-derived imidazolium building blocks as versatile scaffolds, 54*f*
carboxylic functionalities, reactions, 56
conclusion, 66
decarboxylation of ImZw and formation of ionic liquids, 61*f*
elemental analysis of cellulose materials as function of amount, 66*t*
functionalization of cellulose, 64
synthesis of esterified ionic liquid used, 65*f*
glass transition temperatures for adducts, 58*t*
hydrothermal decarboxylation of ImZw, 60*f*
imidazolium zwitterions as building block molecules, 55
ImZw and acidic ILs, acidic properties, 55
ImZw derived from amino acids and carbohydrate derivable dicarbonyl compounds, 57*t*
introduction, 53
pKa values of some ImZw, 58*t*
Approaches to assess alternative solvent systems
biomass derived solvents
glycerol based solvents, 98
limonene, 100
2-methyltetrahydrofuran, 99
design for environment (DfE), 90
objectives, 90
neoteric solvents, 94
ionic liquids, 96
supercritical carbon dioxide (SC-CO_2), 95
water, 98
program for assisting replacement of industrial solvents (PARIS), 89
solvent-less/mechano-chemical mixing, 92
Toxic Substances Control Act (TSCA), 91

B

Biomass conversion processes
bioelectrocatalytic process for cellulosic materials, 230
biomass categories, 220
biorenewables transformation, applications of nanoparticles, 226
conversion of sugars, 227
decarboxylation of fatty acids, 231
decarboxylation of stearic acid to n-heptadecane, 232*f*
deoxygenation of stearic acid, kinetic study, 233*f*
design of fuel cells, 236
direct glucose fuel cell, 237*f*
hydrolysis of cellulose to glucose, 229*f*
nanoparticle-fostered transformation of cellulose into commodity chemicals, 229
polyol hydrogenolysis, proposed mechanism, 227*f*
production of biodiesel, 234
production of hydrocarbons, 228
vanillin (4-hydroxy-3-methoxybenzaldehyde), 235
conclusions, 238
free fatty acids (FFA), catalytic deoxygenation, 221
hydrolysis of cellulose to glucose, 230
hydrotreatment, 221
introduction, 219
nanoscale materials synthesis, 222
bimetallic nanoparticles, 225
metallic nanoparticles, 223
semiconductor nanoparticles, 226
TEM images for Pd NPs, 224*f*
oleaginous materials, 221

C

Carbon dioxide solvent applications in biorefinery
biological extract classes, solubility parameters, 17*t*
biomass, 10
biorefinery, 11

biorefinery applications for CO_2 as solvent, 30*f*
carbon dioxide, 12
CO_2 utilization for products in biorefinery, 25*t*
aldehyde/ketone production, 26
lactic acid and succinic acid, 26
common extraction solvents and CO_2 co-solvents, solubility parameters, 18*t*
conclusion, 29
introduction, 9
solvent uses for carbon dioxide with biological compounds, 14*f*
supercritical carbon dioxide, biomass extraction, 17*t*
sustainable design
economic and environmental impact, 28
integration of CO_2 into biorefinery processes, 27
Chemical alternatives assessment, 108
CO_2 and biorefinery, 14
biorefinery products utilizing CO_2, 24
extraction, 15
CO_2 as anti-solvent, 20
fractionation and refinement, 19
neat scCO_2, 16
reactions, 22
heterogeneous catalysts, 23
selectivity and tuning, 23
transesterification, 23
solubility of oleate species in CO_2 at 323K, 20*f*
solubility parameter, 16
uses of CO_2 as anti-solvent, 21*t*
CO_2 expanded liquid (CXL), 13
CXL. *See* CO_2 expanded liquid (CXL)

D

Deep Eutectic Solvents (DESs), 37
applications as New *Green* and biorenewable reaction media
lewis acid catalyzed dehydration of carbohydrates, 41
metal-catalyzed dehydration of fructose to HMF, 42*s*
protic- or lewis acid-catalyzed dehydration of fructose, 42*s*
conclusions, 47
Cu(I)-catalyzed click chemistry reactions (CuAAC), 44
CuI-catalyzed 1,3-dipolar cycloaddition of azides, 45*s*
general description, 39
hydrogen-bond-acceptors (HBA) and hydrogen-bond-donors (HBD)
interaction, 39*f*
structure, 40*f*
introduction, 38
lewis acid catalyzed production of furfural (furan-2-carbaldehyde), 42
metal-catalyzed cross-coupling reactions, 43
metal-catalyzed hydrogenation of olefins, 46
metal-catalyzed isomerization reactions, 45
Pd-catalyzed Suzuki reaction in different sugar-based DESs, 44*t*
rearrangement of aldoximes, 46*s*
Rh(I)-catalyzed hydrogenation of methyl A-cinnamide, 47*s*
Ru(IV)-catalyzed redox isomerization of allylic alcohols, 46*s*
Designing highly efficient photocatalysts, some strategies, 139
conclusion, 156
degradation rate of acid orange II over various samples, 156*t*
degradation rate over pure $BiVO_4$, Co/$BiVO_4$ and Cu/$BiVO_4$ samples, 154*t*
design of heterojunction to enhance photocatalytic performance, 142
electronic band structures of Ag_2CO_3 and Ag_2O, 144*f*
formation of mesoporous CdS nanocrystals, 152*f*
microspherical photocatalysts, fabrication, 146
$PbWO_4$ microcrystals with different morphologies, 148*f*
phase transformation, 143*f*
photocatalytic recycling tests, 149*t*
progress in fundamentals of environmental photocatalysis, 140
recycling tests of $PbWO_4$ microspheres in acid orange II dye degradation, 149*t*
samples prepared under different conditions, XRD patterns, 155*f*
schematic illustration for sonochemical reactor, 151*f*
semiconductor photocatalysis, 140*s*
separation of photo-excited electrons and holes, 145*f*
sonochemical fabrication photocatalysts with novel nanostructures, 150

sonochemically prepared samples, mesoporous property, 152*t*
TiO$_2$ microsphere sample, 147*f*
UV-Vis diffuse reflectance absorption spectra, 153*f*
DfE methodology, 108

E

EcoScale comparison, mechanochemistry and solution based reactions
comparison of reactions
calculating EcoScale, 132
comparison of solution against mechanochemistry, 134*t*
overall assessment of two methods, 133
phosphonium salt generation, 131
product isolation, 132
solution based Wittig reaction, EcoScale rating, 135*f*
stilbene formation, 131
comparison of soluion based Wittig against mechanochemical Wittig, 130*f*
conclusions, 136
introduction, 129

G

Green and sustainability aspects of alternatives assessment, 104
conclusion, 109
key steps to conducting
assessment of hazard, 106
cost and life cycle aspects, application, 107
decision making process, application of research results, 107
gathering information, 105
identification of possible alternatives, 106
judging possibility, 105
organizing interested parties, 105
Green technologies for environment
conclusion, 4
food waste, 3
greener reactions, 3
introduction, 1
overview, 2
photocatalytic reactions, 3
Greener oxidation of alcohols, glycosides and sulfides
catalytic oxidation of alcohols directly to carboxylic acids, 121
catalytic oxidation of alcohols to aldehydes & ketones, 118
conclusions, 125
experimental
oxidative ring opening of glycosides, 126
preparation of sulfones, 126
preparation of sulfoxides, 126
iminocyclitols, 123*f*
introduction, 117
oxidation of sulfides to sulfides, 124*t*
oxidation of sulfides to sulfones, 125*t*
oxidation of sulfides to sulfoxides, 124*s*
oxidation of sulfides to sulfoxides and sulfones, 123
PCC catalyzed oxidative ring-opening of glycosides with periodic acid, 123*s*
preparation of polyketoester, 122*s*
selective oxidative ring-opening of glycosides, 121

H

High value chemicals and energy production, food waste valorisation
chemicals and biofuels production, food waste based biorefiney, 189*f*
composition of various food wastes from retail and commercial sectors, 188*t*
conclusions and future prospects, 199
fermentative productions, 188
bio-pigments, 192
direct use of food waste, enzymes and bio-colorant production, 189
fermentative succinic acid production, 195*t*
lactic acid, 194
PHB production, 195*t*
poly(3-hydroxybutyrate) (PHB), 193
production of glucoamylase and protease, 190
production of lactic acid, 196*t*
succinic acid production, 193
utilisation of hydrolysed food waste, 192
introduction, 187
production of glucoamylase and protease from agro-industrial, 191*t*
technical challenges, 198
HMF. *See* 5-Hydroxymethyl-furfural (HMF)
5-Hydroxymethyl-furfural (HMF), 41

I

Ionic liquids, 59
applications
cellulose dissolution, 63
Heck reaction, 62
some physicochemical properties of ILs, 62*t*
XRD patterns of MCC, 64*f*
derived from ImZw, 61*t*

L

Lactic acid fermentation, agricultural residues as feedstocks
agricultural residue, 249
comparison of composition, 250*t*
pre-treatment, 251
comparison of lactic acid concentrations, 257*f*, 258*f*
D(-)- and L(+)-lactic acid, chemical structure, 248*f*
different rapeseed residues, analytical figures, 256*t*
fermentation, 253
rapeseed, 255
straw, 254
future perspectives and conclusions, 259
grass, 258
introduction, 247
lab-scale continuous lactic acid fermentations, main results, 259*t*
utilization of agricultural residues, process flow, 252*f*
Lignin based carbon materials for energy storage applications, 203
anode materials, 206
BET surface areas pore volumes, 213*t*
^{13}C NMR spectra of unmodified and modified lignin, 210*f*
challenges with graphite-based anodes in Li-ion batteries, 206
charge capacities as function of cycling and carbonization temperatures, 216*f*
conclusions, 216
electrodes, cyclic performance and coulombic efficiency, 208*f*
electrospun nanofibers from lignin-PEO blends, 207*f*
introduction, 204
lignin carbon fiber, 204
lignin carbon fiber fused mats, 215*f*
lignin monomeric units, 205*f*
possible nucleophilic acyl substitution mechanism, 209*f*
preparation of carbon fiber from lignin, 205*s*
scanning electron microscope (SEM) images of lignin carbon fibers, 214*f*

M

Metal-organic frameworks (MOFs), 161
conclusions and outlook, 179
green innovations in synthesis, 166
renewable linkers for edible MOFs, 168
super-critical water as solvent, 167
industrial manufacture, criteria, 164
complex mixed-ligand bimetallic (Cu and Zn) framework, 166*f*
high boiling organic solvents, 165
selection of metal precursor, 165
solution-based syntheses, 165
introduction, 162
mechano- and thermochemical synthesis of OLED material, 178*f*
metal-organic OLEDs and metallodrugs, 177
processing steps important in industrial MOF manufacture, 164*s*
prospective applications, 163
solvent-free synthesis, 169
accelerated aging reactions, 173
alkyl-ammonium templates, 173
applications of ILAG in synthesis of microporous MOFs, 175*f*
comparison of rates mechanochemical formation of ZIF-8 by LAG and ILAG, 176*f*
input of energy and reagents, 174
mechanochemical assembly of zinc fumarate, 171*f*
mechanochemical reactions, 170
mimicking and accelerating mineral weathering processes, 174
synthesis of metal-organic frameworks by accelerated aging, 177*f*
thermochemical reactions, 170
understanding reaction mechanisms and kinetics, 172

P

PCC catalyzed oxidation
alcohols to aldehydes, 119*s*

alcohols to aldehydes and ketones, proposed mechanism, 120*s*
alcohols to ketones, 119*s*
ortho diol to lactone, 120*s*
primary alcohols directly to carboxylic acids, 121*s*

R

Replacement/minimization of selective hazardous solvents, opportunities
cause for concern
environmental concerns, 88
health concerns, 86
safety concerns, 89
health concerns
carcinogenic, 87
dermatitis, 87
toxicity of glycol ethers, 87
introduction, 69
organic solvents, industry wide
application, 72
adhesives, 80
paints and associated products, 73
pharmaceuticals and fine chemicals, 76
plastics and rubber materials, 82
printing operations, 78
solvent cleaning, conditioning and degreasing, 79
total release (in pounds), for facilities in NAICS 325510, 75*t*
total release (in pounds), for facilities in NAICS 325520, 81*t*
TRI on-site and off-site reported disposed, 74*t*, 77*t*, 80*t*, 83*t*
organic solvents and commonly associated toxicities, 88*t*
production of hazardous solvents, 70
role and function of solvents, physical and chemical properties, 84
some hazardous air pollutants (HAPs), global production, 72*t*
top twenty produced chemicals in U.S. in year 2012, 71*t*

S

scCO_2. *See* Supercritical carbon dioxide (scCO_2)
Semiconductor photocatalytic materials, 141
Solvent replacement systems, specific examples
printed circuit boards, cleaning solvent alternatives, 100
running costs comparison, 102*t*
1,1,1-trichloroethane, 101
vapor degreasing operations, alternative solvent assessment, 103
vapor degreasing solvent substitution at Durex Inc., 102
Supercritical carbon dioxide (scCO_2), 13

U

Unmodified and modified lignin
carbonizing conditions, 212*f*
oxidizing conditions, 211*f*

V

Valorisation of food waste for biofuel production
anaerobic digestion of food waste for biogas production, 197
organic food waste degradation in anaerobic digestion, key steps, 198*f*
production of biodiesel from food waste lipid, 196
trans-esterification of triglyceride for biodiesel production, 197*f*

Printed in the USA/Agawam, MA
January 13, 2020

747850.006